Horticulture: Contemporary Agricultural Practices

Horticulture: Contemporary Agricultural Practices

Editor: Elijah Baxter

RCALLISTO
REFERENCE

www.callistoreference.com

Callisto Reference,
118-35 Queens Blvd., Suite 400,
Forest Hills, NY 11375, USA

Visit us on the World Wide Web at:
www.callistoreference.com

ISBN: 978-1-63239-979-3 (Hardback)

Cataloging-in-Publication Data

Horticulture : contemporary agricultural practices / edited by Elijah Baxter.
    p. cm.
Includes bibliographical references and index.
ISBN 978-1-63239-979-3
1. Horticulture. 2. Agriculture. I. Baxter, Elijah.
SB318 .H67 2018
635--dc23

Table of Contents

Preface

The science of growing, maintaining and harvesting plants is known as horticulture. It is an amalgamation of the elements of art, technology and business used to grow plants. The field also includes arboriculture, landscape design, plant conservation, landscape restoration, etc. This book includes some of the vital pieces of work being conducted across the world, on various aspects related to horticulture. The topics included in it are of utmost significance and are bound to provide incredible insights about the field. Through this text, readers would gain knowledge that would broaden their perspective about horticulture.

This book has been the outcome of endless efforts put in by authors and researchers on various issues and topics within the field. The book is a comprehensive collection of significant researches that are addressed in a variety of chapters. It will surely enhance the knowledge of the field among readers across the globe.

It gives us an immense pleasure to thank our researchers and authors for their efforts to submit their piece of writing before the deadlines. Finally in the end, I would like to thank my family and colleagues who have been a great source of inspiration and support.

Editor

The Effectiveness of Different Rootstocks for Improving Yield and Growth of Cucumber Cultivated Hydroponically in a Greenhouse

Ali Farhadi [1,2,*], Hossain Aroeii [2], Hossain Nemati [2], Reza Salehi [3] and Francesco Giuffrida [4]

[1] Agriculture and Natural Resource Research Center of Isfahan, Iran
[2] Department of Horticulture Sciences, Agriculture Faculty, Ferdowsi University of Mashhad, Iran; aroiee@um.ac.ir (H.A.); nematih@yahoo.com (H.N.)
[3] Department of Horticultural Sciences, Campus of Agriculture & Natural Resources, University of Tehran, Karaj, Iran; salehir@ut.ac.ir
[4] Department of Agriculture, Food and Environment, Catania University, Catania 95100, Italy; francesco.giuffrida@unict.it
*   Correspondence: farhadi_siv@yahoo.com

Abstract: The use of grafted vegetable seedlings has been popular in many countries during recent years, and several *Cucurbita* species and their interspecific hybrids have been tested as rootstocks. Graft-scion incompatibility and lower fruit quality have prevented their commercial use. In this study, the efficacy of grafting the commercial Khassib cucumber hybrid onto various cucurbits, used as rootstocks, was examined in a greenhouse experiment. This experiment was done in a completely randomized design with eight treatments: local landraces of bottle gourd (Ghalyani) and pumpkin (Tanbal) and commercial *Cucurbita* interspecific hybrids (909, 913, Ferro, 64-19, and Shintoza). Ungrafted plants were used as controls. Results indicated that the highest survival rates were using Ferro hybrid (94%), *Cucurbita maxima* (Tanbal) (92%), 64-19 and Shintoza (90%). These results appeared to be related to the stem diameter of the rootstocks and, to a lesser extent, to the scion diameter. Cucumber production was not improved by grafting. The highest yield was obtained with Ferro rootstock, but it was no different compared to ungrafted plants. A significantly lower production than the control was observed with Ghalyani (−44%), 913 (−73%) and 64-19 (−35%) rootstocks. Total soluble solids (TSS) of fruit produced by ungrafted plants was significantly higher than the other treatments. The highest length/diameter ratio was obtained with 909 and Ghalyani rootstocks, whereas more stocky fruit were produced by plants grafted onto Tanbal and 64-19 rootstocks.

Keywords: *Cucumis sativus* L.; soilless; protected cultivation; production; grafting; quality

1. Introduction

Cucumber (*Cucumis sativus* L.) is one of the most popular vegetables in the world, and annually about 71.4 million tons are produced [1]. Cucumber is also an important vegetable crop for greenhouse production in many countries. Continuous cultivation in greenhouses leads to some problems, including soil-borne diseases, soil nutrient imbalances, salty or alkaline soil conditions, spread of weeds, *etc.* To alleviate these problems, grafting is recommended. Grafting is an important technique for vegetable production in many countries where intensive and continuous cultivation is performed. Vegetable grafting has been used to improve yield, fruit quality and disease resistance in *Solanaceae* and *Cucurbitaceae* [2–5]. In recent years, grafting has been found effective in overcoming abiotic stresses.

Numerous studies have reported that different rootstock genotypes could have variable effects on yield and quality of the grafted vegetable, and the overall performance largely depends on the specificity of the rootstock-scion combination [6–8].

Several *Cucurbita* species and their interspecific hybrids have been used traditionally as rootstocks for cucumber. The performance of the grafted plants, including fruit quality, depends on rootstock-scion interactions and on the cultivation method. Some rootstocks have been reported to positively affect yield and fruit quality. However, graft incompatibility and a decrease in quality appears in most cases [6]. Some authors associate this decline in quality with the translocation of fruit quality agents from the *Cucurbita* rootstock to the cucumber scion. The use of rootstocks belonging to the same species (*C. sativus*) has been proposed as a method to overcome these problems [9].

In some developing countries, vegetable grafting is not popular with farmers, and the use of local landraces is common for vegetable crops. The aim of the research was to assess the compatibility of local landraces and interspecific hybrids as rootstocks for cucumber cultivation in terms of yield and quality.

2. Experimental Section

2.1. Plant Material, Treatments and Growth Conditions

The experiment was conducted in autumn–winter 2014 in a glasshouse situated in the Dastgerd Experimental Station of the Agricultural and Natural Resources Research Center of Isfahan, Central Iran (latitude 32°36′N, longitude 51°36′E, 1576 m above sea level). *Cucumis sativus* L. *cv.* Khassib F_1 (Rijk Zwaan, Fijnaart, The Netherlands) was grafted onto five interspecific hybrids (Shintoza, 909, 913, Ferro and 64-19), one local landrace of pumpkin (*Cucurbita maxima* L. (Tanbal)) and one of bottle gourd (*Lagenaria siceraria* (Ghalyani)) using the "splice grafting" described by Lee [10], whereas ungrafted "Khassib" was used as a control. Rootstock and scion seeds were sown on 10 and 14 September 2014, respectively. They were grafted on 22 September 2014. Grafted plants were kept in a small tunnel under semi-controlled environmental conditions. Seedlings grew without light and with a >90% RH (Relative Humidity) for four days, then light was gradually increased and relative humidity was decreased. At the two true-leaf stage (7 October 2014), grafted plants were transplanted to 10 L plastic pots. The substrate was perlite and coco peat (60:40 v/v). The space between plant pots was 35 cm in the rows and 70 cm between rows, obtaining a plant density of four plants/m². Plants were grown under natural light conditions. During the whole crop cycle, air temperatures inside the greenhouse were recorded daily (Figure 1). The experiment was set up in a completely randomized design with three replicates for each treatment. Each experimental unit consisted of eight plants. Throughout the experiment, a modified Hoagland nutrient solution was used: $Ca(NO_3)_2$ (35 g/L); NH_4NO_3 (18 g/L); KNO_3 (62 g/L); FeEDDHA (1.17 g/L); KH_2PO_4 (15 g/L); $MgSO_4$ (32 g/L); $MnSO_4$ (0.062 g/L); $ZnSO_4·2H_2O$ (0.018 g/L); $CuSO_4·5H_2O$ (0.004 g/L); H_3BO_3 (0.077 g/L); Na_2MoO_4 (0.0015 g/L) [11]. The corresponding EC (Electrical Conductivity) of the nutrient solution was 2 ± 0.2 dS/m.

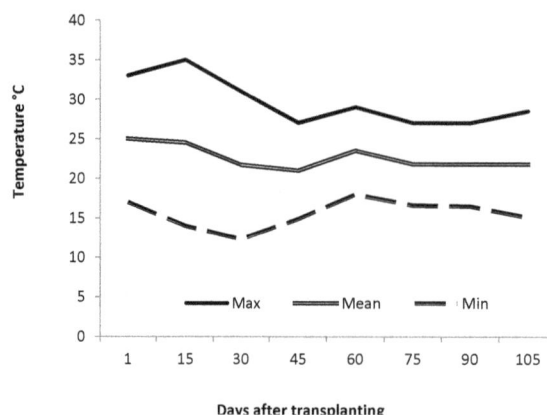

Figure 1. Air temperature inside the greenhouse during the experiment.

2.2. Survival Rate and Plant Growth

The surviving grafted plants were counted 15 d after grafting, and the survival rate was calculated. Plant height, rootstock and scion stem diameter were measured in each treatment. A ruler was used for plant height, with measurement from the substrate surface to the top of the plant. Diameter measurements were made using a digital caliper and were taken 1 cm below and above the graft union. Dry matter production of plant organs (root and shoot g/plant) was determined by drying at 72 °C for 48 h in a thermo-ventilated oven at the end of the growing period. The lengths of the roots and leaf area were measured at the end of the crop.

2.3. Yield and Fruit Quality Parameters

Harvest was started 36 d after transplanting. Harvested fruits were weighed and counted. Total yield, total fruit number, and total marketable and unmarketable fruit (lacking uniformity and/or fruits with physiological disorders) were counted.

2.4. Statistical Analysis

All data were statistically analyzed performing ANOVA with CoStat software. Means were separated using Duncan's Multiple Range Test at $\alpha = 0.05$.

3. Results

3.1. Plant Growth

High compatibility between scion and rootstock and survival was achieved with Ferro hybrid (94%), *Cucurbita maxima* (Tanbal) (92%), 64-19 and Shintoza (Figure 2). The lowest survival rates were observed with *Lagenaria siceraria* (Ghalyani) (72%) and 913 hybrids (73%).

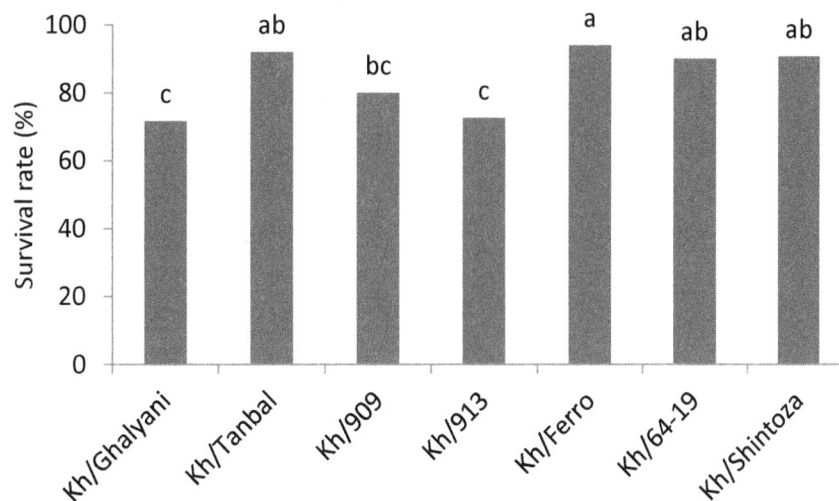

Figure 2. The survival rate of the Khassib cucumber grafted onto different rootstocks (Kh = Khassib scion). Bars with different letters differ statistically at $\alpha = 0.05$ by Duncan's Multiple Range Test.

The rootstock diameter at 15 d after transplanting (DAT) was not affected by grafting, but at the end of the experiment, the diameters showed significant differences (Figure 3). Ferro (8.65 mm), Tanbal and 64-19 hybrid (7.95 mm) had a higher hypocotyl diameter than Shintoza (5.77 mm) and Ghalyani (6 mm). Significant differences were observed in the scion diameter on different rootstocks at 15 DAT and 100 DAT (Figure 4). At 15 DAT, the scion diameter of plants grafted onto 909 was higher than all the other rootstock treatments except for Ferro and 64-19, which were the highest at the end of the experiment (100 DAT).

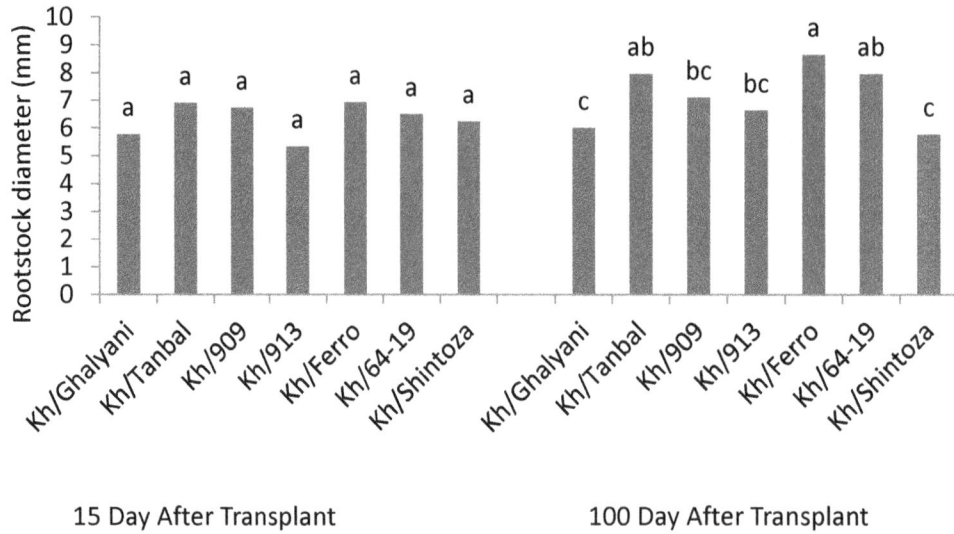

Figure 3. Rootstock diameter at two stages of plant growth (Kh = Khassib scion). Bars with different letters differ statistically at α = 0.05 (Duncan's Multiple Range Test).

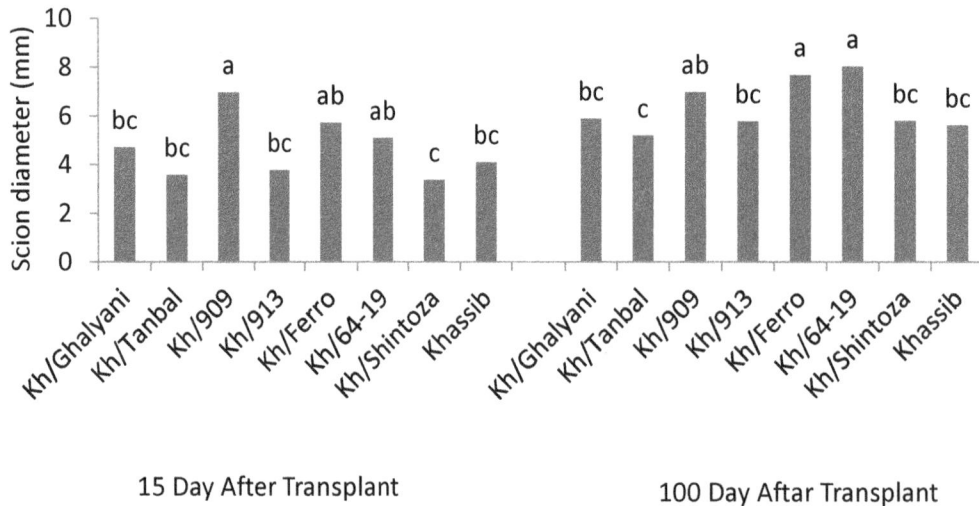

Figure 4. Scion diameter at two stages of plant growth (Kh = Khassib scion). Bars with different letters differ statistically at α = 0.05 (Duncan's Multiple Range Test).

At the end of the experiment, rootstocks significantly influenced plant growth (Table 1). Root dry biomass was the lowest with Ghalyani, reduced by about 80% compared to ungrafted plants. Although greater than Ghalyani, pumpkin and 913 hybrid rootstocks were significantly lower than the other rootstocks. Root length was only partially related to root biomass, highlighting a different ranking for root density among the rootstocks. In particular, 913 and Ferro hybrids had the highest root length but root biomass in 913 was about 55% lower than in Ferro.

Ferro rootstock was the only one with greater growth (+15%) than ungrafted plants (Table 1). The other rootstocks (except 909) produced a lower shoot dry biomass compared to the control, with the greatest reduction with Ghalyani (−60%). Concerning the Root/Shoot (R/S) ratio, Shintoza rootstock showed a higher value compared to the control, whereas the other treatments did not differ from ungrafted plants.

The Shintoza hybrid produced the tallest plant and the highest internode number (Table 1). All other treatments showed values similar or lower than the control. The leaf area showed a high variability among treatments with a lower value with Ghalyani compared to Tanbal, and all others were between these.

Table 1. Effects of rootstocks on cucumber plant growth (*cv.* Khassib).

Treatments	Root Dry Weight (g/plant)	Root Length (cm)	Shoot Dry Weight (g/plant)	R/S	Plant Height (cm)	Node Number	Leaf Area (cm^2/plant)
Kh/Ghalyani	0.52 c	37.5 c	6.4 e	0.11 c	60.7 f	18 d	480.8 b
Kh/Tanbal	1.58 b	52.0 ab	12.7 c	0.12 c	98.0 e	23 cd	904.4 a
Kh/909	2.16 a	48.0 abc	15.3 b	0.14 bc	146.5 cd	31 bc	634.1 ab
Kh/913	1.23 b	57.0 a	8.5 d	0.15 bc	95.0 e	16 d	558.5 ab
Kh/Ferro	2.73 a	59.0 a	18.7 a	0.15 bc	161.5 bc	32 b	612.9 ab
Kh/64-19	2.34 a	47.0 abc	11.9 c	0.20 ab	130.5 d	33 b	555.5 ab
Kh/Shintoza	2.35 a	41.0 bc	10.8 c	0.22 a	241.0 a	43 a	613.5 ab
Ungrafted	2.49 a	48.0 abc	15.9 b	0.16 bc	179.5 b	28 bc	734.2 ab
Significance [1]	***	*	***	*	***	***	*

[1],*,*** = significant. F value at ⩽0.05, 0.01 and 0.001, respectively. Means within columns followed by different letters differ statistically at α = 0.05 (Duncan's Multiple Range Test). Kh = Khassib scion.

3.2. Yield Parameters

Cucumber production was not improved by grafting (Table 2). The highest yield for rootstocks was obtained with Ferro, but it was no different than that for ungrafted plants. A significantly lower production than the control was observed with Ghalyani (−44%), 913 (−73%) and 64-19 (−35%) rootstocks, and these results were mainly due to fruit number. As for marketable fruit, the lowest value was from 913 with 68.7%.

Table 2. Effects of rootstocks on cucumber yield and quality.

Treatments	Yield (g/m^2)	Fruit Number (n/plant)	Marketable Fruit (%)	TSS (Brix)	Fruit Dry Weight (g)	Fruit Diameter (cm)	Fruit Shape (L/D) [2]
Kh/Ghalyani	2954 de	13.7 de	82.0 bc	3.3 cd	3.66 b	2.6 bc	4.2 a
Kh/Tanbal	4096 bc	16.0 cd	96.5 a	3.5 bc	4.98 a	3.1 a	3.2 c
Kh/909	4946 bc	23.7 abc	80.9 cd	3.9 b	4.23 ab	2.5 c	4.7 a
Kh/913	1384 e	5.25 e	68.7 d	3.2 cd	3.39 b	2.9 ab	3.7 ab
Kh/Ferro	6657 a	29.5 a	94.7 ab	3.2 cd	3.39 b	2.9 ab	3.4 bc
Kh/64-19	3407 cd	16.1 cd	91.8 abc	3.9 b	3.69 b	2.6 bc	3.1 c
Kh/Shintoza	4337 bc	20.0 bcd	80.0 cd	2.9 d	4.87 a	2.7 abc	3.5 bc
Ungrafted	5234 ab	27.0 ab	90.5 abc	4.8 a	5.15 a	2.4 c	3.7 ab
Significance [1]	***	***	**	***	**	**	**

[1],**,*** = significant. F value at ⩽0.05, 0.01 and 0.001, respectively. Means within columns followed by different letters differ statistically at α = 0.05 (Duncan's Multiple Range Test). Kh = Khassib scion; [2] L = fruit length, D = fruit diameter.

3.3. Fruit Quality Analysis

Total soluble solids (TSS) of fruit produced by the ungrafted plants were significantly higher than other treatments (Table 2). Fruit from Shintoza had the lowest TSS, 39% lower than the control. TSS content was not directly related to fruit dry weight. In fact, the highest value was observed with ungrafted plants and Shintoza and Tanbal rootstocks. The ungrafted plants had 34% more fruit dry matter than those grafted onto 913 rootstock. The use of different rootstocks led to changes in fruit size and shape. The highest length/diameter ratio was obtained with 909 and Ghalyani, whereas more stocky fruit were produced by plants grafted onto Tanbal and 64-19.

4. Discussion

In this experiment, plants with a thicker diameter stem reflected increased graft compatibility (Figures 2–4). There have been differing opinions on whether a difference in hypocotyl diameter between scion and rootstock affects graft incompatibility. Oda *et al.* [12] and Traka-Mavrona *et al.* [13] reported that smaller differences in hypocotyl diameters between the cucumber scion and squash rootstock may increase compatibility regardless of the quantity of vascular bundles. On the contrary, Edelstein *et al.* [6] did not observe any correlation between the difference of the scion and rootstock hypocotyl diameters or vascular bundles and the survival rate of grafted plants, concluding that the different results were attributed to the grafting techniques. A low survival rate in grafted plants can be due to two main characteristics: (1) the removal of the cotyledons from the rootstock; and (2) the limited number of the vascular bundles that connect the scion to the rootstock [12].

Since the root systems of selected rootstocks are usually much larger and more vigorous, they can absorb water and nutrients much more efficiently as compared to non-grafted plants [3]. However, in this research, we found a positive relation between root growth and biomass and fruit production only using Ferro as a rootstock. In the other scion/rootstock combinations, no clear relation was found between root vigor and scion response. In this regard, plant growth and production could be influenced by hormonal signals from the rootstock that could alter shoot physiology [5,14]. Cucurbit plant stems usually secrete xylem sap when decapitated, which is greatly influenced by the rootstock and contains high concentrations of minerals and plant hormones. This may explain the growth promotion of the cucumbers observed in our experiments when Cucurbita rootstocks were used.

The use of Ferro rootstock significantly increased the total fruit yield compared to the other rootstocks and did not differ from ungrafted plants. Therefore, this rootstock can be used in our conditions against some soil-borne pathogens typical of greenhouse conditions. As regards landraces, *Lagenaria siceraria* gave the lowest cucumber production. Plants grafted onto Tanbal, Ferro and 64-19, as well as ungrafted plants, showed a greater percentage of marketable fruit. In contrast, a higher marketable yield using grafted watermelon was obtained by Colla *et al.* [15] in Mediterranean climatic conditions.

The results of this research also showed that the TSS of fruit produced by ungrafted plants was significantly higher than in other treatments, as was previously found by López-Galarza *et al.* [16] in watermelon.

5. Conclusions

In conclusion, graft compatibility, measured in terms of plant survival after grafting, appeared to be related to the stem diameter of the rootstocks and, to a lesser extent, to scion diameter. Some *C. maxima* × *C. moschata* rootstocks had a similar production to ungrafted plants; thus, these rootstocks could be used for their resistance against soil-borne disease and not for their vigor. After the elimination of methyl bromide, grafting is one of the most effective alternatives worldwide to control root diseases of greenhouse vegetables. However, the high cost of grafted plants is the main problem for the widespread use of vegetable grafting. For these reasons, in developing countries vegetable grafting is still rare compared to the use of grafting in tree crops. The use of local landraces as rootstocks could help decrease the cost of grafted plants. However, our results using local landraces highlighted the importance of checking the grafting compatibility with the most important hybrid used as a scion under biotic and abiotic stresses.

Author Contributions: This work was a product of the combined effort of all authors. Ali Fharadi performed the experiments, gathered and analyzed the data, and wrote the manuscript with the help of Hossain Aroeii, Hossain Nemati and Reza Salehi. Francesco Giuffrida revised and improved the manuscript.

Conflict of Interest: The authors declare no conflict of interest.

References

1. FAO. Statistics at FAO. Available online: www.fao.org/statistics/en (accessed on 25 December 2015).

2. Kubota, C.; McClure, M.A.; Kokalis-Burelle, N.; Bausher, M.G.; Rosskopf, E.N. Vegetable grafting: History, use, and current technology status in north America. *HortScience* **2008**, *43*, 1664–1669.

3. Lee, J.M.; Kubota, C.; Tsao, S.J.; Bie, Z.; Echevarria, P.H.; Morra, L.; Oda, M. Current status of vegetable grafting: Diffusion, grafting techniques, automation. *Sci. Hort.* **2010**, *127*, 93–105. [CrossRef]

4. Savvas, D.; Colla, G.; Rouphael, Y.; Schwarz, D. Amelioration of heavy metal and nutrient stress in fruit vegetables by grafting. *Sci. Hort.* **2010**, *127*, 156–161. [CrossRef]

5. Schwarz, D.; Rouphael, Y.; Colla, G.; Venema, J.H. Grafting as a tool to improve tolerance of vegetables to abiotic stresses: Thermal stress, water stress and organic pollutants. *Sci. Hort.* **2010**, *127*, 162–171. [CrossRef]

6. Edelstein, M.; Burger, Y.; Horev, C.; Porat, A.; Meir, A.; Cohen, R. Assessing the effect of genetic and anatomic variation of *Cucurbita* rootstocks on vigour, survival and yield of grafted melons. *J. Hort. Sci. Biotechnol.* **2004**, *79*, 370–374.

7. Cohen, R.; Burger, Y.; Horev, C.; Koren, A.; Edelstein, M. Introducing grafted cucurbits to modern agriculture: The Israeli experience. *Plant Dis.* **2007**, *91*, 916–923. [CrossRef]

8. Davis, A.R.; Perkins-Veazie, P.; Hassell, R.; Levi, A.; King, S.R.; Zhang, X. Grafting effects on vegetable quality. *HortScience* **2008**, *43*, 1670–1672.

9. Rouphael, Y.; Schwarz, D.; Krumbein, A.; Colla, G. Impact of grafting on product quality of fruit vegetables. *Sci. Hort.* **2010**, *127*, 172–179. [CrossRef]

10. Lee, J.M. Cultivation of grafted vegetables: 1. Current status, grafting methods, and benefits. *HortScience* **1994**, *29*, 235–239.

11. Zhang, Z.; Liu, S.; Hao, S.; Liu, S. Grafting increases the copper tolerance of cucumber seedlings by improvement of polyamine contents and enhancement of antioxidant enzymes activity. *Agr. Sci. China* **2010**, *9*, 985–994. [CrossRef]

12. Oda, M.; Tsuji, K.; Sasaki, H. Effect of hypocotyl morphology on survival rate and growth of cucumber seedling grafted on *Cucurbita* spp. *Jap. Agr. Res. Quart.* **1993**, *26*, 259–263.

13. Traka-Mavrona, E.; Koutsika-Sotiriou, M.; Pritsa, T. Response of squash (*Cucurbita* spp.) as rootstock for melon (*Cucumis melo* L.). *Sci. Hort.* **2000**, *83*, 353–362. [CrossRef]

14. Pérez-Alfocea, F.; Albacete, A.; Ghanem, M.E.; Dodd, I.C. Hormonal regulation of source-sink relations to maintain crop productivity under salinity: A case study of root-to-shoot signaling in tomato. *Funct. Plant Biol.* **2010**, *7*, 592–603. [CrossRef]

15. Colla, G.; Rouphael, Y.; Cardarelli, M.; Salerno, A.; Rea, E. The effectiveness of grafting to improve alkalinity tolerance in watermelon. *Environ. Experim. Bot.* **2010**, *68*, 283–291. [CrossRef]

16. Lopez-Galarza, S.; San Bautista, A.; Perez, D.M.; Miguel, A.; Baixauli, C.; Pascual, B.; Maroto, J.V.; Guardiola, J.L. Effects of grafting and cytokinin-induced fruit setting on colour and sugar-content traits in glasshouse-grown triploid watermelon. *J. Hort. Sci. Biotechnol.* **2004**, *79*, 971–976.

2

Hydraulic Performance of Horticultural Substrates—2. Development of an Evaluation Framework

Uwe Schindler * and Lothar Müller

Leibniz Centre for Agricultural Landscape Research (ZALF), Institute of Landscape Hydrology,
Eberswalder St. 84, Muencheberg D15374, Germany; mueller@zalf.de
* Correspondence: uschindler@zalf.de

Abstract: Sustainable, environmentally friendly and resource-saving water and nutrient management in horticulture requires knowledge of the hydraulic suitability of horticultural substrates for each specific application. The aim of this study was to develop methods and a rating framework to assess the hydraulic performance of horticultural substrates. The hydraulic criteria related to high-quality horticultural substrates were defined as the amount of easily plant-available water (EAW) in the tension range between 10 and 100 hPa, the air capacity and the height of capillary rise. Limiting factors could be water repellency effects and shrinkage. The rating framework consisted of 5 classes between non-satisfactory and very good. The assessment of the hydraulic performance was split for the cultivation into 10-, 20-, and 30-cm-high containers. It was tested on 18 commercial substrates. More than 70% of the tested substrates revealed scores between very good and good. About 30% were evaluated as medium or satisfactory. The most critical aspect was the low air capacity in shallow containers. Shrinkage and water repellency sometimes strongly diminished the score. Both the measurement methods for quantifying substrate hydraulic properties and the evaluation procedure proved applicable. The impact of different ingredients and the composition of substrates of their hydraulic performance should be statistically analysed in further studies.

Keywords: horticultural substrates; hydraulic properties; water retention curve; unsaturated hydraulic conductivity; water repellency; water drop penetration time; shrinkage; extended evaporation method (EEM); HYPROP

1. Introduction

Horticultural substrates, also referred to as growing media, potting soils and gardening or soilless substrates, are widely used as a basis for vegetable and flower production in horticulture and private households, either in greenhouses or under field conditions. They are created as a composition of different ingredients. In most cases, bog peat, mainly consisting of sphagnum moss, is used as the basis for producing horticultural substrates [1,2].

Besides nutrient composition, the hydraulic performance of horticultural substrates is a key issue for evaluating its quality for horticultural purposes. The water and air capacity and the suitability for transporting water are important hydraulic quality indicators [1]. The basic properties are the water retention curve and the hydraulic conductivity function. Shrinkage and water repellency could have negative impacts on storing and transporting water and solutes [1,3]). Hydraulic data for only a few substrates have been reported [4–6]. In most cases the methods and devices used are outdated, the measurements have been time-consuming, the equipment is expensive and the results are subject to uncertainties [7]. Raviv and Lieth [1] concluded that there are a lack of technologies and methods for the effective physical characterization of substrates in horticulture. Many different horticultural substrates are available on the market. The declaration on the package generally provides information on the ingredients and the chemical composition. There is generally no information on physical properties such as the water and air capacity, water and solute transport, shrinkage, and rewettability. However, accurate hydraulic criteria, parameters and measurement data can improve the evaluation of the hydraulic performance of substrates in horticulture [1].

The aim of this study was the development of a rating framework to compare and to assess the hydraulic suitability of substrates used in horticulture.

2. Experimental Section

2.1. Hydraulic Quality Indicators

The hydraulic performance of substrates for plant production in horticulture depends on their capacity to store, transport and supply water, air and nutrients. The higher the water and air capacity and the transport properties of the substrate, the better is its performance for horticultural applications. The important hydraulic variables are the water retention curve and the hydraulic conductivity function [1]. Bulk density and sample preparation are key factors for hydraulic properties of substrates. The sample preparation method was derived from DIN EN 13041 [8] and [2] and guarantees a high reproducibility. It enables comparison of hydraulic properties of growing media even if the substrates are of different basic moistures.

The optimum water supply in horticulture depends on the crop and variety and is achieved in a tension range between 10 and 100 hPa [1,2]. A lower tension limit should guarantee sufficient air [1,9] to avoid any adverse impact on crop growth. According to Raviv and Lieth [1], the easily plant-available water capacity (EAW) should exceed 24% by vol. Higher water capacity is better, but a lower water capacity is defined as under-supplied. According to Raviv and Lieth [1], the air capacity should exceed 10% by vol. to avoid stress due to air limitations. Synonymous with air capacity are air space or air-filled porosity. All water- and air-related parameters in the paragraphs below are expressed as % by vol. The water and air capacities in containers depend on the water retention function and the height of the container. After free drainage, the air capacity is zero at the bottom. That means the substrate is saturated with water. The tension at the surface in hPa corresponds to the height of the container in cm. Heiskanen 1995 [10] Raviv & Lieth [1] concluded that the measurement of hydraulic parameters for describing and quantifying water and solute transport are heavily underrepresented in horticulture. Therefore, the capillary height calculated for a 5 mm·day^{-1} rate (CR$_5$) was used as an additional indicator for estimating the flow resistance and the hydraulic suitability and quality of the substrate reflecting the ease of exchange of water and nutrients in the growing layer. Due to the low transfer distances required, we defined the suitability of horticultural substrates as not limited for capillary heights greater than 30 cm. Limiting factors could be water repellency effects and shrinkage of the horticultural substrates. With reference to Blanco-Canqui and Lal [11] and Poulter et al. [12], the water drop penetration time (WDPT) should not exceed 5 s to avoid negative effects on water infiltration due to water repellency. Longer wetting times could be an indicator of rewetting limitations and preferential flow [13,14]. A shrinkage volume of 5% by vol. should not be exceeded to avoid adverse effects on plant growth and resource management.

2.2. Rating Framework

The hydraulic rating framework of horticultural substrates consisted of two parts: the rating of the basic soil hydrological properties (S_B, Table 1) and the rating of the limitations (S_L, Table 2). The easily plant-available water (EAW) and air capacity (Air) were assessed on a 5-point scale, and capillary rise (CR) on a 2-point scale. The water capacity depended on the storage volume and the thickness of the storage layer. The best score of the basic properties (EAW, Air, CR) was 5, and the poorest rating was 1. The limitations (WDPT and soil shrinkage) were assessed in a 3-point scale. The score of the limitations (S_L) ranged between 0 (no limitation) and 2 (strong limitation). Only the score of the most severe limitation was considered. In the case of a score of 2 for strong rewetting limitation (S_L$_{WDPT}$) and a score of 1 for shrinkage (S_L$_{S100}$), the value 2 had to be considered. The total rating (S_T, Table 3) for the evaluation of the hydraulic performance of the substrates was calculated as the sum of the basic score S_B minus the score of the dominating limitation (S_L). The highest score was 12, the lowest was 1. The hydraulic rating in this study was calculated for horticultural substrates in 10-, 20- and 30-cm high containers (P10, P20 and P30, respectively).

Table 1. Rating scale of basic hydraulic requirements (Basic score (S_B)).

Parameter	Basic Score (S_B)				
	1	2	3	4	5
EAW [1] (% by vol.)	<12	12–16	16–20	20–24	≥24
Air capacity (% by vol.)	<2	2–5	5–8	8–10	≥10
CR$_5$ [2] (cm)	<30	≥30			

[1] Easily plant-available water; [2] CR$_5$—capillary height for a 5 mm·day^{-1} rate.

Table 2. Rating scale of hydraulic quality limitation scores (S_L).

Parameter	Limitation Score (S_L)		
	0	1	2
WDPT (in seconds) [1]	≤5	5–15	>15
S$_{100}$ [2] (% by vol.)	≤2	2–5	>5

[1] WDPT—water drop penetration time at 100 hPa; [2] Shrinkage at 100 hPa.

Table 3. Total rating (S_T) of the hydraulic performance of substrates in horticulture.

Evaluation	Total Score (S_T)
Very good	≥10
Good	<10–>8
Medium	<8–>6
Satisfactory	<6–>4
Non-satisfactory	≤4

2.3. Substrate Samples and Hydraulic Measurements

As examples, the hydraulic quality indicators were determined for 18 commercial horticultural substrates (Table 4).

Table 4. Collection of horticultural substrates (HS).

HS No.	Ingredients, Texture Class	Ash Content (%)
1	Hh (H3–H8), R, G	68.1
2	Hh (H2–H4) H7–H9, G, R	16.8
3	Hh (H2–H5), P, C, K	35.1
4	Hh (H3–H8), R, G, P, C, K	21.4
5	95% Hh (H3–H7), P, Co	24.4
6	R, C, Co, Guano	25.3
7	90% Hh (H4–H8), 10% C, K	41.0
8	Hh (H3–H8), C, P	35.9
9	75% Hh (H3–H5 and H6–H7), Co, C, K	25.1
10	80% Hh (H3–H5 and H6–H7), Co, C	39.9
11	Hh (H2–H5), G, R, P, C, K	48.3
12	Hh (H3–H8), G, R, P, C, C	42.7
13	Hh (H3–H5 and H7–H9)	15.1
14	Hh (H2–H5), G, R, K	35.8
15	Hh (H3–H8), G, P, K	10.8
16	60% Hh (H3–H5 and H6–H7), R, G, Co, K	25.5
17	60% Hh (H3–H5 and H6–H7), Co, C, P	42.8
18	50% Hh (H3–H5), G, R, C	36.2

Abbreviations: Hh—bog peat; H—degree of decomposition; R—Compost of forest residuals; G—Compost of garden residuals; Co—Coir (raw coconut fibre); P—perlite; K—lime; C—clay; S—sand.

The substrates mainly consisted of 30% to 100% bog peat (degree of decomposition between H3 and H7, von Post, 1922 [15], and different portions of organic residuals (garden (G) and forest (F) compost), coir (Co) and mineral additives such as perlite (P), lime (K), clay (C) and sand (S). One of them (No. 6) was totally free of peat.

The water retention curve and the unsaturated hydraulic conductivity function were measured with the extended evaporation method (EEM) and the HYPROP system from saturation to close to the wilting point [7,16,17]. These functions were used for calculating the water (EAW) and air (Air) capacity and for quantifying the capillary rise (CR). The WDPT and shrinkage (S_{100}) were measured during the evaporation experiment at a tension of about 100 hPa [8].

3. Results and Discussion

According to Raviv and Lieth [1], the maximum tension for avoiding water stress was assumed to be 100 hPa. The height of the container markedly influenced the water and air capacity. The range of the easily plant-available water (EAW) and, especially, the air capacity for plant cultivation in the containers of varying heights differed strongly between the substrates. The EAW was evaluated as sufficient for most (more than 90%) substrates in the different containers, with the exception of HS6 which, at 10.7%, revealed an above-average low water storage capacity in P30. However, contrary results relevant to plant stress were measured for the air capacity. In P10, only HS 6 achieved and clearly exceeded the threshold value of 10%. Plants in all other substrates were strongly undersupplied with air. In P20 about half of the substrates revealed no air limitation, and in P30 only HS 13 was undersupplied with air. The capillary water transport was evaluated as sufficient for about half of the substrates. Only the substrates HS 1, 11 and 16 showed a considerable reduction of capillarity. The hydraulic performance (Table 5) of 11 substrates was negatively affected to different extents by shrinkage during drying, and only 3 samples showed an occurrence of water repellency.

Table 5. Hydraulic properties of the horticultural substrates.

HS No.	Θs	Air			EAW			CR_5	WDPT	S_{100}	DBD
		P10	P20	P30	P10	P20	P30				
		% by vol.						cm	sec	% by vol.	g·cm^{-3}
1	71.8	2.5	9.3	17.5	35.1	28.2	20.1	10.1	0.1	2.1	0.43
2	86.0	1.4	6.5	14.2	41.8	36.6	29.0	45.7	1	6.2	0.17
3	79.6	2.3	9.4	17.6	35.0	27.9	19.7	54.7	0.1	5.4	0.30
4	79.0	6.3	11.8	19.4	32.3	26.7	19.2	26.7	15	3.3	0.26
5	86.2	2.3	11.7	21.8	43.8	34.4	24.3	42.9	2	2.1	0.20
6	87.1	20.1	31.7	47.8	38.3	26.7	10.7	17.9	0.1	0.4	0.18
7	84.2	0.7	4.5	10.4	34.8	31.1	25.1	24.4	1	9.1	0.21
8	81.2	1.5	5.7	11.8	33.4	29.1	23.0	45.7	0.1	6.6	0.25
9	80.7	6.4	13.9	23.8	41.0	33.5	23.6	29.3	0.1	0.8	0.22
10	84.4	6.1	13.6	23.3	38.9	31.4	21.7	29.3	0.1	6.2	0.18
11	83.1	1.9	7.2	13.4	31.8	26.5	20.2	13.1	2	6.2	0.31
12	75.8	0.8	6.8	14.4	38.7	32.8	25.2	26.7	6	1.0	0.30
13	84.5	0.6	2.7	7.5	34.6	32.5	27.7	47.7	1	0.6	0.21
14	78.8	3.9	11.6	20.2	36.9	29.2	20.6	15.9	6	0.8	0.28
15	83.4	1.3	6.9	14.0	35.2	29.6	22.4	36.4	2	7.0	0.19
16	81.1	6.1	13.7	25.1	40.1	31.5	21.1	12.7	1	3.3	0.19
17	80.8	5.5	13.8	23.8	40.4	32.1	22.1	29.3	0.1	6.2	0.23
18	81.0	2.6	7.6	14.5	35.6	30.6	23.7	79.9	2	7.0	0.26

Abbreviations: Θs—saturated water content; Air—air capacity in 10, 20 and 30 cm high containers P; EAW—easily plant-available water; CR_5—steady-state capillary height for a 5 mm·day^{-1} rate; S_{100}—shrinkage at 100 hPa; WDPT—water drop penetration at 100 hPa; DBD—Dry bulk density.

The evaluation of the hydraulic performance of the substrates was split for cultivation in different high containers, as described in Tables 1–3 (Table 6). More than 70% of the tested substrates revealed scores between very good and good. About 30% of the substrates were evaluated as medium or satisfactory. The average score in P10 was 8.1 and was 9.0 in P20 and P30. However, there was a great variation between the substrates in the special container and between the containers of different heights. The score ranged from 12 (very good) for HS 5 in P20 and P30 to 5 (satisfactory) for HS 7 and 11 in P10. About 70% of the substrates revealed scores between good and very good. The most critical factor was the low air capacity in shallow containers. However, due to strong shrinkage or water repellency during drying, the score sometimes changed to medium or satisfactory. These processes could lead to reduced water and nutrient supply and solute leaching [11]. The addition of perlite and coir induced positive effects on the hydraulic performance. High amounts of bog peat tended to encourage shrinkage.

Table 6. Rating of the hydraulic performance of the horticultural substrates.

HS No.	Score_Basic Requirement S_B						CR_5	Score_S_L		Score S_T		
	Air			EAW				WDPT	S_{100}	P10	P20	P30
	P10	P20	P30	P10	P20	P30						
1	2	4	5	5	5	4	1	0	0	8	10	10
2	1	3	5	5	5	5	2	0	2	6	8	10
3	2	4	5	5	5	3	2	0	2	7	9	8
4	3	5	5	5	5	3	1	2	1	7	9	7
5	2	5	5	5	5	5	2	0	0	9	12	12
6	5	5	5	5	5	1	1	0	0	11	11	7
7	1	2	5	5	5	5	1	0	2	5	6	9
8	1	3	5	5	5	4	2	0	2	6	8	9
9	3	5	5	5	5	4	1	0	0	9	11	10
10	3	5	5	5	5	4	1	0	2	7	9	8
11	1	3	5	5	5	4	1	0	2	5	7	8
12	1	3	5	5	5	5	1	1	0	6	8	10
13	1	2	3	5	5	5	2	0	0	8	9	10
14	2	5	5	5	5	4	1	1	0	7	10	9
15	1	3	5	5	5	4	2	0	2	6	8	9
16	3	5	5	5	5	4	1	0	1	8	10	9
17	3	5	5	5	5	4	1	0	2	7	9	8
18	2	3	5	5	5	4	2	0	2	7	8	9

4. Conclusions

The hydraulic measurement techniques and methods (EEM, HYPROP, WDPT) used here proved to be suitable for characterizing the hydraulic properties of horticultural substrates as a basis for evaluating their hydraulic applicability for horticultural use. The appropriateness of the methods and the devices were tested and confirmed for 18 commercial horticultural substrates. The proposed evaluation framework provides an opportunity to compare the hydraulic properties of the substrates taking into account threshold values for the plant water and air supply.

Moreover, the gas diffusivity, respiration rate and hysteresis of the hydraulic functions should be taken into account for a better understanding of plant growth in horticulture [1,9]. The latter is of special importance in the low-tension range, because the drainage and the rewetting curves of the hydraulic functions could lead to different results and evaluations [17].

Irrigation of shallow containers should be carried out in a very sensitive manner. Water loss at the bottom of the container induces saturation. This behaviour is associated with lack of air in the container and could lead to growing stress and decreasing yields.

This study provides initial information about the hydraulic properties and the performance of some commercial substrates for horticultural applications. This kind of information is intended to expand our knowledge of substrate hydraulic properties and help evaluate substrate suitability in horticulture. Until recently, substrate buyers were not able to draw any conclusions regarding hydraulic properties on the basis of the declaration and the ingredients of the particular product they had bought. Therefore, statistical analyses of a large number of substrates with different amounts of bog-peat as well as those without peat involving various ingredients should provide real help. The proposed rating framework could be the basis of this evaluation.

Author Contributions: Uwe Schindler was responsible for the soil hydraulic measurements. Lothar Müller and Uwe Schindler developed the rating framework.

Conflicts of Interest: The authors declare no conflict of interest.

References

1. Raviv, M.; Lieth, J.H. *Soilless. Culture*; Elsevier Publications: London, UK, 2008; p. 608.

2. Verdonck, O.; Gabriels, R.I. Reference method for the determination of physical properties of plant substrates. II. Reference method for the determination of chemical properties of plant substrates. *Acta. Hortic.* **1992**, *302*, 169–179. [CrossRef]

3. Dekker, L.W.; Ritsema, C.J.; Oostindie, K.; Boersma, O.H. Effect of drying temperature on the severity of soil water repellency. *Soil Sci.* **1998**, *163*, 780–796. [CrossRef]

4. Fonteno, C.F. Problems & considerations in determining physical properties of horticultural substrates. *Acta Hortic.* **1993**, *342*, 197–204.

5. Bougoul, S.; Ruy, S.; de Groot, F.; Boulard, T. Hydraulic and physical properties of stonewool substrates in horticulture. *Sci. Hortic.* **2005**, *104*, 391–405. [CrossRef]

6. Paraskevi, A.L.; Psychoyou, M.; Valiantzas, J.D. Evaluation of substrate hydraulic properties amended by urea-formaldehyde resin foam. *Hortscience* **2012**, *47*, 1375–1381.

7. Schindler, U.; Durner, W.; Von Unold, G.; Mueller, L.; Wieland, R. The evaporation method—Extending the measurement range of soil hydraulic properties using the air-entry pressure of the ceramic cup. *J. Plant Nutr. Soil Sci.* **2010**, *173*, 563–572. [CrossRef]

8. DIN EN 13041. *Bodenverbesserungsmittel und Kultursubstrate—Bestimmung der Physikalischen Eigenschaften— Rohdichte (Trocken), Luftkapazität, Wasserkapazität, Schrumpfungswert, und Gesamtporenvolumen*; Beuth Verlag GmbH: Berlin, Germany, 2012.

9. Caron, J.; Pepin, S.; Periard, Y. Physics of growing media in green future. *Acta Hortic.* **2014**, *1034*, 309–317. [CrossRef]

10. Heiskanen, J. Physical properties of two-component growth media based on Sphagnum peat and their implications for plant-available water and aeration. *Plant Soil* **1995**, *172*, 45–54. [CrossRef]

11. Blanco-Canqui, H.; Lal, R. Extent of soil water repellency under long-term no-till soils. *Geoderma* **2009**, *149*, 171–180. [CrossRef]

12. Poulter, R.; Duff, A.A.; Bauer, B. Quantifying Surfactant Interaction Effects on Soil Moisture and Turf Quality. Horticulture Australia Ltd., 2009. Available online: http://www.deedi.qld.gov.au/ (accessed on 14 August 2015).

13. Bauters, T.W.J.; Steenhuis, T.S.; DiCarlo, D.A.; Nieber, J.L.; Dekker, L.W.; Ritsema, C.J.; Parlangea, J.-Y.; Haverkampe, R. Physics of water repellent soils. *J. Hydrol.* **2000**, *231*, 233–243. [CrossRef]

14. Ritsema, C.J.; Dekker, L.W. Water repellency and its role in forming preferred flow paths in soils. *Aust. J. Soil Res.* **1996**, *34*, 475–487. [CrossRef]

15. Von Post, L. Sveriges Geologiska Undersöknings torvinventering och nogra av des hittils vunna resultat (SGU peat inventory and some preliminary results). *Svenska Mosskulturforeningens Tidskrift Jonköping* **1922**, *36*, 1–37.

16. Schindler, U.; Doerner, J.; Müller, L. Simplified method for quantifying the hydraulic properties of shrinking soils. *J. Plant Nutr. Soil Sci.* **2015**, *178*, 36–145. [CrossRef]

17. Schindler, U.; Von Unold, G.; Müller, L. Laboratory measurement of soil hydraulic functions in a cycle of drying and rewetting. *Int. J. Emerg. Technol. Adv. Eng.* **2015**, *5*, 281–286.

3

Identification of Differentially Expressed Genes between "Honeycrisp" and "Golden Delicious" Apple Fruit Tissues Reveal Candidates for Crop Improvement

Scott Schaeffer [1,2,†], Christopher Hendrickson [2,‡], Rachel Fox [2] and Amit Dhingra [1,2,*]

[1] Molecular Plant Science Program, Washington State University, Pullman, WA 99164, USA; scott.schaeffer@bcm.edu

[2] Department of Horticulture, Washington State University, Pullman, WA 99164, USA; chendrickson@nu.edu (C.H.); rldaniel4@gmail.com (R.F.)

* Correspondence: adhingra@wsu.edu

† Current address: Department of Agriculture-Agriculture Research Service, Children's Nutrition Research Center, Baylor College of Medicine, Houston, TX 77030, USA.

‡ Current address: Department of Mathematics and Natural Sciences, National University, La Jolla, CA 92037, USA.

Abstract: Cultivars of the same species exhibit a large degree of variation in fruit quality traits, which can be directly influenced by differences in gene expression due to allelic variations and interactions with the environment. For *Malus × domestica* Borkh. (apple), fruit quality traits, including color, texture, aroma, flavor profile, and shelf life, are of utmost economic importance. In order to identify genes potentially influencing these traits, a direct comparative transcriptome profiling approach, based on the differential display technique, was performed using "Golden Delicious" and "Honeycrisp" apple endocarp and peel tissues. A total of 45 differentially expressed sequence tags were identified between the two apple varieties. Reanalysis of a previously published fruit developmental microarray expression experiment revealed that only one of the 45 sequence tags was represented on the array. Differential expression of 31 sequence tags from the peel tissue was validated using quantitative reverse transcription PCR, confirming the robustness of the differential display approach to quickly identify differentially expressed sequence tags. Among these were genes annotated to be involved in ripening, phytohormone signaling, transcription factors, and fruit texture. This work demonstrates yet again the utility of the differential display technique to rapidly identify genes related to desirable traits.

Keywords: differential display; differential expression; fruit quality; *Malus × domestica* Borkh

1. Introduction

Fruit quality traits vary across different cultivars of the same species during development, storage, and/or transit to market. In pome fruit, these traits can be broadly classified as peel or endocarp-specific. Peel-specific traits include fruit color, photo-protective capacity, scalding susceptibility, lenticel injury, aroma, etc. Endocarp-specific traits include propensity for rotting and mold development, hypoxic injury, flesh-sweetness, acidity, and textural qualities. Adding further complexity, these traits are under polygenic control and can vary by season, soil conditions and elevation [1–3].

Development of new tree fruit cultivars remains a laborious and time-consuming process. Juvenility periods and a perennial fruit-bearing nature can require at least five years to study the effect of genes in progeny of a controlled cross. Additionally, desirable traits in tree fruits may be highly dependent on genetic and environmental interactions during the course of fruit development and maturation, which can complicate attempts at performing uniform phenotyping. As a result, identification of causal genetic mechanisms underlying traits of interest in tree fruits can be difficult. Additionally, genomic structure of such species can compound challenges. Theorized duplication events have yielded allotetraploid genome structures in several perennial tree fruits including apple and pear [4,5]. This can result in numerous functionally redundant or interrelated families of genes involved in controlling traits of interest in tree fruit. To facilitate crop improvement in such systems, desirable genes must first be functionally validated prior to them being selectively introgressed into future genetic selections.

Variation in fruit quality traits can be influenced by numerous genetic factors including, but not limited to, allelic variation, gene copy number, and transcript variants. Differential gene and allelic expression may also result in aberrant folding or post-translational modification of proteins within the members of the same species [6,7]. While the causes of the phenotypic differences can vary, approaches used to identify causal genetic elements have been largely uniform. To identify the genetic elements, researchers rely upon segregation probabilities of DNA polymorphisms in populations segregating for the desired traits with the use of molecular markers such as RFLPs (Restriction Fragment Length Polymorphism), SNPs (Single Nucleotide Polymorphism), and microsatellites [8,9]. Environmental cues, polygenic control of quantitative traits, and allelic expression differences can complicate the use of molecular markers in identification of underlying genetic elements in an apple. For these reasons, the exclusive use of marker-based approaches to identify underpinnings of desirable fruit phenotypes can be challenging. While not feasible for all traits, a gene-linked mechanistic understanding of desirable fruit traits wherever possible may provide a reliable and reproducible resource for crop improvement. The Malus research community has developed and deployed molecular markers from densely populated maps of all *Malus* × *domestica* Borkh. linkage groups. Fruit firmness in apple was reported to be linked with the allelic composition of aminocyclopropane-1-carboxylic acid synthase (ACS) and oxidase (ACO) genes that regulate ethylene biosynthesis in climacteric fruit [10,11]. Zhu and Barrett [12] applied this information to nearly 100 apple lines and concluded that those homozygous for ACS1-2 and ACO1-1 had firmer fruit both at harvest and after typical commercial postharvest storage, compared to other genotypes. Additional correlations have been made between a Malus SSR marker and resistance to *Erwinia amylovra* (fireblight) [13]. These signatures of economically-important traits can be used to develop improved germplasm for the commercial market.

Recently, a wealth of additional tree fruit genomic resources have been established, with the release of the genomes from the "Golden Delicious" apple, sweet orange, peach, and Chinese white pear, as well as the recently announced European pear genomes [4,5,14–16]. These resources are enabling the understanding of genetic causes underlying numerous traits in tree fruit such as resistance to pathogens, fruit sweetness, acidity, texture, peel and flesh coloring, and responses to numerous postharvest storage regimes.

One of the less used approaches to identify the genetic basis for phenotypic differences between two varieties is to perform a direct comparison of gene expression in a spatio-temporal context. One such small scale approach for comparative transcriptome profiling is the differential display reverse transcriptase polymerase chain reaction, or DD [17]. The DD technique can amplify all poly-adenylated transcripts allowing for direct capture of actively expressed genes and alleles in a comparative context. The resulting gene fragment amplicons can be isolated based upon either their presence or absence, or based on differential intensity on a polyacrylamide gel that is indicative of differential expression. Identified fragments are cloned, sequenced, and annotated to infer their biological function. This approach has been utilized in many plant systems to identify putative targets involved in many traits or processes including reproductive roles in semigametic Pima cotton (*Gossypium barbadense*), heat responsive genes in peach (*Prunus persica*), hormone and abiotic stress changes in the gene expression of tea (*Camelina sinensis*), and ripening-related genes in banana

(*Musa acuminata* AAA) [18–21]. In each of these instances, DD aided in direct identification of genes potentially involved in the regulation of the phenotypes being studied.

Despite the labor and time-related challenges associated with performing a DD approach, the technique retains some significant advantages over newer next-generation sequencing based approaches. Differential display offers an inexpensive means of identifying candidate genes without the complexities associated with RNA sequencing (RNA-seq) based approaches. The presence or absence of a PCR amplicon in a DD polyacrylamide gel provides a positive result that does not require a reference genome and advanced computational resources associated with comparative RNA-seq workflows. Further, DD can aid in the identification of potentially important regulatory genes between samples which can be overlooked using microarray-based comparisons. The complex data capture, assembly and statistical analysis associated with microarray and next-generation sequencing approaches can result in missing or misidentifying the genes that control traits of interest. Caution is prescribed in carefully examining novel transcripts identified via RNA-seq before proceeding to biological experiments.

Available apple germplasm abounds in diversity of fruit quality traits. The apple cultivar "Honeycrisp" (HC) falls into a category of "JFC high quality" indicating a juicy and crispy-textured flesh, while "Golden Delicious" (GD) is classified as "American/European dessert". HC is typified by a color profile which is 60% orange/red on top of a yellow base color, large size, as well as having an exceptional texture and juiciness [22]. GD, on the other hand, has a completely yellow color and texture that is juicy, firm and crisp. HC apples maintain a crisp texture up to 6 months after harvest even without controlled atmosphere conditions due to maintenance of cell wall integrity and cellular turgor potential [23]. These two apple varieties are popular among consumers, but exhibit highly diverse fruit quality characteristics through ontogeny and ripening. In this study, we investigated the differences in the expression of peel and flesh specific genes of GD and HC apples at two different developmental stages. Results obtained from this work identified several important genes related to fruit quality and demonstrate the utility of an inexpensive method for direct identification of candidate genes that could be utilized for development of improved apple varieties.

2. Materials and Methods

2.1. Sampling Strategy

HC and GD apples were harvested during two growing seasons at 125 or 129 and 150 or 160 days after anthesis (DAA), respectively. The time points sampled in this study were chosen due to their correlation with early and late stages of the ripening process in an apple, which typically begins around 95 days after anthesis (DAA) and concludes at around 150 DAA [1]. Additionally, these samples represent apple fruit at pre-climacteric and climacteric stages of maturity [2]. Tissue processing was performed directly in the orchard where the peel and endocarp samples were obtained from nine apples derived from three trees of the same variety. Samples were immediately frozen in liquid nitrogen. Tissues were stored at −80 °C and subsequently ground for three cycles of four minutes at top speed using a SPEX SamplePrep 6870 Freezer mill (Metuchen, NJ, USA). Total RNA was extracted from ground tissue using Plant RNeasy Minikit according to the manufacturer's instructions (Qiagen, Valencia, CA, USA). RNA was quantified using a Nanodrop 8000 spectrophotometer (Wilmington, DE, USA) and its integrity was confirmed with a denaturing formaldehyde MOPS gel.

2.2. Differential Display

Differential displays were performed on HC endocarp, HC peel, GC endocarp and GC peel samples. Removal of DNA contamination from RNA was performed with the MessageClean Kit (GenHunter, Nashville, TN, USA) according to manufacturer's instructions. Synthesis of cDNA and PCR amplification was performed using the GenHunter RNAimage Kit (GenHunter) according to the manufacturer's instructions. Primers 193 to 200 (GenHunter) in combination with the anchoring poly-T, H-T$_{11}$A, were used for PCR amplification. ^{33}P-labeled cDNA fragments were separated on a 6% acrylamide denaturing gel. After imaging through radiography, bands were excised if differential expression was observed between HC and GD tissues.

2.3. Cloning and Sequencing

Excised bands were boiled for 10 min with 40 μL of deionized water in a 1.5 mL microcentrifuge tube. Re-amplification was performed using 1 μL of the recovered DNA as a template and the differential display primers used for their initial amplification. PCR amplicons were cleaned using the QIAquick PCR Purification Kit (Qiagen) and subsequently ligated into the pGEM-T Easy vector (Promega, Madison, WI, USA) according to the manufacturer's protocol. Ligated fragments were then transformed into *Escherichia coli* XL1-Blue cells via the heat-shock method. Colony PCR was used to confirm the presence of an insert using M13 forward and reverse primers. Plasmids were sequenced using Sanger sequencing with M13 forward primers at Eurofins MWG Operon. Gene IDs as predicted from the apple genome of the sequenced fragments are provided in Table S1.

2.4. Annotation of Differential Display Fragments

BLASTN was performed for all sequenced differential display fragments against the predicted gene sequences from the apple genome release 1.0. As differential display sequences were derived from the 3' end of transcripts and could contain untranslated regions, a second BLASTN was performed against the NCBI EST database for *Rosaceae* sequences (NCBI taxid: 3745). Matching EST sequences were subsequently mapped to full length apple gene sequences with BLASTN which were extracted for subsequent analysis. All full length apple gene sequences were then analyzed by Blast2GO against the NCBI nr database to determine putative functions [3,4]. Localization of each protein sequence was performed with TargetP using plant networks with default parameters [5]. A final BLASTX step was performed against the NCBI protein collections from *Arabidopsis thaliana* and *Solanum lycopersicum* to identify homologs in these systems, with the top result from each species with an E-value of at least 1E^{-20} listed. The description, if more detailed than the original Blast2GO hit, was also retained.

2.5. Quantitative Real Time PCR Validation of Differential Expression

RNA from HC and GD peel tissue collected over two growing seasons was extracted as described above. Validation of differential expression using qPCR was performed with peel tissues. Samples were DNAse-treated using the DNA-free kit (Ambion) according to the manufacturer's instructions. RNA was quantified using Nanodrop 8000 and its integrity was tested on a 1.0% agarose gel. First-strand cDNA was synthesized using the SuperScript VILO cDNA Synthesis Kit (Life Technologies, Carlsbad, CA, USA) following manufacturer's instructions. Concentration was diluted to 50 ng/μL with deionized water. Validation of differential expression was performed for 31 apple genes: MDP0000200646, MDP0000037251, MDP0000875654, MDP0000296716, MDP0000128468, MDP0000712586, MDP0000618650, MDP0000920333, MDP0000152947, MDP0000883284, MDP0000253074, MDP0000213808, MDP0000547450, MDP0000310811, MDP0000232309, MDP0000161275, MDP0000176723, MDP0000304285, MDP0000172863, MDP0000180389, MDP0000523205, MDP0000138340, MDP0000166116, MDP0000220601, MDP0000237908, MDP0000273484, MDP0000286959, MDP0000292888, MDP0000320533, MDP00005772242, and MDP0000697474. Primers for PCR were designed with the assistance of custom scripts based on predicted gene sequences from the apple genome project. These primers were then used in quantitative reverse-transcription PCR validation of DD results using iTaq Universal SYBR Green Supermix (Bio-Rad Laboratories, Hercules, CA, USA) on a Stratagene MX3005P light cycler (Agilent, Santa Clara, CA, USA). All reactions were performed using 50 ng cDNA and carried out in triplicate along with a β-tubulin internal control. β-Tubulin was selected as it has been previously used as an internal control in several prior apple gene expression studies [6–8]. Efficiencies and Cq values were determined using the tool LinRegPCR [9]. Any reactions with an efficiency less than 1.8 or more than 2.2 were discarded from further analysis. Cq values were compared between the three replicates and, in the case that the standard deviation exceeded one, one of the three outliers was omitted from the analysis. The Pfaffl correction was then applied to Cq values of remaining reactions. Fold change was determined by comparing the highest Cq value to a β-tubulin internal control [10]. To validate the identity of the genes, 10 random qPCR products were sequenced. These included MDP0000312808, MDP0000232309, MDP00000320533, MDP0000547450, MDP0000697474, MDP0000166116, MDP0000253074, MDP0000166116, MDP0000253074, MDP0000255887, MDP0000200646, and MDP0000920333. Products were sequenced using the forward and reverse primers that were originally used to amplify them. Sequences were assembled using Lasergene 11 SeqMan Pro (DNASTAR, Madison, WI, USA) and compared to the full length apple gene using a pairwise BLAST.

2.6. Comparison of Expression with Previous Studies

Previously generated "Royal Gala" apple microarray expression data published in 2008 [1] was reanalyzed to identify data from genes recognized in the differential display experiment. All ESTs represented in the experiment were compared to the predicted apple gene sequences generated from the apple genome with BLASTN. The top hits with an e-value lower than $1E^{-10}$ were then assigned the apple gene identifiers. These identifiers were then compared with the genes represented in the differential display experiment.

3. Results and Discussion

3.1. Differential Display

Apple ripening is characterized by the conversion of starches to simple sugars, decrease of total chlorophyll and photosynthetic activity, increase in total carotenoids, flesh softening and cell wall modification as well as accumulation of volatiles and flavor compounds [11–13]. Any differences in these properties would likely be mirrored in differences in gene expression. Differential display analysis of the peel tissues from two developmental stages of GD and HC tissues in apple, generated 115 bands corresponding to potentially differentially expressed genes. Of these, 105 fragments could be isolated which were then cloned into a plasmid and sequenced. The remaining ten could not be re-amplified after extraction from the PAGE (Polyacrylamide Gel Electrophoresis). BLASTN was performed for each sequence against the predicted gene set from the apple genome as well as the NCBI EST sequences available for *Rosaceae* (NCBI taxid: 3745). The apple genome and *Rosaceae* EST sequences similar to the 105 query sequences were extracted and processed through BLAST2GO workflow and the output is available in Table S2. For each matching EST sequence, a final BLAST was performed against the apple predicted gene set. This analysis resulted in the identification of 42 sequences from the differential display experiment, which had significant identity with a full-length apple gene. Three of these sequences, however, had different apple gene matches between the two BLAST methods. In order to account for this discrepancy, all predicted matches were included resulting in 45 sequences which were chosen for further characterization and validation.

Blast2GO analysis was performed on matching full length apple sequences to predict potential functions associated with the sequences, while TargetP was used to predict the cellular localization of each protein (Table 1) [3–5,14]. Of these genes, seven are predicted to be secreted from the cell, while fewer are predicted to be localized to the mitochondria and the chloroplast. The most represented molecular functions in this group include transporter activity (GO:0005215, four sequences), sequence-specific DNA binding transcription factor activity (GO:0003700, three sequences), and protein binding (GO:0005515, three sequences). Relatively more information was available for the biological processes in which these proteins may participate, which include response to stress (GO:0006950, 11 sequences), response to biotic stimulus (GO:0009607, nine sequences), transport (GO:0006810, eight sequences), and catabolic process (GO:0009056, seven sequences). Molecular function and biological process gene ontologies for these genes are summarized in Figure 1. A final BLASTX was performed against the NCBI protein collections from *Arabidopsis thaliana* and *Solanum lycopersicum* to identify the homologs in these systems (Table 2).

Table 1. Identification of fragments isolated from an initial differential display procedure. Differential expression was assessed using "Honeycrisp" and "Golden Delicious" apple fruit peel and endocarp (HCP, HCE, GDP, and GCE, respectively) collected at 129 and 160 days after anthesis (DAA) in the 2007 growing season. Bands from differential display gels were excised, cloned and sequenced. Qualitative abundance was determined for each tissue and noted as either absent (−) present (+) or overexpressed (++). Sequences were matched to predicted apple genes via BLAST and analyzed for predicted localization using TargetP (M—mitochondria, C—chloroplast, S—secreted, _—no localization).

Gene Identifier	Blast2GO Annotation	129 DAA-2007 Season				160 DAA-2007 Season				Cellular Localization
		GDE	GDP	HCE	HCP	GDE	GDP	HCE	HCP	
MDP0000037251	cinnamyl alcohol dehydrogenase-like protein	++	++	+	++	++	++	+	++	−
MDP0000128468	abscisic acid stress ripening protein homolog	++	++	++	+	++	++	++	+	M
MDP0000129664	3-ketoacyl-thiolase	−	−	+	+	−	−	+	+	−
MDP0000138340	NAC domain ipr003441	++	++	+	+	++	++	+	+	−
MDP0000152947	wound-induced protein	+	++	+	+	+	+	+	+	−
MDP0000161275	mitochondrial substrate carrier family protein	+	+	+	++	+	+	+	++	−
MDP0000166116	acyl:CoA ligase acetate-coa synthetase-like protein	+	+	+	+	++	++	+++	+++	S
MDP0000172863	protein	−	−	−	−	+	+	−	−	S
MDP0000176723	acyl:CoA ligase acetate-coa synthetase-like protein	++	++	+	+	++	++	+	+	S
MDP0000180389	disease resistance protein at3g14460-like	++	++	++	+	++	++	+	+	S
MDP0000200646	NAC domain ipr003441	+	+	+	+	+	+	++	++	−
MDP0000213808	probable ubiquitin conjugation factor e4-like	++	++	+	+	++	++	+	+	−
MDP0000220601	zinc finger CCCH domain-containing protein 53-like	+	+	+	+	+	+	+	++	−
MDP0000232309	transmembrane BAX inhibitor motif-containing protein 4	+	+	+	+	++	++	+	+	−
MDP0000233229	Unknown protein	+	+	+	−	+	+	+	++	−
MDP0000234325	WWE protein-protein interaction domain family protein	−	−	+	+	−	−	+	+	−
MDP0000237908	metallothionein-like protein	++	++	+	+	++	++	+	+	−
MDP0000253074	abscisic acid stress ripening protein homolog	−	−	+	−	−	−	+	+	−
MDP0000255887	TIR-NBS-LRR resistance protein	−	+	−	+	+	+	−	−	−
MDP0000273484	SKP1-like protein	−	+	+	−	+	−	+	−	−
MDP0000281279	Unknown protein	++	+	+	++	++	+	+	+	−
MDP0000286959	dentin sialophosphoprotein	+	+	−	+	+	+	−	−	−
MDP0000292888	GYF domain-containing protein	−	−	++	−	−	−	−	+	C
MDP0000296716	ethylene-responsive transcription factor RAP2-7-like	++	+	++	++	++	+	++	++	−
MDP0000304285	xanthine uracil permease family expressed	−	−	−	+	−	−	−	+	−

Table 1. *Cont.*

Gene Identifier	Blast2GO Annotation	129 DAA-2007 Season				160 DAA-2007 Season				Cellular Localization
		GDE	GDP	HCE	HCP	GDE	GDP	HCE	HCP	
MDP0000310811	cysteine protease inhibitor	-	-	-	-	-	-	-	+	M
MDP0000316244	probable ADP-ribosylation factor GTPase-activating protein AGD15-like	+	+	++	++	++	++	+	+	-
MDP0000320533	proteasome assembly chaperone	-	-	-	+	-	-	-	+	-
MDP0000378585	at4g03420 f9h3_4	+	+	++	+	+	+	++	+	C
MDP0000443265	Unknown protein	-	-	-	-	-	-	+	-	-
MDP0000523205	Unknown protein	-	-	-	+	-	+	-	+	-
MDP0000547450	UNC93-like protein	+	++	+	+	+	++	+	+	-
MDP0000572242	probable xyloglucan glycosyltransferase 12-like	+	++	+	+	+	++	+	-	-
MDP0000580900	porin voltage-dependent anion-selective channel protein	+	+	-	-	+	-	-	-	S
MDP0000584042	protein	-	+	-	+	+	+	+	-	-
MDP0000606453	probable nitrite transporter at1g68570-like	-	-	-	+	-	-	+	-	-
MDP0000618650	NAC domain ipr003441	++	-	+	+	++	-	+	+	S
MDP0000689999	protein disulfide isomerase	-	+	-	-	+	+	+	+	S
MDP0000697474	reverse transcriptase	+	+	++	++	++	++	+	+	S
MDP0000712586	protein SCA1-like	++	++	+	+	+	+	+	+	-
MDP0000875654	hydroquinone glucosyltransferase	-	+	-	-	-	-	+	+	S
MDP0000876817	Unknown protein	+	+	-	+	+	+	-	-	S
MDP0000883284	PREDICTED: uncharacterized protein LOC100248602 [Vitis vinifera]	-	-	-	+	-	-	-	+	S
MDP0000901731	Unknown protein	-	-	-	-	-	-	-	+	-
MDP0000920333	putative protein phosphatase 2C 10	-	-	-	-	-	-	-	-	-

Table 2. Linking differentially expressed genes to changes in fruit traits. Functional annotation for differentially expressed genes was performed using Blast2GO, and additional BLASTX was performed against the proteomes of *Arabidopsis thaliana* and *Solanum lycopersicum*. Many of the differentially expressed genes influence important traits within the fruits.

Gene ID	Location	Gene	At Homolog	Sl Homolog	Potential Trait/Fruit Characteristic Influenced	Reference(s)
Transcriptional Regulation						
MDP0000138340	chr1:12,793,865..12,795,081	NAC transcription factor-like 9	AT4G35580	XP_004239709.1	Stress response	[15]
MDP0000200646	chr1:12,796,087..12,797,315	NAC transcription factor-like 9	AT4G35580	XP_004239709.1	Stress response	[15]
MDP0000618650	chr15:14,098,214..14,102,470	NAC domain containing protein 75	AT4G29230	XP_004239596.1	Stress response/general ripening regulation	[15]
MDP0000296716	chr3:9,357,237..9,360,502	AP2 transcription factor SlAP2d	AT2G28550	NP_001234647.1	Ethylene-regulated response	[16,17]
Secondary Biosynthetic Processes						
MDP0000037251	chr1:6,887,281..6,890,898	Putative cinnamyl alcohol dehydrogenase 9	AT4G39330	XP_004250169.1	Fruit firmness	[18]
MDP0000572242	chr15:1,231,253..1,235,097	Probable xyloglucan glycosyltransferase 12-like	AT4G07960	XP_004238064.1	Fruit firmness	[19]
MDP0000875654	chr7:16,340,110..16,341,528	Hydroquinone glucosyltransferase	AT4G01070	XP_004231207.1	Volatile/polyphenol glycosylation	[20,21]
MDP0000523205	chr7:5,077,703..5,080,814	Flavonoid biosynthesis oxidoreductase protein	AT4G10490	NP_001233840.1	Flavanoid synthesis	[22]
Signaling						
MDP0000128468	chr16:4,935,225..4,935,891	Abscisic acid stress ripening (ASR1) protein	No hit	NM_001247208.2	Abiotic stress response, developmental response to sugar levels	[23,24]
MDP0000176723	chr17:23,896,643..23,897,062	Acyl:CoA ligase acetate-CoA synthase-like protein	AT5G16370	XP_004231630.1	Abiotic stress response/ripening	[25–27]
MDP0000920333	chr10:29,789,438..29,789,994	Putative protein phosphatase 2C-10	AT1G34750	XP_004237914.1	Ripening/abiotic stress response	[28,29]
MDP0000213808	chr17:4,103,927..4,109,528	putative ubiquitin conjugation factor E4	AT5G15400	XP_004232186.1	Ethylene production, development, ripening	[30]
Cell/Organ Development						
MDP0000232309	chr16:13,766,655..13,768,223	BAX inhibitor-1 like protein	No hit	No Hit	Cell elongation/fruit size	[31,32]
MDP0000547450	chr7:26,032,235..26,033,783	UNC93-like protein 1-like	AT1G18000	XP_004235041.1	Apogamy	[33]
Stress Response						
MDP0000712586	chr14:27,117,925..27,120,988	SCA1-like protein	No Hit	No Hit	Redox homeostasis/respiration	[34]
MDP0000161275	chr12:24,524,346..24,526,137	Mitochondrial succinate-fumarate transporter 1-like	AT5G01340	XP_004249636.1	Hypoxia-induced fermentation/respiration	[35,36]
MDP0000310811	chr7:3,004,269..3,005,959	Cysteine proteinase inhibitor 6	AT3G12490	XP_004228480.1	Abiotic stress tolerance	[37,38]

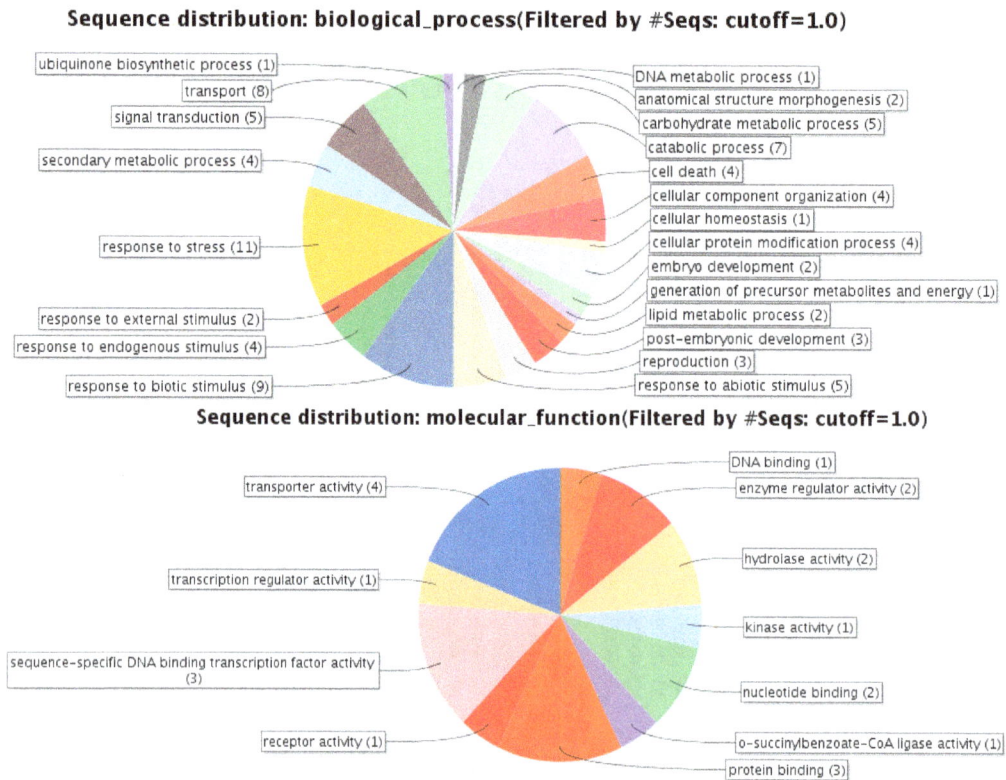

Figure 1. GO-terms associated with excised and sequenced bands from the differential display procedure. Functional analysis of sequences was performed using the Blast2GO annotation suite from full length apple genome (v. 1.0) sequences to which differential display band sequences were matched via BLAST.

3.2. Quantitative Reverse Transcription PCR Validation of Differential Expression

Differential expression observed in the differential display experiment was validated using quantitative reverse-transcription PCR (qPCR). This was performed for all samples when at least one peel tissue exhibited differential gene expression. A second biological replicate from the second growing season was also used to test for seasonal variation. Primer sets were designed based upon predicted gene sequence custom script which selects 26-mer primers, and screens the entire reference genome to ensure the specificity of the oligonucleotide. Primer sequences are provided in Table S3. Expression analysis was performed in triplicate for each gene along with a β-tubulin control which exhibited a global standard deviation of <1.0 across all replicates making it an ideal reference gene. Validation of qPCR specificity was performed by selecting ten differentially expressed genes and sequencing the qPCR products from GD and HC separately. Nine qPCR products aligned with the predicted sequences that were used for the primer design, while product for one, MDP0000255887, failed to yield a consensus sequence or match the original gene.

Analysis of the resulting amplification plots from qPCR revealed that the expression of MDP0000129664 (3-ketoacyl-thiolase) and MDP0000255887 (TIR-NBS-LRR resistance protein) were detected within the set threshold (Cq value lower than 35) in only some of the tested tissues. This could explain why sequences from the qPCR products of MDP0000255887 failed to yield appropriate sequence data. Additionally, no transcripts were detected for MDP0000233229 (protein) and MDP0000901731 (unknown) in any of the samples investigated. Further analysis of the expression for these genes was not pursued.

For the 31 genes whose expression was validated with qPCR, differential expression for each gene was observed, with at least one tested sample having a two-fold increase. Differential gene expression was separated into four categories: high differential expression (greater than five-fold change in expression) (Figure 2), moderate differential expression (three to five-fold change in expression) (Figure 3), low differential expression (zero to three-fold change in expression) (Figure S1), and seasonally-variable differential expression (Figure S2). The largest differences in gene expression were seen in MDP0000618650 (NAC domain containing protein), MDP0000875654 (hydroquinone glucosyltransferase), MDP0000037251 (cinnamyl alcohol dehydrogenase), MDP0000712586 (SCAI-like protein), MDP0000883284 (unknown), MDP0000200646 (NAC domain containing protein), and MDP0000920333 (protein phosphatase 2C). All of these genes were expressed at a higher level in GD compared to HC.

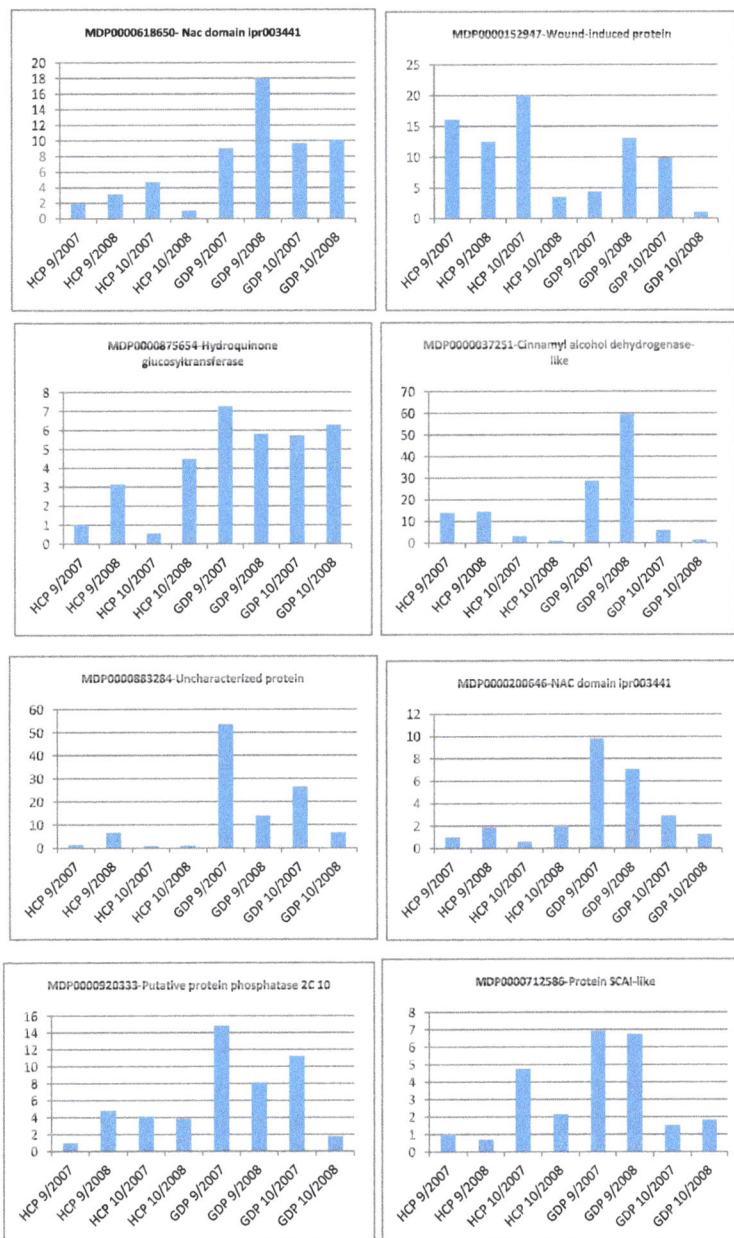

Figure 2. "High" differential expression of genes identified with the differential display procedure. Genes whose expression was greater than or equal to five-fold higher in expression between "Golden Delicious" and "Honeycrisp" peel tissue were classified as highly differentially expressed. Expression is separated by samples tested and plotted against fold expression relative to a β-tubulin internal control.

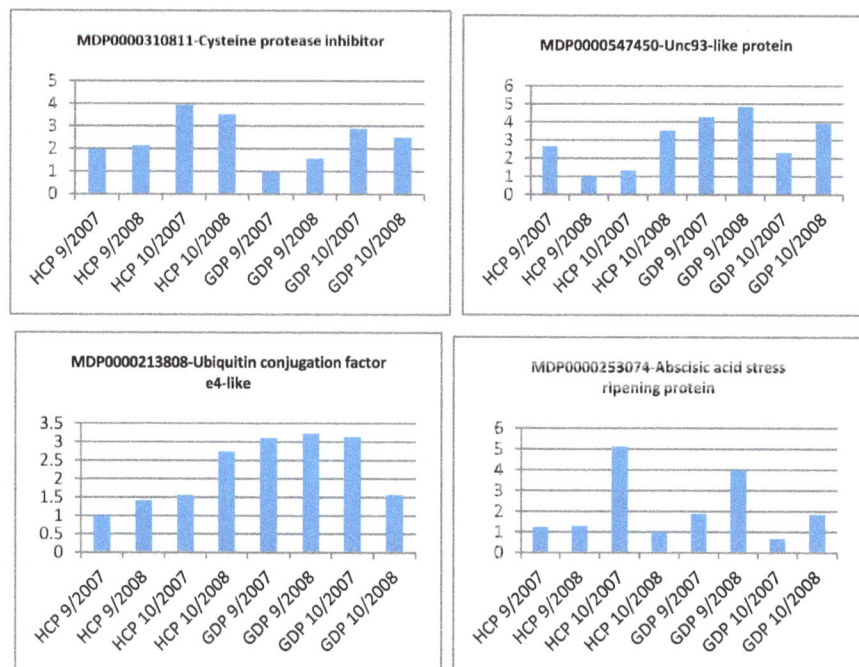

Figure 3. "Moderate" differential expression of genes identified with differential display. Samples with three to five-fold differences in expression relative to the control were classified as moderately differentially expressed. Relative expression relative to a β-tubulin internal control is plotted in each unique biological sample.

3.3. Comparison of Expression with Previous Studies

In order to determine if the expression of the genes identified in this study have been profiled in greater detail in apple fruit, data obtained previously from developing "Royal Gala" tissue was reanalyzed [1]. This study reported the expression of 13,000 unigenes across eight developmental stages. BLASTN was performed for all EST sequences used in the microarray study against the predicted genes from the apple genome project. This study identified 7265 unique genes which were represented in the microarray study. Presence and absence in the DD study was then determined for all sequences represented in the microarray study. Of the 45 genes identified through DD, only MDP0000292888 (GYF domain-containing protein) was represented in the microarray study. This gene was most highly expressed at zero days after anthesis (DAA) with one and a half-fold higher expression compared to the other seven time points, suggesting stable expression over time in "Royal Gala".

When comparing the DD approach with a microarray approach such as that described in the Janssen study referenced earlier, this study showed that the DD approach was quite useful in identifying differentially expressed genes. Only one of the 45 genes represented in the DD dataset was described in the Janssen study, demonstrating that a more thorough transcriptomic profiling of multiple apple genotypes may help in furthering our knowledge of differential gene expression in apple fruit. It appeared that the most significant differences in expression were primarily derived from GD tissues. This could be in fact due to a bias of using the GD predicted gene sequences for designing primer sequences. As genomic and transcriptomic resources for HC increase, additional information from fragments corresponding to the initially isolated differential display bands may come to light.

3.4. Differentially Expressed Homologs in Heterologous Systems

In this study we identified several genes with differential gene expression between HC and GD peel tissues. Differences in the expression of these genes may have important influences upon fruit characteristics throughout the course of development and ripening. Interestingly, some of the differentially expressed genes between GD and HC have been studied in heterologous systems, and are linked or related to major fruit quality traits and regulatory pathways [39,40]. These genes have been linked to various processes including biotic and abiotic stress tolerance, secondary metabolism, signaling, cell or organ development, and transcriptional regulation (Table 2). Such variability may have a significant influence on fruit characteristics and be involved in the phenotypic diversity between HC and GD apples.

Two differentially expressed genes identified in this study, cinnamyl alcohol dehydrogenase (CAD) (MDP0000037251) and xyloglucan transferase (or xyloglycan endotransglycosylase) (MDP0000572242), may have significant effects on cell wall modification. These modifications can have drastic effects upon the texture of fruits either by increasing firmness or promoting loosening of cell walls. Within strawberry fruit, CAD has been previously shown to be localized to lignified cells [41]. Differential expression studies performed with strawberry genotypes varying in fruit firmness found CAD and cinnamyl CoA reductase to have the highest difference in expression in a microarray representing 1701 strawberry ESTs [42]. CADs have also been associated with lignification and consequently increased firmness in loquat (*Eriobotrya japonica*) [18]. Xyloglycan endotransglycosylase (XETs), on the other hand, modify xyloglucans which are hemicelluloses that link together cellulose microfibrils. XETs have been found in multiple isoforms, aiding either in cell expansion during fruit growth or softening during fruit ripening [19]. These enzymes have been associated with fruit softening in kiwi (*Actinidia deliciosa*), tomato (*Solanum lycopersicum*), and other fruits [43,44]. At 129 DAA GD exhibited higher expression of both of these genes compared to HC. At 160 DAA there was a drastic reduction in expression. The expression of CAD decreased in ripening fruit, as it was approximately 12-fold higher at 129 DAA compared to 160 DAA in HC apples, and 28 to 60-fold higher at 129 DAA than at 160 DAA in GD. While the reduction in expression between the time points could correlate with fruit softening associated with ripening, the drastic differences in expression between cultivars could be related to differences in fruit texture. Further studies on the functional analysis of the XET allele in the two cultivars could be used to test this association.

Two differentially expressed genes in this study, MDP0000523205 (homologous to *Solanum lycopersicum* flavonoid biosynthesis oxidoreductase protein) and MDP0000875654 (hydroquinone glucosyl transferase), may be involved in important biosynthetic activities in apple fruit. In apples alone, there are over 300 compounds linked to the aroma profile, with few enzymes involved in their synthesis presently characterized [11]. Hydroquinone glucosyl transferase is a putative target for volatile and phenolic-compound glycosylation. Glucosyl transferases have been previously shown to be involved in the glycosylation of plant volatiles in tomato and have been shown to have relaxed substrate specificity [20]. A hydroquinone glycosyl transferase has also been linked to the synthesis of the phenolic compound arbutin, known to act as an antimicrobial agent and maintain potential anticarcinogenic effects [45]. Flavonoid biosynthesis oxidoreductase proteins (flavanone-3-hydroylases) are important components of the flavonoid biosynthetic pathway in apples which lead to the production of proanthocyanidins [22]. While further research on biochemical activity of these specific enzymes will need to be performed, differential expression of the genes encoding these proteins reveals targets for differences in volatile composition, phenolic compound composition, and flavonoid content between the two cultivars.

Additionally, two auxin-responsive genes were identified that differed between GD and HC tissues. MDP0000176723 (AAE18) is thought to play a role in response to auxin precursors (indole-3-butyric acid, IBA) in the peroxisome, and may modulate available IBA via amino acid conjugation for downstream processing into active IAA [46–49]. With recent focus on the roles of auxin in fruit ripening, this presents a powerful link between peroxisome metabolites and auxin accumulation as apple fruit undergo ripening. Future research may clarify the influence of this relationship in pre-climacteric and ripening apple fruit. Similarly, protein phosphatase 2C-10 (PP2C-10, MDP0000920333) transcripts were found in higher abundance in GD peel than in HC peel. PP2C-10 is reported to function as a serine/threonine phosphatase and is thought to be responsive to active auxin in the cell [28]. While current understanding of these proteins in fruit ripening remains unclear, PP2C proteins have been implicated in MAPK cascades, calcium-calmodulin signaling, and ABA-signaling in the cell which all play a role in numerous ripening-related pathways [29]. More studies of PP2C genotypic and expression variability are needed to determine implications of this gene's expression on fruit quality in pre and postharvest conditions.

Ethylene is a critical plant hormone in modulating many late-stage fruit maturation and ripening processes in apples. Climacteric fruit exhibit an autocatalytic phase of ethylene biosynthesis. This behavior allows ethylene to function both as a trigger and a result of fruit ripening. An ethylene-responsive transcription factor RAP2-7-like transcript (MDP0000296716) was found in higher abundance in GD peels then in HC peel tissues in this study. Containing an AP2/ERF-like domain, RAP2-7 homologs are highly induced during various abiotic stresses and during floral development [16,17]. Following studies implicating floral development genes in later fruit maturation, RAP2-7-like genes may also control expression of genes associated with fruit quality.

Many of the observed differentially expressed genes from this study appear to be induced preferentially at different periods of fruit development, suggesting they may be induced in response to unique cues. A larger scale RNA-seq analysis of GD and HC tissues is expected to provide genetic blueprints for unique responses to the stresses encountered by fruit during the production, storage and postharvest chain. While desirable in the retail market, HC apples exhibit increased frequency of disorders associated with storage conditions relative to other popular cultivars [48]. Identification of these genes may provide an indication of future apple germplasm sensitivity to oxidative stress and frequency of abiotic stress-related storage disorders after harvest. Older varieties such as GD remain popular in the industry due to their unique fruit quality characteristics. In such backgrounds, expression of a senescence-associated gene may mitigate oxidative stress in the fruit at the expense of desirable traits such as acidity, crispiness, and juiciness. Evidence of variability in protein homeostasis activity was also observed among GD and HC tissues from our study. In both sampling time points, GD peel tissues were observed to have higher abundance of SKP1-like and e4-like ubiquitin conjugation factor transcripts (MDP0000213808), than in HC peel and endocarp. As critical components of the 26S-ubiquitin proteolytic pathway, differential abundance of these transcripts may suggest variability in GD and HC post-translational regulation of key points of both ethylene signaling and biosynthesis in fruits [49,50]. We also identified a transcript (MDP0000547450) with homology to UNC93-like genes in higher abundance in GD peel, compared to all HC tissues. Little is known about the role of these proteins in higher plants, or in fruit development. After initial identification in lower plants, UNC93-like genes were found to be involved in regulation of asexual reproduction in *Arabidopsis* [33]. UNC93-like transcripts were present in seed and silique tissue only, suggesting a role in reproductive tissue development. As with MADS-box genes of the ABC model of floral meristem development, UNC93-like genes may have unknown roles in fruit development [50].

In order to identify the components guiding specific apple fruit traits, differences between cultivars at the genomic, transcriptomic, proteomic, and epigenetic states should be investigated. With respect to probing the changes in apple gene expression, several studies have used microarray analysis of gene expression [1,51,52]. With the issues associated with microarray accuracy and reproducibility [53], expanding genomic resources available for Rosaceae, and increased access to next-generation sequence technologies, RNA-seq is emerging as the gold standard to identify transcriptional differences between apple cultivars. RNA-seq has already been shown to be more sensitive in detecting genetic variants, resolving isoforms, detecting transcripts with low abundance, and had higher sensitivity in detecting differential expression [54]. While the DD method enables identification of differentially expressed genes between apple cultivars, it does have the obvious limitations of being labor intensive and time consuming along with a capacity to capture a limited number of gene fragments. A combined DD/next-generation sequencing approach may reduce upfront experimental costs and time, while providing useful insights into differential gene expression.

4. Conclusions

Given the widespread adoption of next-generation sequencing approaches, the differential display technique can often be thought of as dated. Here, we show that differential display effectively identified differential expression between samples, and has the potential to complement next-generation sequencing approaches. Out of a total of 105 extracted bands in this experiment, we confirmed differential expression of 31 genes in the peel tissue of "Golden Delicious" and "Honeycrisp" at two important time points in fruit development. Many of these differentially expressed genes encode for potentially important proteins involved in the fruit traits that differ between the two investigated cultivars. Differential expression appears to occur in developmental, cultivar-specific, and potentially seasonally-variable trends. Of these, an intriguing group of transcripts were identified as potential regulators of tissue redox homeostasis and respiration. In general, the roles of many of the identified proteins in both the peroxisome and mitochondria concerning fruit development, ripening regulation, and fruit quality remain unclear. However, these proteins present interesting groups of genes for future characterization.

Acknowledgments: The authors thank Deb Pehrson for assistance with sample procurement at Washington State University Tukey Orchard, Artemus Harper for assistance with qPCR primer design, Tyson Koepke for assistance in performing differential display, and Salma Tariq, Michela Quickstad, Kiara B. Little, and Emily Rose Rodgers for assistance in cloning differential display fragments. The publication was in part supported by NIH/NIGMS through institutional training grant award T32-GM008336. Its contents are solely the responsibility of the authors and do not necessarily represent the official views of the NIGMS or NIH. USDA-NIFA grant 2008-35300-04676 and ARC Hatch grant WNP00677 to AD are gratefully acknowledged for partial support of the work in this publication.

Author Contributions: S.S. and A.D. designed the study. Samples were obtained by S.S. Samples were prepared by S.S. Sequencing was performed by S.S. and R.F. qRT-PCR was performed by S.S., C.H. and R.F. Data analysis and interpretation was performed by S.S., C.H., R.F. and A.D. The manuscript was prepared by A.D., S.S., C.H. and R.F. All authors reviewed and approved of the final manuscript.

Conflicts of Interest: The authors declare no conflict of interest.

References

1. Janssen, B.J.; Thodey, K.; Schaffer, R.J.; Alba, R.; Balakrishnan, L.; Bishop, R.; Bowen, J.H.; Crowhurst, R.N.; Gleave, A.P.; Ledger, S.; et al. Global gene expression analysis of apple fruit development from the floral bud to ripe fruit. *BMC Plant Biol.* **2008**. [CrossRef] [PubMed]

2. Kupferman, E. The Role of Ethylene in Determining Apple Harvest and Storage Life. Available online: http://postharvest.tfrec.wsu.edu/pages/N4I1C (accessed on 4 March 2016).

3. Conesa, A.; Gotz, S.; Garcia-Gomez, J.M.; Terol, J.; Talon, M.; Robles, M. Blast2GO: A universal tool for annotation, visualization and analysis in functional genomics research. *Bioinformatics* **2005**, *21*, 3674–3676. [CrossRef] [PubMed]

4. Götz, S.; García-Gómez, J.M.; Terol, J.; Williams, T.D.; Nagaraj, S.H.; Nueda, M.J.; Robles, M.; Talón, M.; Dopazo, J.; Conesa, A. High-throughput functional annotation and data mining with the Blast2GO suite. *Nucleic Acid Res.* **2008**, *36*, 3420–3435. [CrossRef] [PubMed]

5. Emanuelsson, O.; Brunak, S.; von Heijne, G.; Nielsen, H. Locating proteins in the cell using TargetP, SignalP and related tools. *Nat. Protoc.* **2007**, *2*, 953–971. [CrossRef] [PubMed]

6. Nosarzewski, M.; Archbold, D.D. Tissue-specific expression of SORBITOL DEHYDROGENASE in apple fruit during early development. *J. Exp. Bot.* **2007**, *58*, 1863–1872. [CrossRef] [PubMed]

7. Zhang, S.; Chen, W.; Xin, L.; Gao, Z.; Hou, Y.; Yu, X.; Zhang, Z.; Qu, S. Genomic variants of genes associated with three horticultural traits in apple revealed by genome re-sequencing. *Hortic. Res.* **2014**, *1*, 1–14. [CrossRef] [PubMed]

8. Zsolt, A.; Deák, C.; Miskó, A.; Tóth, M.; Papp, I. Development of cDNA normalization system and preliminary transcription analysis of KCS genes in apple tissues. *Acta Univ. Agric. Silvic. Mendel. Brun.* **2014**, *59*, 9–12.

9. Ruijter, J.M.; Ramakers, C.; Hoogaars, W.M.H.; Karlen, Y.; Bakker, O.; van den Hoff, M.J.B.; Moorman, A.F.M. Amplification efficiency: Linking baseline and bias in the analysis of quantitative PCR data. *Nucleic Acid Res.* **2009**, *37*, 1–12. [CrossRef] [PubMed]

10. Pfaffl, M.W. A new mathematical model for relative quantification in real-time RT-PCR. *Nucleic Acid Res.* **2001**, *29*, 1–6. [CrossRef]

11. Dixon, J.; Hewett, E.W. Factors affecting apple aroma/flavour volatile concentration: A Review. *N. Z. J. Crop Hortic. Sci.* **2000**, *28*, 155–173. [CrossRef]

12. Goulao, L.F.; Santos, J.; de Sousa, I.; Oliveira, C.M. Patterns of enzymatic activity of cell wall-modifying enzymes during growth and ripening of apples. *Postharvest Biol. Technol.* **2007**, *43*, 307–318. [CrossRef]

13. Solovchenko, A.E.; Merzlyak, M.N.; Pogosyan, S.I. Light-induced decrease of reflectance provides an insight in the photoprotective mechanisms of ripening apple fruit. *Plant Sci.* **2010**, *178*, 281–288. [CrossRef]

14. Emanuelsson, O.; Nielsen, H.; Brunak, S.; von Heijne, G. Predicting subcellular localization of proteins based on their N-terminal amino acid sequence. *J. Mol. Biol.* **2000**, *300*, 1005–1016. [CrossRef] [PubMed]

15. Crifò, T.; Puglisi, I.; Petrone, G.; Recupero, G.R.; Piero, A.R.L. Expression analysis in response to low temperature stress in blood oranges: Implication of the flavonoid biosynthetic pathway. *Gene* **2011**, *476*, 1–9. [CrossRef] [PubMed]

16. Dietz, K.J.; Vogel, M.O.; Viehhauser, A. AP2/EREBP transcription factors are part of gene regulatory networks and integrate metabolic, hormonal and environmental signals in stress acclimation and retrograde signalling. *Protoplasma* **2010**, *245*, 3–14. [CrossRef] [PubMed]

17. Chung, M.Y.; Vrebalov, J.; Alba, R.; Lee, J.; McQuinn, R.; Chung, J.D.; Klein, P.; Giovannoni, J. A tomato (*Solanum lycopersicum*) APETALA2/ERF gene, SlAP2a, is a negative regulator of fruit ripening. *Plant J.* **2010**, *64*, 936–947. [CrossRef] [PubMed]

18. Cai, C.; Xu, C.-J.; Li, X.; Ferguson, I.; Chen, K. Accumulation of lignin in relation to change in activities of lignification enzymes in loquat fruit flesh after harvest. *Postharvest Biol. Technol.* **2006**, *40*, 163–169. [CrossRef]

19. Rose, J.K.C.; Bennett, A.B. Cooperative disassembly of the cellulose–xyloglucan network of plant cell walls: Parallels between cell expansion and fruit ripening. *Trends Plant Sci.* **1999**, *4*, 176–183. [CrossRef]

20. Louveau, T.; Leitao, C.; Green, S.; Hamiaux, C.; van der Rest, B.; Dechy-Cabaret, O.; Atkinson, R.G.; Chervin, C. Predicting the substrate specificity of a glycosyltransferase implicated in the production of phenolic volatiles in tomato fruit. *FEBS J.* **2011**, *278*, 390–400. [CrossRef] [PubMed]

21. Puupponen-Pimiä, R.; Nohnek, L.; Meier, C.; Kähkönen, M.; Heinonen, M.; Hopia, A.; Oksman-Caldentey, K.M. Antimicrobial properties of phenolic compounds from berries. *J. Appl. Microbiol.* **2001**, *90*, 494–507. [CrossRef] [PubMed]

22. Flachowsky, H.; Halbwirth, H.; Treutter, D.; Richter, K.; Hanke, M.V.; Szankowski, I.; Gosch, C.; Stich, K.; Fischer, T.C. Silencing of flavanone-3-hydroxylase in apple (*Malus × domestica* Borkh.) leads to accumulation of flavanones, but not to reduced fire blight susceptibility. *Plant Physiol. Biochem.* **2012**, *51*, 18–25. [CrossRef] [PubMed]

23. Çakir, B.; Agasse, A.; Gaillard, C.; Saumonneau, A.; Delrot, S.; Atanassova, R. A grape ASR protein involved in sugar and abscisic acid signaling. *Plant Cell* **2003**, *15*, 2165–2180. [CrossRef] [PubMed]

24. Golan, I.; Guadalupe, P.; Dominguez, Z.K.; Shkolnik-Inbar, D.; Carrari, F.; Bar-Zvi, D. Tomato ABSCISIC ACID STRESS RIPENING (ASR) gene family revisited. *PLoS ONE* **2014**, *9*, e107117.

25. Eubel, H.; Meyer, E.H.; Taylor, N.L.; Bussell, J.D.; O'Toole, N.O.; Heazlewood, J.L.; Castleden, I.; Small, I.D.; Smith, S.M.; Millar, A.H. Novel Proteins, putative membrane transporters, and an integrated metabolic network are revealed by quantitative proteomic analysis of Arabidopsis cell culture peroxisomes. *Plant Physiol.* **2008**, *148*, 1809–1829. [CrossRef] [PubMed]

26. Kamada, T.; Nito, K.; Hayashi, H.; Mano, S.; Hayashi, M.; Nishimura, M. Functional differentiation of peroxisomes revealed by expression profiles of peroxisomal genes in *Arabidopsis thaliana*. *Plant Cell Physiol.* **2003**, *44*, 1275–1289. [CrossRef] [PubMed]

27. Wiszniewski, A.A.G.; Zhou, W.; Smith, S.M.; Bussell, J.D. Identification of two Arabidopsis genes encoding a peroxisomal oxidoreductase-like protein and an acyl-CoA synthetase-like protein that are required for responses to pro-auxins. *Plant Mol. Biol.* **2009**, *69*, 503–515. [CrossRef] [PubMed]

28. Goda, H.; Sawa, S.; Asami, T.; Fujioka, S.; Shimada, Y.; Yoshida, S. Comprehensive comparison of auxin-regulated and brassinosteroid-regulated genes in Arabidopsis. *Plant Physiol.* **2004**, *134*, 1555–1573. [CrossRef] [PubMed]

29. Schweighofer, A.; Hirt, H.; Meskiene, I. Plant PP2C phosphatases: Emerging functions in stress signaling. *Trends Plant Sci.* **2004**, *9*, 236–243. [CrossRef] [PubMed]

30. Sharma, B.; Joshi, D.; Yadav, P.K.; Gupta, A.K.; Bhatt, T.K. Role of ubiquitin-mediated degradation system in plant biology. *Front. Plant Biol.* **2016**. [CrossRef] [PubMed]

31. Ascencio-Ibanez, J.T.; Sozzani, R.; Lee, T.J.; Chu, T.M.; Wolfinger, R.D.; Cella, R.; Hanley-Bowdoin, L. Global analysis of Arabidopsis gene expression uncovers a complex array of changes impacting pathogen response and cell cycle during geminivirus infection. *Plant Physiol.* **2008**, *148*, 436–454. [CrossRef] [PubMed]

32. Sierla, M.E.; Feys, B.J. Characterization of *Arabidopsis thaliana* orthologues of GAAP, a Golgi-localized anti-apoptotic protein. In Proceedings of the 20th International Conference on Arabidopsis Research, Edinburgh, UK, 30 June–4 July 2009.

33. Cordle, A.R.; Irish, E.E.; Cheng, C.L. Gene expression associated with apogamy commitment in *Ceratopteris richardii*. *Sex. Plant Reprod.* **2012**, *25*, 293–304. [CrossRef] [PubMed]

34. Foyer, C.H.; Noctor, G. Oxidant and antioxidant signalling in plants: A re-evaluation of the concept of oxidative stress in a physiological context. *Plant Cell Environ.* **2005**, *28*, 1056–1071. [CrossRef]

35. Catoni, E.; Desimone, M.; Hilpert, M.; Wipf, D.; Kunze, R.; Schneider, A.; Flügge, U.I.; Schumacher, K.; Frommer, W.B. Expression pattern of a nuclear encoded mitochondrial arginine-ornithine translocator gene from Arabidopsis. *BMC Plant Biol.* **2003**. [CrossRef]

36. Picault, N. Plant mitochondria: From genome to function. In *Plant Mitochondrial Carriers*; Day, D.A., Millar, A.H., Whelan, J., Eds.; Kluwer Academic Publisher: Dordrecht, The Netherlands, 2004; Volume 17.

37. Solomon, M.; Belenghi, B.; Delledonne, M.; Menachem, E.; Levine, A. The involvement of cysteine proteases and protease inhibitor genes in the regulation of programmed cell death in plants. *Plant Cell* **1999**, *11*, 431–443. [CrossRef] [PubMed]

38. Krüger, J.; Thomas, C.M.; Golstein, C.; Dixon, M.S.; Smoker, M.; Tang, S.; Mulder, L.; Jones, J.D.G. A tomato cysteine protease required for Cf-2-dependent disease resistance and suppression of autonecrosis. *Science* **2002**, *296*, 744–747. [CrossRef] [PubMed]

39. Costa, F.; Alba, R.; Schouten, H.; Soglio, V.; Gianfranceschi, L.; Serra, S.; Musacchi, S.; Sansavini, S.; Costa, G.; Fei, Z.; et al. Use of homologous and heterologous gene expression profiling tools to characterize transcription dynamics during apple fruit maturation and ripening. *BMC Plant Biol.* **2010**, *10*, 1–17. [CrossRef] [PubMed]

40. Gosch, G.; Halbwirth, H.; Schneider, B.; Hölscher, D.; Stich, K. Cloning and heterologous expression of glycosyltransferases from *Malus × domestica* and *Pyrus communis*, which convert phloretin to phloretin 2′-O-glucoside (phloridzin). *Plant Sci.* **2010**, *178*, 299–306. [CrossRef]

41. Blanco-Portales, R.; Medina-Escobar, N.; López-Ráez, J.A.; González-Reyes, J.A.; Villalba, J.M.; Moyano, E.; Caballero, J.L.; Muñoz-Blanco, J. Cloning, expression and immunolocalization pattern of a cinnamyl alcohol dehydrogenase gene from strawberry (*Fragaria × ananassa* cv. Chandler). *J. Exp. Bot.* **2002**, *53*, 1723–1734. [CrossRef] [PubMed]

42. Salentijn, E.M.J.; Aharoni, A.; Schaart, J.G.; Boone, M.J.; Krens, F.A. Differential gene expression analysis of strawberry cultivars that differ in fruit firmness. *Physiol. Plant.* **2003**, *118*, 571–578. [CrossRef]

43. Redgwell, R.J.; Fry, S.C. Xyloglucan endotransglycosylase activity increases during kiwifruit (*Actinidia deliciosa*) ripening (Implications for fruit softening). *Plant Physiol.* **1993**, *103*, 1399–1406. [PubMed]

44. Cutillas-Iturralde, A.; Zarra, I.; Fry, S.C.; Lorences, E.P. Implication of persimmon fruit hemicellulose metabolism in the softening process. Importance of xyloglucan endotransglycosylase. *Physiol. Plant.* **2006**, *91*, 169–176. [CrossRef]

45. Gutiérrez-Larraínzara, M.; Rúaa, J.; Carob, I.; de Castroa, C.; de Arriagaa, D.; García-Armestob, M.R.; del Valle, P. Evaluation of antimicrobial and antioxidant activities of natural phenolic compounds against foodborne pathogens and spoilage bacteria. *Food Control* **2012**, *26*, 555–563. [CrossRef]

46. Osorio, S.; Scossa, F.; Fernie, A.R. Molecular regulation of fruit ripening. *Front. Plant Sci.* **2013**, *198*, 1–8. [CrossRef] [PubMed]

47. Pattison, R.J.; Csukasi, F.; Catalá, C. Mechanisms regulating auxin action during fruit development. *Physiol. Plant.* **2014**, *151*, 62–72. [CrossRef] [PubMed]

48. Watkins, C.B.; Nock, J.F.; Weis, S.A.; Jayanty, S.; Beaudry, R.M. Storage temperature, diphenylamine, and pre-storage delay effects on soft scald, soggy breakdown and bitter pit of "Honeycrisp" apples. *Postharvest Biol. Technol.* **2004**, *32*, 213–221. [CrossRef]

49. Cherian, S.; Figueroa, C.R.; Nair, H. "Movers and Shakers" in the regulation of fruit ripening: A cross-dissection of climacteric versus non-climacteric fruit. *J. Exp. Bot.* **2014**, *65*, 4705–4722. [CrossRef] [PubMed]

50. Ireland, H.S.; Yao, J.-L.; Tomes, S.; Sutherland, P.W.; Nieuwenhuizen, N.; Gunaseelan, K.; Winz, R.A.; David, K.M.; Schaffer, R.J. Apple SEPALLATA1/2-like genes control fruit flesh development and ripening. *Plant J.* **2013**, *73*, 1044–1056. [CrossRef] [PubMed]

51. Soria-Guerra, R.E.; Rosales-Mendoza, S.; Gasic, K.; Wisniewski, M.E.; Band, M.; Korban, S.S. Gene expression is highly regulated in early developing fruit of apple. *Plant Mol. Biol. Rep.* **2011**, *29*, 885–897. [CrossRef]

52. Zhu, Y.; Zheng, P.; Varanasi, V.; Shin, S.; Main, D.; Curry, E.; Mattheis, J.P. Multiple plant hormones and cell wall metabolism regulate apple fruit maturation patterns and texture attributes. *Tree Genet. Genomes* **2012**, *8*, 1389–1406. [CrossRef]

53. Draghici, S.; Khatri, P.; Eklund, A.C.; Szallasi, Z. Reliability and reproducibility issues in DNA microarray measurements. *Trends Genet.* **2006**, *22*, 101–109. [CrossRef] [PubMed]

54. Zhao, S.; Fung-Leung, W.-P.; Bittner, A.; Ngo, K.; Liu, X. Comparison of RNA-seq and microarray in transcriptome profiling of activated T cells. *PLoS ONE* **2014**, *9*, e78644. [CrossRef] [PubMed]

4

Organic Horticulture in India

Sisir Mitra [1,]* and Hidangmayum Devi [2]

[1] Section Tropical and Subtropical Fruits, International Society for Horticultural Science,
Faculty of Horticulture, Bidhan Chandra Krishi ViswaVidyalaya, Mohanpur, B-12/48, Kalyani, Nadia,
West Bengal 741252, India
[2] Indian Council of Agriculture Research complex for NEH Region, Tripura Centre, Lembucherra,
West Tripura 799210, India; lembihort@gmail.com
* Correspondence: sisirm@vsnl.net

Abstract: During the previous three decades, organic produce has attracted the attention of a growing health-conscious population across the globe. Both international and domestic communities are becoming aware of issues like agrochemical residues, produce quality, and food safety. Worldwide, over 37.5 million ha of land (0.87% of total agricultural land) is being managed organically by 1.9 million producers in 164 countries. In addition, there is another 31 million ha certified for wild harvest collection. Global sales of organic products have reached U.S. $75 billion, with the U.S. and Europe as the largest consumers. The concept of organic farming is not new to the Indian farming community. Several forms of organic farming are successfully practiced in diverse climates, particularly in rain-fed, tribal mountains, and hilly areas of the country. Many of the forest products of economic importance, such as herbs and medicinal plants are in this category by default. The report of the Task Force on Organic Farming appointed by the Government of India noted the vast areas of the country where limited amounts of synthetic chemicals are used, although they have low productivity, but also which could have unexploited potential for organic agriculture. As of March 2014, India had 4.72 million ha under an organic certification process, including 0.6 million ha of cultivated agricultural land and 4.12 million ha of wild harvest collection forest area. During 2012–2013, India exported 165,262 million tons of organic products across 135 commodities valued at $312 million. The domestic market for organic commodities is also growing at an annual growth rate of 15%–20%. The crops grown organically include cashew nut, spices, cotton, rice, sugarcane, pineapple, passion fruit, groundnut, sunflower, millet, vegetables, wheat castor, mustard, walnut, tea, coffee, banana, and mango. Institutional support for organic exports from India was created by the launch of the National Program for Organic Production (NPOP) by the Agriculture and Processed Food Export Development Authority (APEDA), Ministry of Commerce. The NPOP supports promotional initiatives, accreditation by inspection and certification agencies, and offers support to agri-business enterprises to facilitate export. India now has 26 accredited certification agencies to facilitate the certification of growers.

Keywords: food safety; environmental health; sustainability; organic

1. Introduction

In India, historical farming systems were by and large organic, where crop rotation, choice of cultivar for region, and utilization of solar radiation for soil sterilization were used, and soil fertility was maintained through organic manure and symbiotic soil microflora. Food security is not simply a function of production or supply, but also of availability, accessibility, stability of supply, affordability, the provision of adequate quantity and quality, and safe nutritious food at all times [1,2]. In a world where consumers are increasingly aware of what they eat and consume, quality and safety have become key issues and unique selling propositions.

Though India experienced the green revolution from a "begging bowl" status to greater abundance with the increased use of synthetic agrochemicals such as fertilizers and pesticides, the adoption of nutrient-responsive high-yielding varieties of crops, and greater exploitation of irrigation potential, the continuous use of these high energy inputs has indiscriminately led to the deterioration of soil health and the environment, and food safety has become a major concern.This fertilizer and pesticide overuse has put forth a question regarding the sustainability of agriculture/horticulture in the long term, calling attention to sustainable production practices which address social, ecological, and economic issues together. Recognizing the impact of the excessive use of chemical fertilizers on soil health and pesticides on human health, there has been an effort to develop integrated management systems. Organic farming addresses soil, human, and environmental health, and is eco-friendly, and thus may be one of the options for sustainability [3].

2. Organic Farming

In a very simplistic form, organic farming is the practice of growing crops without the use of synthetic pesticides, herbicides, and fertilizers. Rather, it relies mainly on crop rotation, use of crop residues by soil incorporation of cover crops, and biomass recycling such as composting, animal integration, and biofertilizers to maintain soil health without disturbing the ecological environment in order to obtain sustainably high yields. The Codex Alimentarius commission—a joint body of FAO/WHO—defined organic agriculture as a unique production management system which promotes and enhances agro-ecosystem health, including biodiversity, biological cycles, and soil biological activity, accomplished by the use of on-farm agronomic, biological, and mechanical methods and the exclusion of all synthetic off-farm inputs [4]. The International Federation of Organic Agriculture Movements (IFOAM) has formulated four broad principles of organic farming, which are the basic roots for the growth and development of organic agriculture in a global context: (1) health (to sustain and enhance the health of soil, plant, animal, human, and planet as one and indivisible); (2) ecology (attaining ecological balance through the design of farming systems, establishment of habitats, and the maintenance of genetic and agricultural diversity); (3) fairness (to ensure that fairness characterized by equity, respect, justice, and stewardship of the shared world, with regard to the common environment and life opportunities); and (4) care (managing in a precautionary and responsible manner to protect the health and well-being of current and future generations and the environment by adopting appropriate technologies and rejecting unpredictable ones, such as genetic engineering) [4].

3. Indian Initiatives

Historically in India, agriculture utilized organic-like practices. Even presently, much of the forest produce of economic importance, such as herbs, medicinal plants, etc., is in this category. India has brought 4.72 million ha under organic certification processes, including 0.6 million ha of cultivated agricultural land and 4.12 million ha for wild harvest collection in forests, as of March 2014 [5]. Sikkim has become India's first fully organic state by implementing organic practices on approximately 75,000 ha of agricultural land [6]. There has been a consistent increase in the number of certified farmers in the country every year. Organic farmers' associations have taken the lead in adopting and spreading technology. Institutional support for organic production was created by the launch of the National Programme for Organic Production (NPOP) by the Agriculture and Processed Food Export Development Authority (APEDA), Ministry of Commerce. The NPOP supports promotional initiatives, accreditation of inspection and certification agencies, and offers support to agri-business enterprises to facilitate export. APEDA has been interacting with the European Union (EU), the United States Department of Agriculture (USDA), Japan, and IFOAM for recognition of equivalence of the Indian quality assurance system. India was recognized for the purpose of equivalence since June 2006 for an unspecified duration, as reported in the Annex III of Regulation

(EC) 1235/2008. However, the European Union's organic production rules are not formally recognized as equivalent by India.

In 2004, India's Ministry of Agriculture launched a nationwide initiative—the National Project on Organic Farming (NPOF)—with the objective of promoting organic farming in India by facilitating access to organic inputs, streamlining production, facilitating certification, and developing domestic markets for organic commodities. In 2004–2005, a national Centre for Organic Farming was established under the Ministry of Agriculture at Ghaziabad to provide institutional support and to facilitate moving farmers into organic crop production by providing suitable logistics. The National Horticulture Mission launched by India's Department of Agriculture and Cooperation in 2005 offers assistance for transitioning to organic farming of horticultural crops. As a result of these interventions, organic agriculture has seen unexpectedly high growth. By the combined effort of farmers, the government, NGOs, and market forces, the Indian organic movement has reached a stage where it can swiftly move to occupy a desired space in Indian agriculture. Today, India is alongside the EU, the U.S., Japan, Brazil, Argentina, and Switzerland, which have adopted organic standards and put in place an inspection and certification mechanism [7].

4. Current Status

India has the largest number of organic producers in the world most with small holdings. During 2013–2014, India exported 135 products valued at $403 million. Major destinations for organic products from India were the U.S., the EU, Canada, Switzerland, Australia, New Zealand, South-East Asian countries, West Asia, and South Africa. Soybean comprised 70% of the commodities, and products exported followed by cereals and millets other than basmati rice (4%), sugar (3%), tea (2%), pulses and lentils (1%), dry fruits (1%), and spices (1%) [8].

Organic farming is growing rapidly among Indian farmers and entrepreneurs, especially in low productivity areas, rain-fed zones, hilly areas, and the northeastern states, where fertilizer consumption is less than 25 kg/ha/year. Nine states in India have promoted policies and programs on organic farming. Uttrakhand has made organic a thrust for improving its mountain agriculture farm economy and livelihood. Mizoram and Sikkim declared their intentions to move to total organic farming. Karnataka has formulated organic policies, and Maharashtra, Tamil Nadu, and Kerala have supported public–private partnerships for the promotion of organic farming.

5. Quality Regulation

Organic agriculture systems and products are required to be certified by accredited agencies to indicate that they have been produced, stored, processed, handled, and marketed in accordance with technical specifications. The organic label is a production process claim, as opposed to a product quality claim. India now has 26 accredited certification agencies to facilitate the certification of growers. In India, the Tea Board, Coffee Board, Spices Board, and Coconut Development Board have developed guidelines for production and certification, and have encouraged the production and export of organic produce and products. Recognizing the need for organic farming, the Ministry of Agriculture has also taken the major initiative that production and certification for domestic markets shall be the responsibility of the Ministry of Agriculture.

There are now over 150 countries in the world exporting certified organic products. Organic trade is expanding at the rate of 15%–20% per year. Over 500 public and private certification bodies now operate in the global organic market place [9]. The many governmental and private standards and technical regulations governing organic production and certification have placed a burden on producers and traders and created barriers for trade on many levels. There is a need for tools to be in place to support equivalency and harmonization in the global organic trade. By cooperation within and among governments and the private sector, trade barriers can be reduced.

6. Key Opportunities in India

India is one of the leading fruit producers in the world, producing about 10% of the world's fruit production [10]. Most of the produce is consumed fresh and domestically. The main destinations are the Middle East, Europe, and Southeast Asia. India is the largest mango producer in the world; however, a negligible amount of fresh (42,998.31 MT) and processed mangoes are exported due to huge domestic demand [11]. The UK, Netherlands, and Germany have a high demand for organic mangoes, which could be exploited by India. Indian organic banana exports are negligible in relation to the world trade. India needs to follow a two-pronged strategy for increasing organic banana exports. First, it should target the processed organic banana market (pulp, purees, and concentrates), and second, it should focus on the geographically closer Japanese market and the EU [9].

India has good potential for the export of organic pineapples, as three major importing markets are the U.S., EU, and Japan. As is the case with most other fruit exports from India, the prime export destination for Indian grapes is the Middle East, but it offers limited opportunities for organic grapes. The main target destination market for Indian organic grapes is the EU, especially the UK and the Netherlands. Moreover, there is a current consumption trend increasingly favoring organic wine, further increasing the demand for organic grapes. Other organic fruits which could be successfully exported include litchi, passion fruit, pomegranate, sapota, apple, walnut, and strawberry [9].

India is the second largest producer of vegetables in the world after China, followed by the Middle East, Singapore, Malaysia, Sri Lanka, Bangladesh, Nepal, the EU, and Australia [12]. Traditional vegetables like onion, potato, okra, bitter gourd, and green chilies, and non-traditional vegetables like asparagus, celery, paprika, sweet and baby corn, and cherry tomato are all exported. Global demand is increasing for organic vegetables, and Indian organic vegetable producers would be in a position to expand their market in the EU, Australia, and Singapore. India is also the largest producer and exporter of organic tea. With the European Commission having granted "equivalence" status to Indian organic certifying agencies, Indian organic tea producers are in a position to expand their markets in Europe, one of the leading tea consuming regions. Organic coffee is mainly consumed by the developed countries; namely, the U.S., Germany, France, Italy, Japan, and the EU. India accounts for 1% of the estimated world organic coffee market, so there is exceptional potential to increase its exports in the near future. India currently accounts for over 12% (in terms of quantity) of the world spice market. The main consumers of organic spices are Germany, the UK, France, Japan, and the U.S. However, organic spices in India represent a very negligible part of total spice production. Organic spices produced by India and having export potential include pepper, ginger, turmeric, cloves, mace, nutmeg, vanilla, cardamom, chili, mustard, tamarind, camboge, thyme, rosemary, oregano, marjoram, parsley, and sage (fresh, dehydrated, and oil). India is a significant supplier of certified organic ingredients to the global organic cosmetics and health care industries, and has a vast area under herbal and aromatic plant production. India is also a significant producer of essential oils in world. Given these advantages, India could become one of the leading suppliers of organic ingredients to the global organic cosmetic and health care industry [13].

7. Indian Organic Food Market

India's organic industry is just beginning to gain attention in the market place in urban centers, primarily with small shops and groceries. However, the retail scene in India is beginning to see dramatic changes with the recent development of hyper-markets in most metropolitan cities. Today, every supermarket has an organic food section, and every large city in India has numerous organic food stores and restaurants. This is a huge change considering that the first organic food store in Mumbai was started in 1997. The pattern of organic food consumption in India is much different than in developed countries. In India, consumers prefer organic fruits, vegetables, spices, strawberry, tea, organic marmalade, organic honey, organic butter, and various organic flours. However, there are many consumers who are unaware of the difference between natural and organic food. Many people purchase products labeled as "Natural" thinking that they are organic. However, consumers are not aware of the certification system, since certification is not compulsory for domestic retail in India. The overall growth in the market of organic products was estimated at 14% for spices, 15% for banana, 14% for tea, 11% for rice, 8% for fruit, 8% for herbal extracts, 8% for cotton, and 4.5% for turmeric [13]. Organic production is not limited to the foods sector, but also applies to significant amounts of organic cotton fiber, garments, cosmetics, functional food products, and body care products.

8. Conclusions

India's organic export markets would grow with the support of the industry, the government, and NGOs coming together to work with farmers. The future for markets for organic foods is definitely bright, as it is growing rapidly in the EU, in the U.S. and Canada, and in Japan and Australia, as well as in some developing countries. With growing consumer awareness of food safety, health, and environmental issues, the organic food sector has become an attractive opportunity for export from developing countries.

Acknowledgments: This review article was compiled by collecting information's from university library; hence no fund was required for writing the manuscript.

Author Contributions: Sisir Mitra conceived the idea of reviewing and writing an article on organic horticulture in India. Sisir Mitra and Hidangmayum Devi together collected the information and compiled the article.

Conflicts of Interest: The authors declare no conflict of interest.

References

1. Boon, E.K. *Food Security in Africa: Challenges and Prospects*; Encyclopaedia of Life Support Systems (EOLSS): Oxford, UK, 2007.
2. Mclntyre, D.B.; Herren, H.R.; Wakhungu, J.; Watson, R.T. *International Assessment of Agricultural Knowledge, Science and Technology for Development—IAASTD, Agriculture at a Cross Road, Global Report*; Island Press: Washington, DC, USA, 2009; p. 590.
3. Singh, H.P. Organic Horticulture—Retrospect and Prospect. In Proceedings of the National workshop on Organic Horticulture—Its Production, Processing, Marketing and Export for Sustainability, Bidhan Chandra Krishi Viswavidyalaya, Mohonpur, West Bengal, India, 8–10 June 2007; Bidhan Chandra Krishi Viswavidyalaya: West Bengal, India, 2007; pp. 1–8.
4. FAO. *Organic Agriculture*; Food and Agriculture Organization of the United Nations: Rome, Italy, 1999.
5. Yadav, A.K. Organic Agriculture at a Glance 2015. Available online: http://krishijagran.com/farm/scenario-in-india/2015/03/Organic-Agriculture-At-a-Glance (accessed on 2 April 2015).
6. Sikkim Becomes India's First Organic State. Available online: http://www.thehindu.com/news/national/sikkim-becomes-indias-first-organicstate/article8107170.ece (accessed on 14 January 2016).
7. Chadha, K.L. Organic Farming: Concept, Initiatives, Status and Implication. In Proceedings of the Padmanav Panda Memorial Oration at Orissa University of Agriculture & Technology, Bhubaneswar, India, 25 July 2008.
8. Why Organic Farming has not Caught Up Yet in India. Available online: http://www.thehindubusinessline.com/markets/commodities/why-organic-farming-has-not-caught-up-yet-in-india/article6933518.ece (accessed on 25 February 2015).
9. Mitra, S.K. Organic tropical and subtropical fruit production in India—Prospects and challenges. *Acta Hort.* **2013**, *975*, 303–307. [CrossRef]
10. Indian Horticulture Database, 2014. National Horticulture Board, Ministry of Agriculture, Government of India: Gurgaon, India, 2015. Available online: http//www.nhb.gov.in (accessed on 25 October 2015).
11. Palanivel, V.; Muthukumar, C.M.; Gurusamy, M. A study on cultivation and marketing of mangoes in Krisnagiri district. *Int. J. Adv. Eng. Recent Tech.* **2015**, *2*, 31–43.

12. Ingh, V.B.; Kanaujia, S.P. Vegetable Production Scenario in India. In Proceedings of the National Seminar on Sustainable Horticulture vis-à-vis Changing Environment, Dimapur District, India, 26–28 February 2016; The Horticultural Society of North-East India: Medziphema, India, 2015; pp. 73–82.

13. Garibay, S.V.; Jyoti, K. Market Opportunities and Challenges for Indian Organic Products. Available online: http://www.orgprints.org/2684/1/garibay-2003-Market-Study-India.pdf (accessed on 25 October 2016).

5

Vermiculture for Sustainable Organic Agriculture in Madagascar

Holy Ranaivoarisoa [1,*], Solofoniaina Ravoninjiva [1], Sylvain Ramananarivo [1] and Romaine Ramananarivo [2]

[1] Agro Management, Développement Durable et Territoires, Ecole Doctorale Gestion des Ressources Naturelles et Développement, Ecole Supérieure des Sciences Agronomiques, University of Antananarivo, Antananarivo 101, Madagascar; aina.njiva@yahoo.com (S.R.); sylaramananarivo@yahoo.fr (S.R.)

[2] Ecole Supérieure de Management et d'Informatique Appliquée, Antananarivo 101, Madagascar; agromanagement1@yahoo.fr

[*] Correspondence: rholy1@yahoo.fr

Abstract: Despite the possession of arable land, Malagasy farmers do not have sufficient access to capital and equipment and invest little to improve their low agricultural productivity in Madagascar. Vermicomposting is the result of research on the culture of earthworms to overcome problems with fertilization and provide benefits to farmers, including improvement in crop performance and yield, and preservation of the environment. Each farmer can practice vermicomposting because of its simple technology. Our concern has been how to develop the production and use of vermicompost at the household level in rural areas in order to solve problems of soil fertility, improve agricultural productivity, and increase farmers' incomes with this organic technique. Thus, the objective of this research was to propose a model for the development of production and use of vermicompost in rural areas in order to minimize the costs of agricultural inputs, improve soil fertility and increase long-term household incomes. A typology of operators was carried out according to defined factors of production and activities performed, followed by a socio-economic analysis and a comparative analysis based on the types obtained. It will be essential to clearly define a national policy on organic farming by supporting private sector groups, NGOs or associations, and encouraging farmers to produce their own fertilizer. Technical and financial support will be needed for the development of concrete visual references which can demonstrate the technical and economic value that organic farming brings using vermicompost.

Keywords: vermicompost; household income; sustainable agriculture

1. Introduction

Malagasy farmers do not have sufficient access to materials and capital, despite the possession of arable land. They use very few modern inputs, such as fertilizers and improved seeds, or modern technologies. They engage in few activities for improvement of agricultural land: 16% of cultivated land is fertilized, 2% with mineral fertilizers and 14% with organic manure [1]. Vermicompost is a fertilizer that farmers can produce themselves by using their land and recycling household waste. Its use is considered organic, and greatly reduces inputs excluding expenditures for synthetic chemicals [2]. However, the number of farmers who currently use vermicompost is very small. In the Rural Municipality of Ambohimanambola with 2228 farmers, 0.85% use vermicompost, while others use either chemical fertilizers (NPK, urea), animal manure (cattle, rabbits, poultry), or ash, and only in small quantities. Development of vermicomposting in this community has been led by the association Tanora Andrin'ny tontolo Ambanivohitra, or TATA. TATA is a peasant association created in December 1997 and legalized in 2004, and is ranked among the first producers of vermicompost in Madagascar. The goals of TATA have been to support achieving sustainable improvements in agricultural productivity and increase the income of rural households in the community.

The overall objective of this research was to propose a developmental model of manufacture and use of vermicompost to minimize spending on agricultural inputs, improve the fertility of the soil in the long term, and increase household incomes. The corresponding specific objectives were (i) to characterize the agricultural practices of farmers in the Rural Municipality of Ambohimanambola; (ii) to determine the socio-economic benefits of using vermicompost; and (iii) to determine the conditions for developing the use of vermicompost among farmers.

The hypotheses to be tested were: (i) farmers using vermicompost have better crop production than by fertilization with other modes; (ii) the use of vermicompost reduces spending on agricultural inputs and increase farmers' incomes; and (iii) the lack of information and awareness of farmers inhibits further developing the use of vermicompost.

2. Experimental Section

2.1. Materials

2.1.1. Study Area

The research was carried out in the Rural Municipality of Ambohimanambola, the District of Antananarivo Avaradrano, Analamanga Region, Madagascar. It included 2228 households that are in 10 villages, each having their own characteristics. The majority of people (85%) are devoted mainly to agriculture. The leading producers of vermicompost in Madagascar are in this area, grouped in the TATA association, whose headquarters is located in this rural municipality.

2.1.2. Data Collection

The documentation collected revolved around the following themes: (i) analysis of farmers' fertilization practices; (ii) all work addressing the practice of vermicomposting; and (iii) all of the work already carried out on the use of vermicompost. Interviews were conducted with: (i) the President and the trainers of the TATA association; (ii) the authorities responsible for the Rural Municipality of Ambohimanambola; and (iii) 5 leaders of the respondents in each village. A survey of farmers produced all of the data for the study through a questionnaire. It was conducted among 60 of 928 households in 5 villages of the 10 selected according to their fertilization practices. A focus group was also conducted with households who are members of the TATA association. It was designed to obtain information about their opinions, attitudes and experiences or to explain their expectations about the intervention.

2.2. Methods

2.2.1. Typology

A typology of farms was used to identify and differentiate them, using data from the surveyed farms about their practices and procedures. Successful differentiation criteria were based on the production system, inputs, production activities, and non-agricultural activities. The results of the questionnaire from the 60 households were used to perform a multivariate analysis with XLSTAT software (Addinsoft, Paris, France). Four methods of analysis were selected: Agglomerative Hierarchical Clustering (AHC), k-means clustering, Discriminant Analysis (DA) with an analysis of variance, and Multiple Correspondence Analysis (MCA).

2.2.2. Identification of Socio-Economic Performance of Rural Vermicomposting

The profitability of fertilizer use was analyzed in two stages. First, a socio-economic analysis of the production of vermicompost focused on evaluation of the time required for making vermicompost, assuming that farmers made their own. It was important to identify the cost in Malagasy ariary (1$ = 2000 ariary (2012 average)) per kg vermicompost, and the Internal Rate of Return, or IRR, of investment using a Simulation Test, or TSIM, in XLSTAT. Second, a comparative analysis between the treatment group, composed of farmers using vermicompost, and other types of fertilization and/or other modes was performed. In the Discriminant Analysis, a Wilks' Lambda test was done to determine if the averages of the types statistically differed. Comparative criteria were based on fertilizer prices in ariary/kg, comparing costs of fertilizers and the yields acquired, costs of agricultural inputs, and the impact of the fertilization methods with total income of the operator.

2.2.3. Multiple Correspondence Analysis

A Multiple Correspondence Analysis (MCA) in XLSTAT was used to study the variables according to membership in a defined type. The variables that had strong correlations with major constraints farmers faced with the use or development of vermicompost were analyzed based on the responses of each household in the investigative survey and following the typology.

3. Results and Discussion

3.1. Differentiation of Farms

The factor analyses method classified the operators into three types (Table 1). Households using vermicompost as fertilizer were categorized as Type 2, and were considered the minority (25% of the total). Households using other types of fertilization were classified as either Type 1, judged as the majority class (38%) using more fertilizer, or as Type 3 (37%) using less fertilizer. The number of households using vermicompost was low compared to all households, while 75% of surveyed operators used other types of fertilization, such as chemical fertilizer and/or cattle manure.

Table 1. Summary of household typology.

Types of Households	Type 1	Type 2	Type 3
Denomination	Type using more fertilizer	Type using vermiculture	Type using less fertilizer
Fertilization	mixed fertilizers	vermiculture	mixed fertilizers
Percentage	38.3%	25.0%	36.7%
People employed	>4	1 to 2	3 to 4
Total harvest area	>0.3 ha	0.15–0.3 ha	≤0.15 ha
Area reserved for rice	>0.15 ha	0.05–0.15 ha	≤0.05 ha
Area for vegetables crops	>0.2 ha	0.1–0.2 ha	<0.1 ha
Area for cassava	>0.1 ha	0.02–0.1 ha	<0.02 ha
Area reserved for making vermicompost	0	$\geq 2 \text{ m}^2$	0
NPK	>15 kg	0	≤15 kg
Urea	>8 kg	0	≤8 kg
Cattle manure quantity purchased	>500 kg	<200 kg	200 to 500 kg
Insecticides used	>2 L	0	≤2 L
Cattle	≥2	0	0
Pigs	1 to 2	0	0
Poultry	≤4	0	>4
Practice other activities	No	Yes	Yes

3.2. Socio-Economic Analysis of Vermicompost Production

Based on the practical experiences of households using vermicompost, and comparing three methods for installing a worm bin (in basement, ground framed by brick lined, floor framed by boards), the time involved was not a constraint, as the amount of labor required in persons/day was low and did not exceed 1 person/day. In addition, with a cost of about 60 ariary/kg, vermicompost was produced at a low cost (Figure 1). By investing more to get more fertilizer, the IRR was high, assuring the profitability of the investment. The low cost incurred for the production of one kg of vermicompost was explained, first, by the tools of construction of the worm bin, which were not expensive, reducing the costs incurred [3]. In addition, the raw materials needed were readily available in the area at a very low price or were even free. Furthermore, vermicompost provided a waste recycling opportunity for households. Thus, the overall advantage was that this system facilitated the farmers' respect of the environment.

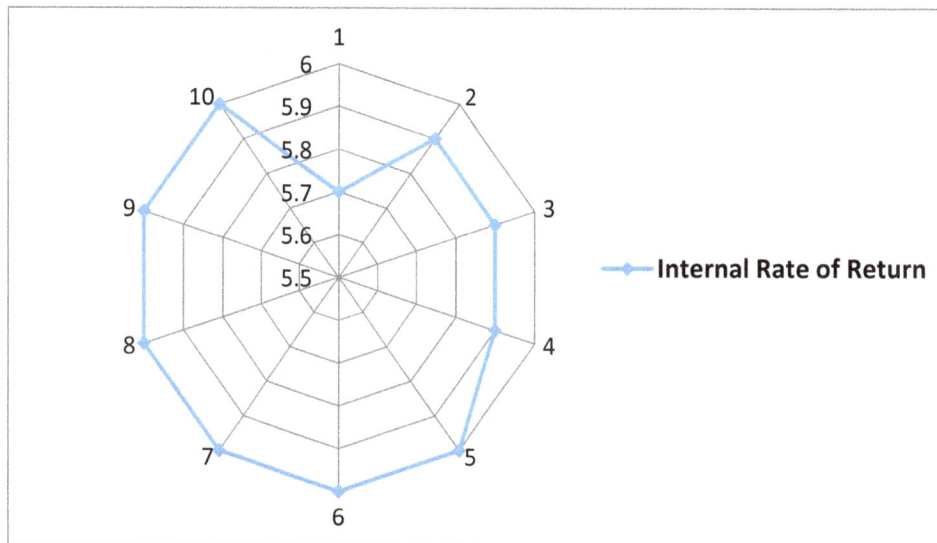

Figure 1. Internal Rate of Return of the manufacture of vermicompost (in increments of 100).

Figure 1 shows that the IRR increased every year during 10 years, and had a high value, which assured a good return on investment when making vermicompost.

3.3. Fertilizers Costs Impact on Crop Yields

There were some significant differences between household Types 1 and 3 using mixed fertilizers with respect to their fertilizer costs, and between these two types compared to Type 2 that used vermicompost (Figure 2A). High fertilizer costs were not associated with greater yields (Figure 2B). The low costs of fertilizers for Type 2 households were due to making their own manure, and suggested that high yields were a result of good quality vermicompost. The use of vermicompost reduces production costs by reducing expenses in chemical inputs, particularly fertilizers and pesticides [4].

Type 1 households had the highest expenses for fertilizers for all crops except zucchini and cassava, while their yields were in the middle among the three types, and they did not differ from Type 3 households. Type 2 had lower fertilizer costs and also generally higher returns. In the case of cassava cultivation, Type 1 and 3 households did not use fertilizer while a Type 2 household paid 600 ariary/are. Type 2 obtained a yield of 47 kg/are, but Type 1 had 22 kg/are and Type 3 had 20 kg/are. Logically, when fertilizer expenses decrease, and yield increases, the value of production also increases, from 10% to 50% in the present study.

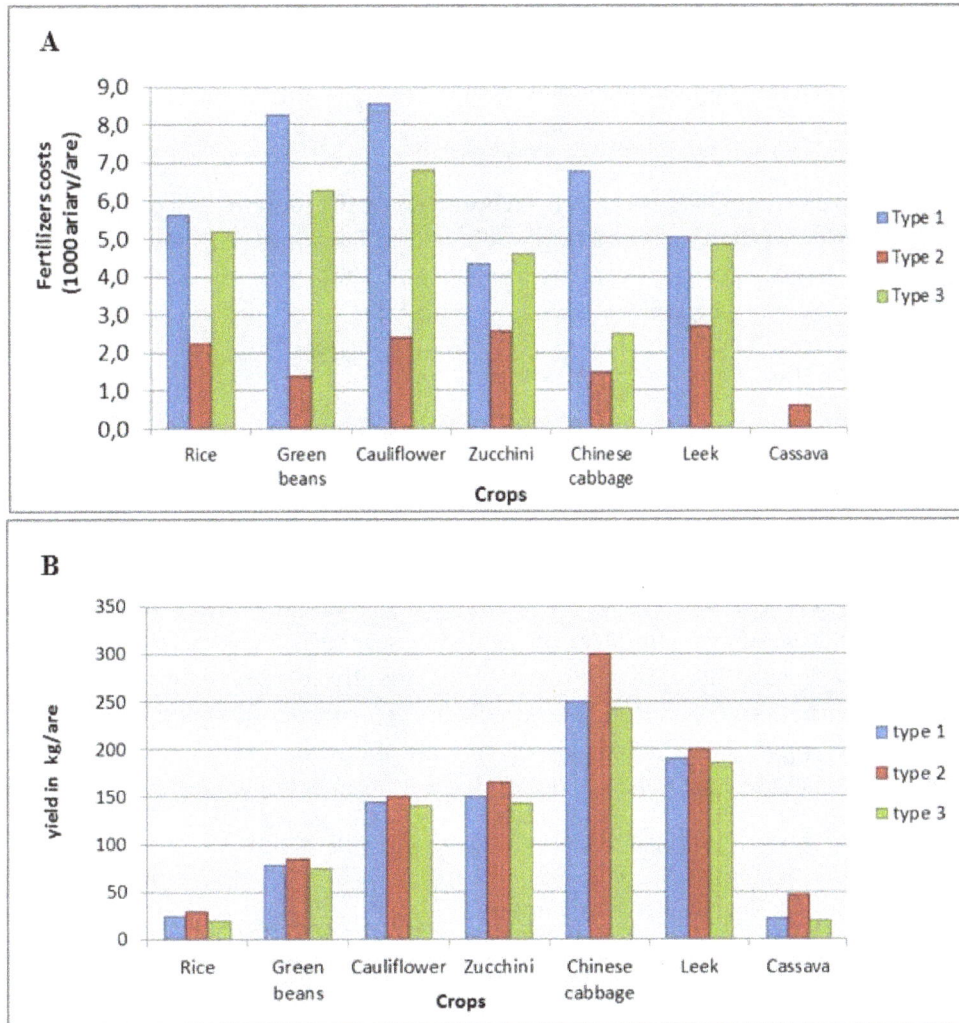

Figure 2. Comparison of (**A**) fertilizer costs and (**B**) yields for several crops by household type using Wilks' Lambda test. Area units are "are", or 100 m^2 or 0.01 ha.

3.4. Impact of the Use of Vermicompost on Total Household Income

In addition to the benefits obtained from the use of vermicompost on the farm, the manufacturing process creates a new source of income. The TATA Association offers its members an opportunity in terms of a market, but considers requests from entities or individuals with combined vermicompost production of 500 kg to 1 ton per month, depending on the growing season. In addition to the advantages of selling and using vermicompost, households would reduce their spending on agricultural inputs and improve soil fertility while increasing long-term yields [4].

3.5. Factors Affecting Development of Vermicomposting

3.5.1. Analysis of Households' Opinions by MCA

Survey results refer to different perspectives and expertise of households on the use and the manufacture of vermicompost. Type 1 households did not know vermicompost manufacturing techniques, or had not fully mastered the methods. Some did not understand the value of vermicomposting, while others consider it a more or less good fertilizer according to information circulating in their locality. Type 2 households knew that the manufacture of fertilizer from worms was feasible, and had mastered the manufacturing techniques in general, so for them the value of vermicomposting was appreciable. Type 3 households only knew that it was possible to produce vermicompost with earthworms, although some were unable to control its manufacture, and others had very few production skills. They knew that vermicompost had good qualities and allowed good yields, but some hesitation was noticed among them.

3.5.2. Constraints for Those Producing Vermicomposting

Failures due to behavioral factors have been observed by the founder and president of the TATA trainers association, given the amount that a household could produce. The number of active people was not a constraint, but insufficient observation, anticipation, and the lack of development of planning the manufacturing of vermicompost could be constraints [5].

3.5.3. Constraints for Those Not Producing Vermicompost

In analyzing the situation of Type 1 and Type 3 households, one of the major problems facing the use of vermicompost was insufficient information. Indeed, in the Rural Municipality of Ambohimanambola, vermicomposting training has been done only by the TATA association, but there has been no specific budget for public awareness campaigns and/or information sessions. In short, the lack of information has led to a lack of awareness of the benefits of using vermicompost. Moreover, farmers have not mastered the steps to make it. Many operations have succeeded over time in promoting the use of agricultural inputs in Madagascar such as chemical fertilizers, but efforts promoting vermicomposting have been unsuccessful. Currently, in most of the rural world, the only information on fertilizers has been for the use of chemical fertilizers. Farmers believe that the best solutions for improving their operations are by the use of chemical fertilizers. Thus, insufficient information has been an important factor blocking the development of the use of vermicompost.

3.6. Valuation and Maximizing the Benefits of Vermicomposting on the Farm

3.6.1. Optimization of the Benefits of Vermicompost

To make the most of the advantages of using vermicompost, more needs to be used. Farmers should be encouraged to consider making vermicomposting a substantial part of their operation. With an increase in production of vermicompost, the farmer/producer could receive more income by selling unused quantities. To better maximize the benefits of vermicompost, it would also be important to restore and maintain soil fertility and ecological balance by crop rotation, vegetation cover, natural fertilization and minimum tillage [6].

3.6.2. Valuation of Organic Products and Local Market Development

Currently, the possibility of disposal of agricultural products for all growers is through local distributors who offer the same price irrespective of the type of fertilization. It would be important to organize marketing by selling organic products at a separate price. It would be useful to establish some regulations concerning product quantity and quality. The creation of a common marketing center would be necessary, which would mean uniting producers in cooperative-type associations authorized to act as strong traders in managing the flow of quality product. Finally, it would be recommended to induce consumers to buy organic products, eating food that balances human health.

3.6.3. Strategies for Developing the Use of Vermicompost

State Intervention

The development of organic farming, in direct relation to the use of organic fertilizers like vermicompost, is not included fully in national politics in Madagascar [7]. It would be important to clearly define a national policy on organic agriculture by supporting the private sector, NGOs and associations working in this field such as the TATA association, and encouraging farmers to produce their own fertilizer as vermicompost.

Creating Network or Movement that Brings Together All Stakeholders

Currently, the only group in Madagascar which brings companies operating in the field of organic farming together is PRONABIO (Produits Naturels et Biologiques), comprised of 31 companies working in the field of organic farming. Members have the support, activities, and ambitions of its members as a permanent source of advice and information, promotion of activities and products (participation in fairs, exhibitions, conferences, seminars and international exhibitions, creation of a permanent office with a technical secretariat), and the defense of the interests of its members. Most of these companies establish contracts with smallholder farmers, collect their products, and export them. By creating a network, farmers could be motivated to use vermicompost and also produce more for the international organic market. The network should include all the players in this sector at a regional level: research, production, training, supervision, technical assistance, and export financing. Members could share their experiences, lobby to defend the rights of players in the industry, represent the sector at national and international levels, and encourage professionalisation of the sector for better quality, improved competitiveness in the marketplace, and sustainability of organic farming in Malagasy [8]. To establish the development of the use of vermicompost, it would also be important to establish a website that would bring together farmers associations, private entities, and/or the public who may be applying vermicompost.

4. Conclusions

This research focused on the identification of the mode of development of vermicomposting by farmers in the Rural Municipality of Ambohimanambola to resolve fertility problems, improve organic agriculture and agricultural productivity, and increase household incomes. The analyses indicated that the agricultural practices of farmers using vermicompost were different from those of households using other types of fertilization according to predefined criteria. About 25% of the households surveyed were classified as Type 2 that practiced vermicomposting, while the other two types used mixed fertilizers and were differentiated by other factors. The results indicated that the use of vermicompost does not yet hold an important place for most farmers. The study also showed that the use of vermicompost reduced average spending on agricultural inputs from 45% to 60%, and resulted in yield increases of 10% to 50%. Sale of unused vermicompost could provide another source of income. The analysis confirmed that the lack of information and awareness inhibited growth in the use of vermicompost, but strategic points for promoting development of vermicomposting were identified. The survey revealed a good knowledge of vermicomposting techniques for improving crop yields, for avoiding unnecessary expenses by the use of chemical fertilizers, and for protecting the environment. A trend towards sustainable organic agriculture and its principles was observed during this research, but it remains to be verified if the presence of the TATA association working in the field of vermicomposting has really expanded its use. For other regions or towns, the question is: "How can vermicomposting be introduced in areas where there are no organizations or associations working in organic agriculture?"

Acknowledgments: Recognition to TATA Association for its material and human support in the accomplishment of this research.

Author Contributions: This work was a product of the combined effort of all of the authors.

Conflicts of Interest: The authors declare no conflict of interest.

References

1. Rasoarimalala, S.; Andriamialijaona, H.; Rabemanantsoa, M. *Support for the Establishment of a System of Production of Inputs (Fertilizers and Pesticides) in Areas of Intervention PARECAM*; Study Report; PARECAM: Antananarivo, Madagascar, 2001; p. 122.
2. BIMTT; MAMABIO; MINAGRI; RTM; TATA. *Manual on Basic Techniques in Organic Farming: For Technicians*; RTM (Reggio Terzo Mondo): Antananarivo, Madagascar, 2011; p. 60.

3. MB Bouche, M.B. *The Earthworm for the Treatment of Waste*; Periodically Biofutur 7; Lavoisier: Paris, France, 1982; pp. 43–46.

4. Kouassi, K.J. Technical and Socioeconomic Analysis on a Farm in Attiégouakro (Ivory Coast). Available online: http://www.memoireonline.com (accessed on 23 December 2012).

5. Rajaonarison, A. Organic Farming in Madagascar. Master's Thesis, Laulanié Green University, Isoraka, Madagascar, 2009.

6. CTA; UNCTAD; WTO. *Global Markets for Organic Fruit and Vegetables*; FAO: Rome, Italy, 2001; pp. 301–309.

7. Vermicompost. Available online: http://www.erails.netconsulted (accessed on 5 November 2011).

8. Andriamady, T.V. Agricultural Experimentation Based Manufacturing Vermicompost for Soil Fertility. Master's Thesis, Superior School of Agronomy, Antananarivo, Madagascar, 2011.

6

Organic and Conventional Produce in the U.S.: Examining Safety and Quality, Economic Values, and Consumer Attitudes

Amarat Simonne [1,*], Monica Ozores-Hampton [2], Danielle Treadwell [3] and Lisa House [4]

[1]   Food Safety and Quality Research Laboratory, Department of Family, Youth and Community Sciences, Institute of Food and Agricultural Sciences, University of Florida, Gainesville, FL 32603, USA

[2]   Horticultural Sciences Department at SWFREC, Institute of Food and Agricultural Sciences, University of Florida, Immokalee, FL 34142, USA; ozores@ufl.edu

[3]   Horticultural Sciences Department, Institute of Food and Agricultural Sciences, University of Florida, Gainesville, FL 32611, USA; ddtreadw@ufl.edu

[4]   Food and Resource Economics Department, University of Florida, Gainesville, FL 32611, USA; lahouse@ufl.edu

*   Correspondence: asim@ufl.edu

Abstract: Organic agriculture is an industry sector that has been experiencing steady global growth in recent years. The United States is ranked first in organic food consumption, followed by Germany and France. In 2014, the estimated market value of organic foods in the U.S. was $42 billion; 43% of this total was attributed to produce (fruits and vegetables). Organic production systems in the U.S. must adhere to National Organic Program (NOP) standards that integrate cultural, biological, and mechanical practices. These standards promote the recycling of resources and ecological balance while conserving biodiversity. While the U.S. organic produce sector is steadily expanding, many questions related to price, safety, nutritional quality, and consumer preference remain. This paper will provide comparisons and insights in the following areas: (1) the economic contribution and impact of the organic produce market; (2) the U.S. National Organic Standards and requirements, as well as the certification process; (3) the nutritional quality and safety of organic produce; (4) consumer attitudes and preferences regarding organic produce; and (5) future research directions and developments for the organic produce industry.

Keywords: food safety; organic produce; nutritional quality; microbial safety

1. Introduction

Organic agriculture is an industry sector that has been experiencing steady global growth in recent years. The difference between organic and conventional products is how they are grown (produced) and processed; specific regulations as to what is considered organic or nonorganic (conventional) can vary by region. For example, the regulations for the European Union (EU) can be found in Council Regulation (EEC) 834/2007 and 889/2008 and 1235/2008 on organic production of agricultural products [1]; for the U.S., pertinent regulations can be found at [2]. The Research Institute of Organic Agriculture (FiBL) in collaboration with the International Federation of Organic Agricultural Movements (IFOAM) recently released a comprehensive review of global organic agricultural statistics and emerging trends. The document provides a comprehensive overview of standards and regulations in various regions of the world as well as current statistics on organic production areas, producers of organic crops, organic markets, and data on specific crops such as citrus, fruits (temperate, tropical and subtropical), and vegetables [3]. As with conventional production systems, organic production systems include a wide range of practices deemed acceptable under various organic rules, laws, and regulations. This particular manuscript will focus on organic and conventional produce (fruits and vegetables) in the United States, with a particular focus on economic value, safety and quality, and consumer attitudes.

2. Economic Impact of Organic Produce in the United States

Globally, the United States is ranked first in organic food consumption, followed by Germany and France. Organic food sales in the United States totaled $24.8 billion in 2012, accounting for 3.7% of total U.S. food sales and 11.4% of U.S. fruit and vegetable sales [4]. Organic production of various crops and animal-derived products has been one of the fastest-growing segments of U.S. agriculture in the last decade (United States Department of Agriculture Agricultural Marketing Service [2,5,6]. In 2014, the estimated market value of organic foods in the U.S. was $42 billion; 43% of this total was attributed to produce (fruits and vegetables) [6]. In 2008, there were 168,776 acres (approx. 68,300 ha) dedicated to certified organic vegetable production in the U.S. [6,7]. California continues to lead the nation in certified organic vegetable production, with more than 104,076 acres (42,118 ha) under cultivation.

3. The U.S. National Organic Standards and Requirements and Certification Process

In the United States, organic vegetable and fruit production must be in compliance with the National Organic Program (NOP) standards [2], which are designed to create production systems that "respond to site-specific conditions by integrating cultural, biological, and mechanical practices that foster cycling of resources, promote ecological balance, and conserve biodiversity" (USDA/AMS, 2009, para. 6) [2]. To be classified (or certified) as organic growers or producers, producers must develop and submit an organic system plan. The plan must cover all aspects of vegetable and fruit production, ranging from soil fertility to postharvest processing and handling; this includes weed and insect control plans, disease management, and crop rotations. (Vegetables and fruit must be grown in rotation with other crops.) In developing an organic system plan, producers need to consult the NOP Generic National List (GNL) which brings together in one location a generic list of materials that can be utilized for organic production. Alternatively, the Organic Materials Review Institute (OMRI) also provides a list of brand name products that can be utilized in organic production (www.omri.org) [8].

The OMRI is a nonprofit organization that has no connection with the NOP, and the OMRI Products List (OPL) represents that organization's recommendations regarding the acceptability of brand-name products for organic production, processing, and handling. The products included in the OPL have been reviewed against the standards developed by OMRI for assessing compliance with the NOP Rule (NOP/7 CFR Part 205) [2]. Only those products that have passed this review are included in the OPL and can display the OMRI Listed™ seal on labels and in advertising and promotions. Because participation in the OMRI program is voluntary, and there is cost involved, the OPL (OMRI Products List) is not a comprehensive listing of products suitable for organic production, processing, and handling. Therefore, a product's absence from the OPL does not imply its failure to comply with the NOP Rule.

Certifying land for organic production requires that no prohibited substances be applied for a minimum of three years prior to harvest. The NOP certifying agencies carry out detailed annual inspections to verify that production and handling processes on organic farms comply with NOP standards. Currently there are approximately 80 certification agencies in the U.S., all accredited by the USDA, and applicants have the freedom to select any accredited U.S. certification agency. As a result, in the U.S., when consumers buy Certified Organic farm products, they can be certain that they are supporting farmers who take their stewardship roles so seriously that they have voluntarily accepted strict federally regulated standards [9]. In addition, the USDA National Organic Program also provides strategic plans, which are open and transparent for 2015–2018 [10].

4. Nutritional Quality and Safety of Organic *vs.* Conventional Produce

Overall, according to the recent literature, consumers continue to perceive that organic produce is more nutritious and safer. However, while the organic production of fruits and vegetables has experienced rapid growth, very few well-controlled studies have been published to prove or disprove these beliefs. This section of the manuscript will focus on traditional nutrients and selected phytonutrients; for the most part, the literature will cover recent major reviews from 2009 to present. Only limited data are available for specific produce in limited geographical areas and for specific nutrients.

A recent review of studies (published in 2011) suggests that ingestion of organic produce leads to a higher intake of phenolic compounds and vitamin C, as well as lowering the amount of nitrates and pesticides consumed [11–14]. According to a review by Hunter *et al.* [13], organically grown produce has about a 12% higher concentration of secondary metabolites. If these secondary metabolites positively influence health, organic foods may possess advantages over conventional foods. According to the latest major review (a meta-analysis) by Barański *et al.* [15], on average, organic crops have higher concentrations of antioxidants, lower concentrations of cadmium, and a lower incidence of pesticide residues than nonorganic crops, across regions and production seasons.

However, other recent systemic reviews of nutrition and health effects of organic foods conducted in the US have revealed few health benefits from consuming organic foods [16,17]. Another recent study by the American Academy of Pediatrics examining the potential health and environmental advantages and disadvantages of organic foods revealed the following: (a) due to a lack of control as well as other issues with measurement, nutritional differences between organic and conventional produce appear to be minimal; (b) organic produce contains fewer pesticide residues, but it remained unclear how clinically relevant this is [18].

In terms of safety, there seems to be a consensus in literature that organically-grown fruits and vegetables tend to have lower incidences of pesticide residue contamination and lower levels of nitrates [11,12,14]. In the United States, according to the latest (2013) report from the USDA, more than 99% of all products sampled by the program had pesticide residues below the EPA tolerance. However, this report did not distinguish between organic or conventional production; it simply designated the sample as domestic, imported, or unknown. It is notable that fresh and processed fruit and vegetables accounted for 84% of the 10,104 collected samples for pesticide analyses. More than 40% of the samples had *no* detectable pesticide residue. Produce with detectable pesticide residues (5%) included bananas, broccoli, carrots, cauliflower, celery, green beans, mushrooms, nectarines, peaches, plums, raspberries, and summer and winter squashes [5].

When it comes to bacterial contamination, the most recent comprehensive review concluded that there is no statistically significant difference between organic and conventional produce [17]. In addition, in terms of fungal toxins and heavy metal contamination, the latest reviews revealed mixed results. These mixed results were also consistent with some earlier major published reviews that are not cited in this manuscript.

Experts agree that quality (including nutritional quality) and safety of fresh produce is a complex issue which is affected by a variety of factors, such as environmental conditions, production practices, postharvest processing and handling, and bioavailability after consumption. A body of research reveals that the compositional differences between organic and conventional produce depends on production practices and surrounding environments for both systems. In addition, postharvest processing, storage, handling, preparation, cooking, and processing prior to consumption all affect nutrition, bioavailability, and functionality (Table 1). Once produce is consumed, only certain amounts of nutrients are absorbed into the human body, bloodstream, or targeted tissue and reach clinically relevant levels. It seems that comparing the absolute difference in the nutrients of fruits and vegetable grown under different production methods may not be sufficient. Many experts have suggested conducting further research, looking at a broader range of nutrients and potential impact on human health outcomes [11,12,14,17,18]. Based on current knowledge, the question becomes: are we asking appropriate research questions?

Table 1. Summary of factors affecting quality and nutritional quality of produce from production to consumption and absorption in the body.

Production Stage	Harvest, Postharvest Handling, Packaging and Storage	Preparation and Cooking and Processing	In Human Body (After Consumption)
-Immediate environment (e.g., geographic zones) -Production conditions -Fertilization -Pest control -Season -Water -Light/day length -Soil types/growing mediums -Plant genotypes -Analytical methods for assessment of selected nutrient components and the sample preparation methods	-Harvest methods -Harvest maturity -Plant parts -Postharvest handling and processing techniques -Time and temperature of storage -Analytical methods for assessment of selected nutrients components and the sample preparation methods	-Preparation methods -Cooking and processing methods -Consumer handling and storage -Analytical methods for assessment of selected nutrients components and the sample preparation methods	-Age, gender and genetic of individual -Enzymatic levels -Manner of eating (e.g., chewing) -Stage of health -Analytical methods for assessment of selected nutrients components and the sample preparation methods

Source: [11–19].

5. Consumer Attitudes and Preference Regarding Organic Produce

According to recent consumer surveys, 84% of American consumers buy organic food. Certified organic produce is common in most grocery chains and fetches substantial price premiums in the U.S. Customers' top motivations for purchasing organic produce include health reasons and avoiding genetically modified organisms (GMOs) [4]. Nevertheless, consumers purchase organic foods for many different reasons, including environmental factors [13,17] and the perception that organic food tastes better [17]. According to a review by Yiridoe *et al.* [20], a clearer picture of consumer attitudes and willingness to pay will help the organic industry in the future. Even if consumers are willing to pay, knowledge and awareness of organic foods may not be consistent, so future consumer research will be helpful to better understand trends.

6. Future Research

Does conventional and organic produce differ when it comes to safety and quality? Are there compositional differences between the two? Current data offer few clear answers. Much depends on production practices and the surrounding environment, as well as postharvest processing, storage, handling, preparation, cooking, and processing prior to consumption. Furthermore, bioavailability after consumption and functionality also depend on the biological condition of the end user. In order to formulate meaningful research in this area, many factors must be considered. For example, with available GPS and geospatial technology, mapping of soil information throughout production regions can be incorporated and adjusted in the comparison studies. Development of specific biomarkers for tracking the consumption of specific secondary metabolites and assessing the benefits of consumption as they relate to different groups of consumers could come into play. In addition, understanding consumers and the knowledge and motivations that affect current and future behaviors will help the industry. Last but not least, critical reviews of analytical procedures will help provide a better understanding of the situation. These are but a few examples of future themes for research in these areas.

7. Conclusions

There is no doubt that the organic produce market provide significant economic impact in the U.S., but when it comes to quality and safety, current data offer few clear answers. Much depends on production practices and the surrounding environment, as well as postharvest processing, storage, handling, preparation, cooking, and processing prior to consumption. While consumers are willing to pay, knowledge and awareness of organic foods may not be consistent.

Acknowledgments: This work is partially funded by the NIFA HATCH Fund (FLA-FYC-005476).

Author Contributions: Amarat Simonne conducted the safety and quality issues, Monica Ozores-Hampton and Danielle Treadwell persued organic standards and production and Lisa House conducted the consumer preference issues.

Conflicts of Interest: The authors declare no conflict of interest.

References

1. International Federation of Organic Agriculture Movements, European Union Regional Group. IFOAM EU Group. Available online: www.ifoam-eu.org/sites/default/files/page/files/ifoameu_reg_regulation_dossier_201204_en.pdf (accessed on 8 March 2016).

2. National Organic Program standards. Available online: www.ams.usda.gov/nop/NOP/standards (accessed on 28 July 2015).

3. The World of Organic Agriculture: Statistics & Emerging Trends. Available online: www.fibl.org/fileadmin/documents/shop/1663-organic-world-2015.pdf (accessed on 1 August 2015).

4. The Organic Trade Association's 2014 Organic Industry Survey. Available online: ota.com/sites/default/files/indexed_files/StateOfOrganicIndustry_0.pdf (accessed on 4 August 2015).

5. Pesticide Data Program Annual Summary, Calendar Year 2013. Available online: www.ams.usda.gov/AMSv1.0/getfile?dDocName=STELPRDC5110007 (accessed on 7 August 2015).

6. 2007 Census of Agriculture. Available online: www.agcensus.usda.gov/Publications/2007/Online_Highlights/Organics/ORGANICS.pdf (accessed on 1 August 2015).

7. Organic Market Overview. Available online: www.ers.usda.gov/topics/natural-resources-environment/organic-agriculture/organic-market-overview.aspx (accessed on 7 June 2015).

8. OMRI Lists. Available online: omri.org/OMRI_about_list.html (accessed on 1 August 2015).

9. What is Organic Farming? Available online: www.extension.org/article/18655 (accessed on 7 August 2015).

10. National Organic Program Strategic Plan 2015–2018. Available online: www.ams.usda.gov/AMSv1.0/getfile?dDocName=STELPRDC5086491&acct=nopgeninfo (accessed on 7 June 2015).

11. Brandt, K.; Leifert, C.; Sanderson, R.; Seal, C.J. Agroecosystem management and nutritional quality of plant foods: The case of organic fruits and vegetables. *Crit. Rev. Plant Sci.* **2011**, *30*, 177–197. [CrossRef]

12. Huber, M.; Rembialkowaska, E.; Srednicka, D.; Bugel, S.; van de Vijver, L.P.L. Organic food and impact on human health: Assessing the status quo and prospects of research. *Life Sci.* **2011**, *58*, 103–109. [CrossRef]

13. Hunter, D.; Foster, M.; McArthur, J.O.; Ojha, R.; Petocz, P.; Samman, S. Evaluation of the micronutrient composition of plant foods produced by organic and conventional agricultural methods. *Crit. Rev. Food Sci.* **2011**, *51*, 571–582. [CrossRef] [PubMed]

14. Lima, G.P.P.; Vianello, F. Review on the main differences between organic and conventional plant-based foods. *Int. J. Food Sci. Technol.* **2011**, *46*, 1–13. [CrossRef]

15. Baranski, M.; Srednicka-Tober, D.; Volakakis, N.; Seal, C.; Sanderson, R.; Stewart, G.B.; Benbrook, C.; Biavati, B.; Markellou, E.; Giotis, C.; *et al.* Higher antioxidant and lower cadmium concentrations and lower incidence of pesticide residues in organically grown crops: A systematic literature review and meta-analyses. *Br. J. Nutr.* **2014**, *112*, 794–811. [CrossRef] [PubMed]

16. Dangour, A.D.; Lock, K.; Hayter, A.; Aikenhead, A.; Ellen, E.; Uauy, R. Nutrition-related health effects of organic foods: A systematic review. *Am. J. Clin. Nutr.* **2010**, *92*, 203–210. [CrossRef] [PubMed]

17. Smith-Sprangler, C.; Brandeau, M.L.; Hunter, G.E.; Bavinger, C.; Pearson, M.; Eschbach, P.; Sundaram, V.; Liu, H.; Schirmer, P.; Stave, C.; *et al.* Are organic foods safer or healthier than conventional alternatives?: A systematic review. *Ann. Intern. Med.* **2012**, *157*, 348–366. [CrossRef] [PubMed]

18. Forman, J.; Silverstein, J. Organic foods: Health and environmental advantages and disadvantages. *Pediatrics* **2012**, *130*, e1406–e1415. [CrossRef] [PubMed]

19. Terry, L.A. *Health-Promoting Properties of Fruit & Vegetable*; CABI: Willingford, UK, 2011.

20. Yiridoe, E.K.; Bonti-Ankomah, S.; Martin, R.C. Comparison of consumer perceptions and preference toward organic *versus* conventionally produced foods: A review and update of the literature. *Renew. Agric. Food Syst.* **2005**, *20*, 193–205. [CrossRef]

7

Hydraulic Performance of Horticultural Substrates—3. Impact of Substrate Composition and Ingredients

Uwe Schindler *, Gunnar Lischeid and Lothar Müller

Leibniz Centre for Agricultural Landscape Research (ZALF), Institute of Landscape Hydrology,
Eberswalder St. 84, Muencheberg D15374, Germany; lischeid@zalf.de (G.L.); mueller@zalf.de (L.M.)
* Correspondence: uschindler@zalf.de

Abstract: Horticultural substrates, also referred to as growing media, potting soils and gardening or soilless substrates, are widely used as a basis for vegetable and flower production in horticulture. They are created as a composition of different ingredients (bog peat, organic residuals, coir, perlite and other components). Hydraulic properties such as water storage capacity, air capacity, shrinkage behaviour, wettability or hydraulic conductivity are important variables for a comprehensive evaluation of the performance of horticultural substrates. A set of 36 commercial potting soils and substrates was selected and the hydraulic properties (water retention curve, unsaturated hydraulic conductivity function, capillary rise and shrinkage) were measured using the extended evaporation method (EEM). Additionally, the water drop penetration time was determined as a measure of wettability. The hydraulic performance of the horticultural substrates was evaluated. Generally, bog peat is the main component of horticultural substrates. Additionally, coir (raw coconut fibre), bark, different composts and mineral ingredients such as perlite, pumice, vermiculite, sand and others are used. The growing medium with the best hydraulic performance in this study revealed substrates composed of bog peat with added coir, perlite and organic residuals. Mineral ingredients in general decreased the content of easily available water but did not exhibit any significant effect on the other properties studied. However, the risk of a lack of air can be increased by the addition of clay. The presence of perlite had positive effects on the air content and the re-wettability. The presence of organic materials had significant and detrimental effects on the height of the capillary rise. We also found that some products declared as preferable for use in containers were better suited as substrates for bed cultivation. However, a comprehensive evaluation of the eligibility of horticultural substrates in horticulture requires not only hydraulic measurements but also growing experiments and an assessment of their chemical, biological and technological suitability.

Keywords: water retention curve; unsaturated hydraulic conductivity; water repellency; water drop penetration time; shrinkage; extended evaporation method (EEM); HYPROP

1. Introduction

Horticultural substrates, also referred to as growing media, potting soils and gardening or soilless substrates, are widely used as a basis for vegetable and flower production in horticulture and private households, either in greenhouses or under field-grown conditions [1–3]. They are created as a composition of different ingredients. In most cases bog peat, mainly consisting of sphagnum moss, is used as the basis for producing horticultural substrates [2,3].

There are many different substrates for horticultural applications on the market. The declaration on the package generally provides information on the ingredients and the chemical composition. Generally, water storage evaluations and water budget declarations on substrate packages are based on assumptions or are missing. However, accurate substrate hydraulic criteria, parameters and measurement data can improve the evaluation of the hydraulic performance of horticultural substrates in horticulture [3].

The aim of this study was to compare and evaluate the hydraulic properties of some commercially available substrates for horticultural applications. Some important questions were: Do commercially available substrates meet hydraulic requirements (sufficient easily available water and air, high capillarity, low shrinkage, good wettability) for gardening? Are there correlations between the ingredients and basic properties of horticultural substrates with hydraulic properties? Does the package labelling provide conclusions on the hydraulic quality of horticultural substrates? The results of evaluation of a set of horticultural substrates are presented.

2. Experimental Section

2.1. Horticultural Substrates

A set of 36 commercial horticultural substrates (HS) for different horticultural applications was analysed (Table 1). The samples varied in their bog peat and ash content (x), the added ingredients, their price and other properties. Most substrates consisted of 80% or more bog peat (HS No. 1, 2, 4, 5, 7, 9, 10, 11, 12, 13, 14, 15, 16, 17, 18, 19, 20, 21, 22, 23, 24, 26, 27, 28, 29, 30, 31, 32 and 35) with added mineral and/or organic ingredients (garden residual and compost, forest residual, clay, sand, perlite, coir, lime, guano). In some substrates there was much less bog peat (No. 3, 6, 25, 36) and two substrates (33, 34) were peat-free. The Chrysal active substrate package (No. 8) provided no information about the ingredients.

2.2. Hydraulic Criteria

The hydraulic properties were evaluated as an example for plant growth in 20-cm-high containers and for bed cultivation with free drainage. Hydraulic criteria were the easily plant-available water capacity at a tension of 100 hPa (EAW_{100}, growth in container, difference of water content at air capacity and at 100 hPa) and 800 hPa (EAW_{800}, bed cultivation, water content difference between field capacity and water content at 800 hPa), the air (Air) capacity, and the capillary rise. High-quality horticultural soils were designed to provide 24% by vol. or more easily plant-available water. The air capacity (Air) was expected to exceed 10% by vol. [3] to avoid stress due to air limitations. The capillary height calculated for a 5 mm·day^{-1} rate (CR_5) was used as an additional indicator to estimate the flow resistance and the hydraulic suitability and quality of the substrate regarding the easy exchange of water and nutrients in the growing layer. A capillary height of 30 cm was defined as the threshold value. The limiting factors were water repellency effects and shrinkage. With reference to Blanco-Canqui and Lal [4], the water drop penetration time (WDPT, Letey [5]) was not to exceed 5 s to avoid negative effects on water infiltration due to water repellency. Longer wetting times could be an indicator for rewetting limitations and preferential flow [6]. The shrinkage volume of 5% by vol. was not to be exceeded to avoid adverse effects on plant growth and resource management.

2.3. Hydraulic Measurements

The water retention curve and the unsaturated hydraulic conductivity function were measured using the extended evaporation method (EEM) and the HYPROP system from saturation to close to the wilting point [7,8]. If hydraulic information in the dry range is not required, the measurement could be stopped at 100 hPa or at any other tension. These functions were used to calculate the water capacities EAW_{100} and EAW_{800}, to measure the Air capacity and to quantify the CR_5. The shrinkage at 800 hPa was derived from the oven-dried (105 °C) sample. Horticultural substrates with measured oven-dry shrinkage (S_{dry}) of less than 25% by vol. generally provide shrinkage lower than the 5% by vol. at 800 hPa (shrinkage threshold value). The WDPT, as a measure for the rewetting behaviour, was measured after 4 h of free evaporation from the fresh substrate sample.

Table 1. List of commercial horticultural substrates.

HS	Product	Price [x]	Application [y]	Ingredients [z]
1	Falkena	M	P	90% Hh (H4–H8), 10% C
2	Plantop	M	P	Hh (H2–H5), G, F
3	Plantop, for grass	M	F	35% Hh (H2–H8), 30% F, 15% G, 20% L, C, S
4	Falkena, rhododen	M	F	Hh (H3–H9)
5	Falkena, potting soil	M	P	Hh (H3–H8), P, C
6	Bodengold, bio.	M	P/F	40% Hh (H2–H8), 20% F, 40% G, P, C
7	Bodengold, premium	M	P	100% Hh (H2–H5), P, C
8	Chrysal, active soil	H	P	no information
9	Cuxin, balcony plants	H	P	Hh (H3–H7), C, Co
10	Cuxin for turf rolls	H	F	Hh, G, Co
11	Falkena, balcony	M	P/F	Hh (H2–H9)
12	Mecklenburger	kA	F	Hh (H3–H5)
13	Treff_Jiffy Products	kA	F	Hh (H5–H8)
14	Plantop, substrate I	M	P/F	80% Hh (H2–H8), 15% F, 5% G
15	Thomas Phillips	M	P	80% Hh (H3–H7), 15% F, 5% G, S
16	Netto supermarket	M	P	Hh (H3–H8), F, G
17	Blumenrisse	M	P	100% Hh (H2–H8), C
18	Gartenkrone	M	P	80% Hh (H4–H8), F, P, Gu
19	Compo Sana	H	P	96% Hh (H3–H8), P, Gu
20	Floragard	H	P	100% Hh (H2–H8), Gu
21	Fleurelle	M	P	Hh (H2–H6), F, G, Gu
22	Hewita Flor	M	P/F	Hh (H2–H6), G, C, S
23	Grüne Welle bio soil	M	P/F	Hh (H2–H5), G, W
24	Compo Bio.	H	F	Hh (H2–H5, H6–H8), G, Gu, Ca
25	Cuxin, for vegetables	H	F	60% Hh (H3–H5, H5–H7), G, C, L
26	Stender potting soil	M	P	100% Hh (H3–H5, H5–H7), C
27	Frux with natural clay	H	P	Hh, C
28	Euflor, plantahum	H	P/F	Hh (H3–H5), S, C
29	Kuhlmann potting soil	M	F	82% Hh, 10% G, 5.5% S, 2.5% C
30	Grüne Welle	M	P	Hh (H3–H6), C
31	Raiffeisen Gartenkraft	M	P	97% Hh (H3–H8), C, Ca, 0.07% Gu
32	Cuxin, for container	H	P	Hh (H3–H4, H5–H6), Co, C
33	DCM Cuxin, peat-free	H	P	100% Co
34	Neudohum, peat-free	H	P	F, CO, C
35	Uniflor, Schohmaker	M	P	100% Hh
36	Pro-green-BK	kA	P/F	30% Hh, 40% Co, 30% P

[x] Price: M—medium; H—high; kA—no information; [y] Recommended use: P—container; F—bed cultivation; P/F—container and bed; [z] Ingredients: Hh—bog peat; H—degree of decomposition; G—garden residual and compost; F—forest residual; T—clay; S—sand; L—loam; P—perlite; Co—coir; Ca—lime; Gu—guano.

2.4. Statistical Analysis

The statistical analysis aimed to identify predictors for the hydraulic properties of the different substrates. Different approaches were used for metric and nominal candidate predictor variables. In total, 22 predictor variables were investigated. The level of significance was $p < 0.05$. To test for significant relationships between nominal predictor variables and the hydrological properties of the substrates, non-parametric Spearman rank correlation was used to account for the non-Gaussian distribution of the values. Out of all the candidate predictor variables, 19 were nominal variables, and 15 of these were in only two groups. Due to the different numbers of replicates and non-Gaussian bivariate distribution, the non-parametric Mann-Whitney test was used to check for significantly different median values between the respective two groups. The statistical analyses and some of the figures were produced using the R software, version 2.12.0 (R Foundation for Statistical Computing, Vienna, Austria) [9].

3. Results and Discussion

3.1. Hydraulic Properties

Table 2 presents all the relevant hydraulic properties of the substrates under study in detail. The results showed great variability among the substrates. The saturated water content varied between 77.2% and 90% by vol. (an average 84.9% by vol.), and the permanent wilting point reached values between 7.9% and 20.9% by vol. (an average 14.2% by vol.). These values were comparable with slightly decomposed natural peat soils [10] and with values for horticultural substrates [3,11–15]. The EAW_{100} (25.4%–44.1% by vol., average 32.4%) in 20-cm-high containers was much higher than under bed cultivation (18.3%–36.0%, average 23.4%). However, in some cases, substrates with high amounts of easily plant-available water showed limitations for air, shrinkage and/or rewetting. The rewetting was limited in more than 30% of the substrates.

Table 2. Hydraulic properties of the horticultural substrates.

HS No.	Θs	FC	pWP	Air_{p20}	Air_{Bed}	EAW_{p20}	EAW_{Bed}	S_{dry}	CR_5	$WDPT_4$
				% by vol.					cm	s
HS 1	86.2	48.4	11.9	9.4	37.8	32.7	20.0	29.2	24.4	5
HS 2	77.2	43.0	18.1	8.5	34.2	31.2	19.8	19.1	10.1	20
HS 3	78.8	42.2	16.2	8.4	36.6	32.9	19.0	20.2	26.7	12
HS 4	88.8	55.0	10.9	7.9	33.8	30.1	25.3	35.9	47.7	13
HS 5	86.3	47.8	14.4	13.1	38.5	27.9	18.3	25.3	45.7	0.1
HS 6	80.7	46.7	13.5	12.8	34.0	25.4	19.7	22.2	13.1	1
HS 7	87.2	52.1	15.3	10.6	35.2	29.2	23.8	27.1	54.7	0.1
HS 8	88.9	55.2	16.3	9.2	33.8	29.9	25.7	23.6	45.3	0.1
HS 9	90.0	54.6	19.0	6.0	35.4	25.9	27.5	27.2	29.3	0.1
HS 10	87.3	45.0	14.5	12.4	42.4	34.9	21.2	22.1	30.6	0.1
HS 11	84.3	46.7	11.6	6.3	37.6	34.2	19.0	29.8	30.4	32
HS 12	88.3	39.5	18.3	10.7	48.8	44.1	20.8	22.8	40.5	3
HS 13	87.6	47.2	16.8	8.3	40.5	37.2	24.9	24.1	18.3	57
HS 14	80.3	44.8	18.7	8.7	35.5	32.3	20.1	24.1	21.8	17
HS 15	86.2	51.7	13.6	7.5	34.5	32.1	25.2	31.3	29.0	7
HS 16	85.5	44.4	17.1	9.1	41.1	36.9	20.3	23.8	15.9	10
HS 17	87.7	54.3	15.2	10.5	33.4	26.4	26.0	35.2	35.3	1
HS 18	88.1	50.8	13.3	13.5	37.3	27.6	24.5	29.5	52.7	0.1
HS 19	88.3	51.1	7.9	10.9	37.2	30.5	26.0	27.1	36.4	1
HS 20	86.5	48.0	16.7	12.2	38.4	30.5	23.1	27.7	46.4	2
HS 21	89.3	48.7	13.2	13.9	40.6	32.1	23.5	19.1	21.5	1
HS 22	86.6	49.2	16.1	6.0	37.5	37.6	24.7	26.3	29.8	0.1
HS 23	77.9	41.4	11.3	8.0	36.5	33.1	18.4	18.4	37.2	19
HS 24	85.9	58.2	14.2	5.0	27.7	30.5	32.2	23.4	87.9	0.1
HS 25	82.9	44.6	16.7	9.8	38.3	34.3	23.9	24.7	12.7	3
HS 26	84.7	47.5	15.9	8.0	37.2	36.2	26.4	23.4	37.2	32
HS 27	83.6	51.9	11.4	4.2	31.7	34.6	36.0	32.1	79.9	240
HS 28	83.6	44.7	20.9	6.5	38.8	36.9	21.0	27.8	53.4	5
HS 29	81.6	47.4	11.1	6.7	34.3	32.3	24.0	24.9	69.6	7
HS 30	79.3	43.3	11.3	8.0	36.0	32.8	19.9	24.0	57.7	1
HS 31	84.5	50.7	10.4	8.7	33.8	30.0	23.9	28.4	53.0	1
HS 32	83.2	49.4	10.8	6.1	33.8	33.8	27.4	20.6	42.9	62
HS 33	89.0	38.1	9.1	17.0	50.9	38.2	25.1	14.5	76.3	0.1
HS 34	84.1	44.3	10.0	6.0	39.9	25.5	19.6	13.0	17.9	0.1
HS 35	79.9	53.2	15.1	23.9	26.7	25.4	22.7	27.6	26.0	10
HS 36	83.4	40.8	12.3	17.5	42.7	37.7	24.7	8.0	40.1	0.1
MW	84.8	47.8	14.1	9.8	37.0	32.6	23.4	24.5	38.8	15.6

Abbreviations:Θs—saturated water content; FC—field capacity at 60 hPa tension; pWP—permanent wilting point; Air_{p20}—air capacity in a 20-cm-high container; Air_{Bed}—air capacity at free drainage; EAW_{p20}—easily plant-available water in a 20-cm-high container; EAW_{Bed}—easily plant-available water fixed between FC and water content at 800 hPa; S_{dry}—shrinkage volume of the oven-dry sample; CR_5—capillary height of a 5 mm·day^{-1} rate; $WDPT_4$—water drop penetration time after 4 hours' free evaporation; MW—mean value.

3.2. Quality Scores

Taking into account the evaluation scales developed by Schindler and Mueller [16], the suitability of the substrates was evaluated for containers and for a bed (Table 3). The "total" score in Table 3 stands for an average evaluation with no direct link to any special application and crop. We could not find substrates which were evaluated as satisfactory or non-satisfactory for horticultural use. Twenty horticultural substrates (HS: 1, 5, 7, 8, 10, 12, 17, 18, 19, 20, 21, 24, 25, 28, 29, 30, 31, 33, 36) did meet all requirements for a very good hydraulic evaluation for horticultural use in both containers and beds. They provided sufficient easily plant-available water and air, and they were not or not markedly limited by shrinkage and water repellency. Two of them, HS 33 and HS 36, provided the best hydraulic performance of all the substrates under study with a score of 11.5. One noteworthy fact was that HS 33 was peat-free and consisted of 100% coir, while HS 36 was composed of 30% bog peat, 40% coir and 30% perlite. Eight of the substrates (HS: 3, 4, 6, 9, 22, 23, 32, 35) were evaluated as very well-suited or well-suited either for containers or for bed cultivation, and 7 (HS: 2, 11, 13, 14, 15, 16, 35) were evaluated as good for both kinds of cultivation. Only one substrate (HS 27) was of lower hydraulic quality than the others. This substrate was medium- to well-suited. It provided the highest level of easily plant-available water but was strongly affected by shrinkage and water repellency. Furthermore, the air volume after free drainage in 20-cm-high containers was strongly limited, only 4.2% by vol., and was the lowest of all the test substrates.

Table 3. Evaluation of the horticultural substrates.

Sub-Strate	Score_Basic Requirement				Score Limitation			Score		
	Container		Bed							
	EAW_{p20}	Air_{p20}	EAW_{Bed}	Air_{Bed}	CR_5	S_{dry}	$WDPT_4$	Bed	Cont.	Total
PS 1	5	4	4	5	1	1	0	9	9	9
PS 2	5	4	3	5	1	0	2	8	7	7.5
PS 3	5	4	3	5	1	0	1	9	8	8.5
PS 4	5	3	5	5	2	2	1	7	9	8
PS 5	5	5	3	5	2	1	0	11	9	10
PS 6	5	5	3	5	1	0	0	11	9	10
PS 7	5	5	4	5	2	1	0	11	10	10.5
PS 8	5	4	5	5	2	0	0	11	12	11.5
PS 9	5	3	5	5	1	1	0	8	10	9
PS 10	5	5	4	5	2	0	0	12	11	11.5
PS 11	5	3	3	5	2	1	2	7	7	7
PS 12	5	5	4	5	2	0	0	12	11	11.5
PS 13	5	4	5	5	1	0	2	8	9	8.5
PS 14	5	4	4	5	1	0	2	8	8	8
PS 15	5	3	5	5	1	2	1	6	8	7
PS 16	5	4	4	5	1	0	1	9	9	9
PS 17	5	5	5	5	2	2	0	10	10	10
PS 18	5	5	5	5	2	1	0	11	11	11
PS 19	5	5	5	5	2	1	0	11	11	11
PS 20	5	5	4	5	2	1	0	11	10	10.5
PS 21	5	5	4	5	1	0	0	11	10	10.5
PS 22	5	3	5	5	1	1	0	8	10	9
PS 23	5	4	3	5	2	0	2	9	8	8.5
PS 24	5	3	5	5	2	0	0	10	12	11
PS 25	5	4	4	5	1	0	0	10	10	10
PS 26	5	4	5	5	2	0	2	9	10	9.5
PS 27	5	2	5	5	2	2	2	5	8	6.5
PS 28	5	3	4	5	2	1	0	9	10	9.5
PS 29	5	3	5	5	2	0	1	9	11	10
PS 30	5	4	3	5	2	0	0	11	10	10.5
PS 31	5	4	4	5	2	1	0	10	10	10
PS 32	5	3	5	5	2	0	2	8	10	9
PS 33	4	5	5	5	2	0	0	12	11	11.5
PS 34	5	3	3	5	1	0	0	9	9	9
PS 35	5	5	4	5	1	1	1	9	8	8.5
PS 36	4	5	5	5	2	0	0	12	11	11.5

Abbreviations: EAW_{p20}—easily plant-available water in a 20-cm-high container; Air_{p20}—air capacity in a 20-cm-high container; EAW_{Bed}—easily plant-available water fixed between FC and water content at 800 hPa; Air_{Bed}—air capacity at free drainage; CR_5—capillary height of a 5 mm·day^{-1} rate; S_{dry}—shrinkage volume of the oven-dry sample; $WDPT_4$—water drop penetration time after 4 hours' free evaporation; Cont.—container.

3.3. Correlations of Hydraulic Ratings with Basic Properties and Statistical Grouping

For four different ingredients of the substrates, the content was declared on the package. Among these, neither the ash content nor the content of organic residuals was significantly correlated with any of the physical substrate properties studied or with the quality scores. The substrates spanned a wide range from 0 to 100% bog peat material, although most of them contained more than 70%. The bog peat content had a significantly positive effect on the scores (S) for the height of capillary rise (S_B$_{CR5}$). In addition, it significantly increased two unfavourable properties: the drop infiltration time (WDPT4) and the extent of shrinking during desiccation (S$_{dry}$ and the score S_L$_{Sdry}$) (Figure 1). No significant effects were found on any of the remaining 11 properties or quality scores.

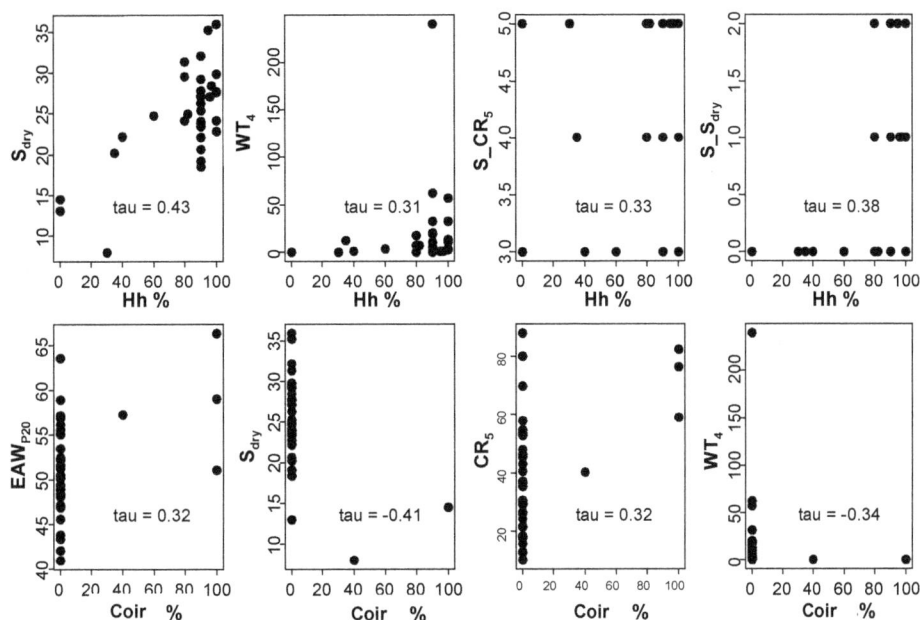

Figure 1. Significant relationships between metric predictor variables and dependent variables.

Only four out of the 32 substrates studied contained coir. Among these, one consisted exclusively of coir. Coir correlated positively with two favourable properties: the height of capillary rise (CR$_5$) and the easily available water content at a container height of 20 cm (EAW$_{p20}$)—and inversely with two detrimental properties: the extent of shrinking during desiccation (S$_{dry}$) and the drop infiltration time (WDPT$_4$) (Figure 1).

As there was no quantitative declaration on numerous packages, the effects of mineral ingredients in general, perlite, and organic residuals could only be checked based on their presence or absence.

Mineral ingredients in general decreased the content of easily available water in the bed (EAW$_{Bed}$) (Figure 2), but did not exhibit any significant effect on any of the remaining properties studied or the quality scores. In contrast, the presence of perlite had significantly positive effects on the mean air content in the case of 20-cm-high containers (Air$_{P20}$, S_B$_{Air\,p20}$) and for total scores for suitability in containers (S_T$_{container}$) and in general (S_T). In addition, perlite increased wettability, indicated by a negative correlation with the infiltration time (WDPT$_4$). The presence of organic residuals had significant and detrimental effects on the height of capillary rise (CR$_5$, S_CR$_5$).

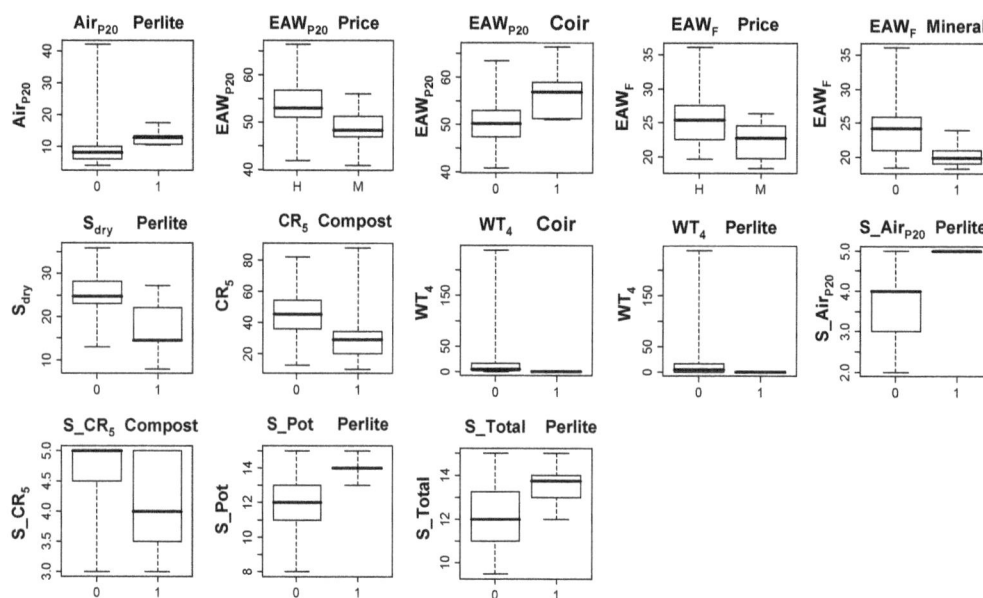

Figure 2. Significant effects of nominal predictor variables on the dependent values (range and quartiles).

The declaration of suitability for different purposes was not significantly correlated with any of the properties studied or the quality scores. High-priced horticultural substrates only differed significantly from less expensive substrates with respect to the content of easily available water both in the bed and in the container (EAW_{p20}, EAW_{Bed}).

Generally, bog peat is the main component of horticultural substrates. However, the best evaluation of the substrate hydraulic suitability in this study revealed substrates composed of bog peat with addition of coir, perlite and organic residuals. Heiskanen's findings [11] agreed with these results, showing the positive impact of perlite on reducing shrinkage and increasing unsaturated hydraulic conductivity in the low tension range. The content of bog peat had a significantly positive effect on the height of capillary rise. In addition, it significantly increased two unfavourable properties: the drop infiltration time and the extent of shrinking during desiccation. The presence of organic residuals had significant and detrimental effects on the height of capillary rise. Coir correlated positively with two favourable properties: height of capillary rise and easily available water content with a container height of 20 cm; and inversely with two detrimental properties: extent of shrinking during desiccation and drop infiltration time.

Results found by Pittenger [17] revealed the importance of bark, bog peat and vermiculite for high-quality substrate mixtures. The results of this study showed that mineral ingredients in general decreased the content of easily available water but did not exhibit any significant effect on the remaining properties studied or the quality scores. However, the risk of a lack of air can be increased by adding clay. In contrast, the presence of perlite had significantly positive effects on air content in the case of 20-cm-high containers. These results fit with the findings by Regand et al. [18], Özkaynak and Samanci [19], and Fazeli et al. [20], who concluded that the best mixture for plantlet growth was peat moss, perlite or sand, respectively. In addition, perlite increased wettability, indicated by a negative correlation with infiltration time. These findings agree with results by Raviv and Lieth [3] and Al Naddafa [14].

4. Conclusions

This study provides some initial information about the hydraulic properties and performance of some commercial substrates for horticulture. This kind of information is intended to expand our knowledge of substrate hydraulic properties and help in the evaluation of substrate suitability in horticulture. The hydraulic performance of totally peat-free substrates was not worse than those containing peat. However, other substrates and substrate mixtures, with bog peat from different parts of the world (Canada, Estonia, Finland and other), and with various alternative ingredients, should be analysed. The proposed method provides a simple and practicable solution.

We found major differences between the tested substrates and substrate compositions. The hydraulic values differed markedly between production in containers versus in beds with free drainage. Lack of air, especially in shallow containers, and water repellency were the relevant deficits. Of the 36 horticultural substrates under study, the great majority earned a good to very good hydraulic rating.

Buyers of horticultural substrates cannot make any assumptions based on the hydraulic properties of the particular product they have bought. However, from the test results above, the probability is very high that the product will perform well or very well hydraulically if the substrate is largely free of clay.

A comprehensive evaluation of the eligibility of horticultural substrates in horticulture, however, requires not only hydraulic measurements but also production studies and an assessment of their chemical, biological and technological suitability. These will be topics for future studies.

Author Contributions: Uwe Schindler was responsible for the hydraulic measurements. The basic evaluation was carried out by Lothar Müller and Uwe Schindler. Gunnar Lischeid was responsible for the statistical evaluation.

Conflicts of Interest: The authors declare no conflict of interest.

References

1. Yogev, A.; Raviv, M.; Hadar, Y. Plant-waste based compost suppressive to deseases caused to pathogenic Fusarium oxysporum. *Eur. J. Plant Pathol.* **2006**, *116*, 267–278. [CrossRef]

2. Verdonck, O. Status of soilless culture in Europe. *Acta Hortic.* **2007**, *742*, 35–39. [CrossRef]

3. Raviv, M.; Lieth, J.H. *Soilless Culture*; Elsevier Publications: London, UK, 2008; p. 608.

4. Blanco-Canqui, H.; Lal, R. Extent of soil water repellency under long-term no-till soils. *Geoderma* **2009**, *149*, 171–180. [CrossRef]

5. Letey, J. Measurement of contact angle, water drop penetration time, and critical surface tension. In *Water-Repellent Soils*; DeBano, L.F., Letey, J., Eds.; University of California: Berkeley, CA, USA, 1968; pp. 43–47.

6. Bauters, T.W.J.; Steenhuis, T.S.; DiCarlo, D.A.; Nieber, J.L.; Dekker, L.W.; Ritsema, C.J.; Parlangea, J.-Y.; Haverkampe, R. Physics of water repellent soils. *J. Hydrol.* **2000**, *231*, 233–243. [CrossRef]

7. Schindler, U.; Durner, W.; von Unold, G.; Mueller, L.; Wieland, R. The evaporation method—Extending the measurement range of soil hydraulic properties using the air-entry pressure of the ceramic cup. *J. Plant Nutr. Soil Sci.* **2010**, *173*, 563–572. [CrossRef]

8. Schindler, U.; Mueller, L.; Eulenstein, F. Hydraulic performance of horticultural substrates—1. Method for measuring the hydraulic quality indicators. In Proceedings of the International Symposium on Quality Management of Organic Horticultural Produce, Ubon Ratchathani, Thailand, 7–9 December 2015.

9. R Development Core Team. *R: A Language and Environment for Statistical Computing*; R Foundation for Statistical Computing: Vienna, Austria, 2010.

10. Wösten, J.H.M.; Lilly, A.; Nemes, A.; Le Bas, C. Development and use of a database of hydraulic properties of European soils. *Geoderma* **1999**, *90*, 169–185. [CrossRef]

11. Heiskanen, J. Physical properties of two-component growth media based on Sphagnum peat and their implications for plant-available water and aeration. *Plant Soil* **1995**, *172*, 45–54. [CrossRef]

12. Richards, D.; Lane, M.; Beardsell, D.V. The influence of particle size distribution in pinebark, sand, brown coal potting mixes on water supply aeration and plant growth. *Sci. Hortic.* **1986**, *29*, 1–14. [CrossRef]

13. Nowak, J.S.; Strojny, Z. Changes in physical properties of peat-based substrates during cultivation period of gerbera. *Acta Hortic.* **2004**, *644*, 319–323. [CrossRef]

14. Al Naddafa, O.; Livieratos, I.; Stamatakisa, A.; Tsirogiannisb, I.; Gizasb, G.; Savvasc, D. Hydraulic characteristics of composted pig manure, perlite, and mixtures of them, and their impact on cucumber grown on bags. *Sci. Hortic.* **2011**, *129*, 135–141. [CrossRef]

15. Caron, J.; Pepin, S.; Periard, Y. Physics of growing media in a green future. International Symposium on Growing Media and Soilless Cultivation. *Acta Hortic.* **2014**, *1034*, 309–317. [CrossRef]

16. Schindler, U.; Mueller, L. Hydraulic performance of horticultural substrates—2. Development of an evaluation framework. In Proceedings of the International Symposium on Quality Management of Organic Horticultural Produce, Ubon Ratchathani, Thailand, 7–9 December 2015.

17. Pittenger, D.R. Potting soil label information is inadequate. *Calif. Agric.* **1986**, *40*, 6–8.

18. Regand, R.V.; Machado, S.; Alam, M.; Ali, A. Greenhouse production of potato (*Solanum tuberosum* cv. *Desiree*) seed tubers using in-vitro plantlets and rooted cutting in large propagation beds. *Potato Res.* **1995**, *38*, 61–68.

19. Özkaynak, E.; Samanci, B. Yield and yield components of greenhouse, field and seed bed grown potato (*Solanum tuberosum* L.) plantlets. *Akdeniz Univ. Ziraat Fak. Derg.* **2005**, *18*, 125–129.

20. Fazeli Sabzevar, R.; Mirabdulagh, M.; Zarghami, R.; Pakdaman, B. Mini-tuber production as affected by planting bed composition and node position in tissue cultured plantlet in two potato cultivars. *Int. J. Agric. Biol.* **2007**, *9*, 416–418.

8

Breeding Trends of Fruit and Vegetable Crops for Organic Production in Lithuania

Audrius Sasnauskas *, Vidmantas Bendokas, Rasa Karklelienė, Danguolė Juškevičienė, Tadeušas Šikšnianas, Dalia Gelvonauskienė, Rytis Rugienius, Danas Baniulis, Sidona Sikorskaitė-Gudžiūnienė, Ingrida Mažeikienė, Audrius Radzevičius, Nijolė Maročkienė, Eugenijus Dambrauskas and Vidmantas Stanys

Institute of Horticulture, Lithuanian Research Centre of Agriculture and Forestry, Babtai LT-54333, Lithuania; v.bendokas@lsdi.lt (V.B.); r.karkleliene@lsdi.lt (R.K.); d.juskeviciene@lsdi.lt (D.J.); t.siksnianas@lsdi.lt (T.Š.); d.gelvonauskiene@lsdi.lt (D.G.); r.rugienius@lsdi.lt (R.R.); d.baniulis@lsdi.lt (D.B.); s.sikorskaite@lsdi.lt (S.S.-G.); i.mazeikiene@lsdi.lt (I.M.); a.radzevicius@lsdi.lt (A.R.); n.marockiene@lsdi.lt (N.M.); e.dambrauskas@lsdi.lt (E.D.); v.stanys@lsdi.lt (V.S.)
* Correspondence: a.sasnauskas@lsdi.lt

Abstract: At the Institute of Horticulture, Lithuanian Research Centre of Agriculture, Babtai, Lithuania, 27 fruit cultivars and 25 vegetable cultivars suitable for organic production were included in the National Plant Variety List and into the European Union Common Catalogue of Agricultural Plant and Vegetable Varieties. The cultivar development programs have focused on improving cold hardiness, resistance to diseases and pests, and yield, as well as on shipping, shelf life qualities, plant development, and biotechnological applications.

Keywords: breeding; cultivar; orchard plants; vegetables; biotechnology; genetics

1. Introduction

Climatic conditions and the socio-economic environment have given rise to unique horticultural traditions characteristic to Lithuania. Plant breeding for organic production has been an important activity at the Institute of Horticulture, Lithuanian Research Centre of Agriculture and Forestry (IH-LRCAF) in Babtai, Lithuania. The cabbage breeding program was started in 1924, onion in 1926, legume crops in 1929, pear in 1940, blackcurrant in 1946, root crops in 1948, apples and stone fruits in 1952, and strawberry in 1953. The efficiency of horticultural plant breeding depends on scientific progress in the following areas: speeding up generation times, creation of screening methods for valuable characters in the earliest plant development stages, investigation of genetic control mechanisms of plant characters and valuable properties, and development of methods for creation of initial breeding material. In breeding efforts directed toward interspecific crossing, selection of acceptable hybrids, male sterility and creation of stable lines are important issues [1]. The breeding program is focused on pest and disease resistance, quality traits, productivity, biochemical composition, tolerance of unfavorable environmental factors, and adaptability to local agro-climatic conditions [2–6]. Successful cultivar development depends on the availability of genetic material, knowledge of important plant traits, and methods of breeding. Biotechnological methods, based on the newest achievements in genetics, physiology and biochemistry provide new possibilities for breeding. Biotechnology and molecular techniques may guarantee production of high quality fruit and vegetable cultivars [7,8].

The aim of this paper was to review fruit and vegetable crop breeding, investigations of gene pools, and creation of new cultivars and hybrids at IH-LRCAF in Lithuania.

2. Breeding Efforts

Crossbreeding and individual or family group selection have been used for fruit and vegetable breeding. Mass selection has been employed in primary seed production for maintenance of valuable family features as long as possible. Primary breeding material and first generation hybrids have been obtained. A description of the cultivars that were created is reported below, including 7 blackcurrant (*Ribes nigrum* L.), 2 strawberry (*Fragaria* × *ananassa* Duch.), 3 apple (*Malus domestica* Borkh.), 3 pear (*Pyrus communis* L.), 4 sweet cherry (*Prunus avium* L.), 2 sour cherry (*Prunus cerasus* L.), 5 plum (*Prunus domestica* L.), 5 carrot (*Daucus carota* L.), 3 red beet (*Beta vulgaris* L.), 2 radish (*Raphanus sativus* L.), 3 cucumber (*Cucumis sativus* L.), 4 tomato (*Solanum lycopersicum* L.), 2 sweet pepper (*Capsicum annuum* L.), and 2 garlic (*Allium sativum* L.) cultivars, and 1 each white cabbage (*Brassica oleracea* L.), onion (*Allium cepa* L.), chive (*Allium schoenoprasum* L.), and bean (*Phaseolus vulgaris* L.) cultivar. The study was carried out in the experimental fields for organic and conventional horticultural plant production at the IH-LRCAF.

3. Results

3.1. Fruit Crops

3.1.1. Pome Fruits

The genetic origin of resistance to apple scab, one of the most damaging diseases of apple worldwide, caused by the fungal pathogen *Venturia inaequalis,* was tested at the embryonic apple development stage using isolated cotyledons. Application of the method allows a 92%–100% recovery of progeny and provides a valuable tool for efficient assessment of disease resistance [9]. Screening methods of scab-resistant genotypes by using cotyledons in vitro were employed for estimation of donor characteristics and potential for breeding and preliminary estimation of interaction of nuclear and cytoplasmic genes. Donors of monogenic V_f and V_m resistance to *Venturia inaequalis* were introduced into the breeding program. Marker-assisted selection (MAS) has been extensively used in apple breeding to identify inherited resistance genes and to reduce the time required for selection of resistant seedlings, and it is also an important tool for searching for resistance donors. The most commonly used markers were based on SSRs and SNPs. This led to new scab-immune apple cultivars. Each of these cultivars was suitable in yield, storage life and fruit quality. DNA analysis using a *Vfa*1 gene-like sequence-specific marker was performed.

Our results showed that the cultivars "Aldas", "Skaistis" and "Rudenis" were immune to *Venturia ineaqualis* races 1 to 5, and had a DNA fragment of 500 bp specific for the *Vfa*1 gene. The evolving virulence of the apple scab pathogen stimulated efforts to adapt new resistance breeding approaches, such as pyramidization of resistance genes. A total of 159 apple hybrids with pyramidic resistance were obtained: 133 seedlings carried *Rvi5* and *Rvi6* genes, and 26 hybrids inherited *Rvi5* and *Rvi10* genes [10,11].

Fruit production and quality and scab resistance are important directions of the pear breeding program. Crossing of local with introduced cultivars has resulted in new pear cultivars, such as "Lukna", "Gaisra" and "Liepona" [12].

3.1.2. Stone Fruits

Crossing local, cold-resistant and disease-resistant cultivars developed abroad has led to the creation of the sweet cherry cultivars *"Jurgita"*, *"Mindaugė"*, *"Jurga"*, *"Meda"* [13], the sour cherry cultivars *"Vytėnų žvaigždė"* and *"Notė"* [14], and the plum cultivars *"Gynė"*, *"Jūrė"*, *"Rausvė"*, *"Kauno vengrinė"*, and *"Aleksona"* [15]. A testing protocol for differentiation of resistance to brown rot (*Monilinia fructicola*) in sweet and sour cherry genotypes was developed, and is being applied for development of new sweet and sour cherry cultivars resistant to the disease.

3.1.3. Small Fruits

Organic blackcurrant production requires cultivars resistant to fungal diseases and pests. The limiting factor is lack of resistance in *R. nigrum*, as plants of other *Ribes* species are often more resistant to key pests and diseases. Therefore, interspecific hybridisation has been used to introduce resistance to pests and diseases into commercially-acceptable blackcurrant cultivars [16]. An investigation of inheritance of disease resistance in interspecific hybrids of *R. americanum*, *R. aureum*, *R. janczewskii*, *R. nigrum*, *R. pauciflorum*, *R. petraeum*, *R. sanguineum*, *R. ussuriense* and *R. uva-crispa* resulted in a large number of the hybrids displaying resistance to powdery mildew, a small percentage showing resistance to *Septoria* leaf spot, and no plants demonstrating any resistance to anthracnose.

Large scale hybridizations of the progeny were made with hybrids from the *R. nigrum* spp. scandinavica line. Breeding with this line for resistance to fungal diseases led to the development and release of a number of powdery mildew-resistant blackcurrants. An improvement in blackcurrant quality achieved during past decades has included an increase in average fruit size from 0.9 to 1.5 g, and a 5 to 7 days advancement of flowering time [17]. The new cultivars include:

"Gagatai" (*"Minaj Shmyriov"* × *"Öjebyn"*). The fruit are large (1.3 g), and very rich in ascorbic acid (197 mg/100 g) and other chemical compounds. Plants are highly resistant to frost, powdery mildew, and gall mite (*Cecidophyopsis ribis*).

"Pilėnai"(*"Minaj Shmyriov"* × *"Öjebyn"*). The fruit are large (1.0 g), and contain 153.1 mg/100 g ascorbic acid. The cultivar is highly resistant to powdery mildew, of medium resistance to leaf spot, and slightly affected by gall mite.

"Tauriai" (*"Titania"* × *"Beloruskaja Sladkaja"*). The fruit are large (1.5 g), and contain 161 mg/100 g ascorbic acid. Plants are highly resistant to powdery mildew and gall mite.

"Blizgiai" (*"Öjebyn"* × *"Minaj Shmyriov"*). The flowering and ripening seasons are earlier than of the cv. *"Minaj Shmyriov"* (1–4 days). The fruit are large (1.5 g), and contain 176 mg/100 g ascorbic acid. The cultivar is resistant to powdery mildew and gall mite in blackcurrant cultivars in Lithuania started in the 1960"s [18]. To overcome the gall mite problem, a new direction in development of genetic lines was initiated by hybridization with the Siberian blackcurrant lineage and selection for field resistance to gall mite. This led to the release of the complex gall mite resistant cultivars *"Vakariai"* and *"Dainiai"*. The main purpose of the breeding program for resistance to gall mite was to introduce the *P* gene from *R. nigrum* spp. *sibiricum* into *R. nigrum*. At the beginning, the agronomic traits of hybrids resistant to gall mite were poor, but these hybrids were backcrossed, and later three commercial cultivars, *"Dainiai"*, *"Kriviai"* and *"Kupoliniai"* [19,20], were released. *"Dainiai"*, released in 2001, is a cultivar with high fruit quality and complete resistance to gall mite, and was also used as a donor of resistance to gall mite. Progeny of *"Dainiai"* were used to map the *P* gene. A 107 bp AFLP marker was identified and its potential in screening progeny for resistance to gall mite hybrids was evaluated [21].

The breeding program for strawberry involved hybridization with European cultivars. PCR-based markers for the *Rpf1* gene were developed and used for screening strawberries for red stele resistance. Two cultivars, *"Dangė"* and *"Saulenė"*, were created, distinguished by their good resistance to fungal diseases, good size and taste, and high yield [22]. A concentrated and high yield with excellent tasting fruits are characteristic of *"Saulene"* (*"Shuksan"* × *"Senga Sengana"*), suitable for fresh use. The mid- to late-season cultivar *"Dange"* (*"Venta"* × *"Redgauntlet"*) distinguishes itself by good fruit size, attractiveness, and stable, high yield every year. New wild strawberry cultivars were developed and submitted for DUS testing (Distinctness, Uniformity and Stability) in Poland.

3.2. Vegetable Crops

3.2.1. Root Vegetables

Investigations of carrots and red beets under different production techniques showed that carrot production was less dependent on technology. The total yield of all cultivars of carrots was higher in a conventional system. The greatest differences were obtained with the cultivar "Šatrija BS", with yield of 48.3 t/ha grown conventionally and 41.0 t/ha grown organically (Table 1). The remaining varieties were less variable. Marketable production was also less varied.

Table 1. Carrots and red beet productivity in Lithuania, Babtai, 2010–2014.

Cultivars	Organic Production		Conventional Production	
	Total Yield (t/ha)	Marketable Production (%)	Total Yield (t/ha)	Marketable Production (%)
Carrot				
"Ieva" H	65.5	88.9	70.0	90.0
"Rokita" H	50.7	88.0	65.3	89.8
"Svalia" H	53.2	87.1	57.7	88.4
"Garduoles"	57.0	87.5	60.0	88.5
"Šatrija BS"	41.0	84.2	48.3	85.4
LSD_{05}	3.186		4.224	
Red beet				
"Joniai"	34.1	75.9	47.2	79.9
"Rikiai"	41.3	82.8	55.5	86.8
"Ilgiai"	40.0	79.4	51.0	80.3
LSD_{05}	7.272		2.469	

Carrot cultivars of the Nantes type are the most popular in Lithuania. Therefore, attention was given to root shape. Roots of the new varieties ("Ieva" H, "Rokita" H, "Svalia" H, "Garduolės", "Šatrija BS") are medium in size and diameter, with a cylindrical, pointed shape. Carrots in Lithuania grow well in a light soil, but the hybrid "Rokita" grows well in a heavier soil. Carotene accumulation ranges from 20–22.5 mg/100 g, dry soluble material from 11%–12.5%, and total sugar from 7.5%–8.5%. Carrots are suitable for harvesting in autumn and for storing [23,24].

Red beet cultivars of different types were created at the Institute [25,26]. The cultivar "Joniai" is distinguished by its small foliage, round and oval-round shape of the root, and not clearly concentrated rings [27]. "Ilgiai" is of medium size with emerald green leaves, with medium-long cylindrical roots. "Rikiai" is a medium early two-seeded (75%) cultivar. The colour of the foliage is dark green with a light yellow tone. The shape of the root is elliptical to round-elliptical with a thin main root with regular skin. "Rikiai" are tolerant to the drought [28]. All new cultivars are suitable for fresh use, storing, and processing.

Cultivars of radish were distinguished by their earliness. Extra earliness is characteristic of "Babtų žara". Its root is round to oval-round, the colour is bright red, the flesh is white, juicy, and delicate in taste. The radish "Liliai" represented medium earliness and has a longer growing cycle. They are suitable for growing in the open field and in greenhouses. Sowing time is possible early in spring or the second and third weeks of August. Roots are round, massive, and red in colour with a purple tone.

3.2.2. Fruit Vegetables

The new cucumber hybrids "Gintai" and "Krukiai BS" are suitable for growing in the open field and in greenhouses [29,30] "Gintai" is a mid-season heterotic hybrid with high yield. It has external attractiveness, is 7–10 cm long, and is light green in colour. "Krukiai BS" is a mid—to late-season hybrid. The fruit is green with white stripes, and does not turn yellow for a long time. The fruit is 8–10 cm long with a diameter of 3.5–4.0 cm. Bees are necessary for pollination. "Daugiai" is distinguished with early ripening and intensive yield. The fruit is 10–13 cm long, dark green, and does not turn yellow for a long time. They are suitable for the fresh market and pickling. Production is optimal in spring and summer in greenhouses.

The new tomato cultivars "Svara", "Viltis", "Balčiai" are suitable for growing in unheated greenhouses and in the open field. "Svara" is a late-season cultivar distinguished with high yield and fungal disease resistance. Fruit are red and weigh approximately 70 g. The fruit are round, and taste somewhat acidic. The cultivar "Viltis" is early ripening. Plants are of medium height and have a short duration of production. Fruit weight at first harvest reach 120–150 g, and later weights decrease to 80–90 g. Fruit are red with a sweet taste, and are flat-round. "Balčiai" is a mid-season cultivar with fruit having red colour, good taste, medium size (75–95 g), with a flat-round shape. "Skariai" is a mid-season cultivar suitable for growing in unheated greenhouses. Fruit are tasty, red in colour, thin skinned, are oblong-oval, cylindrical, and massive (90–150 g). It is sensitive to lack of fertilizer. "Rutuliai" is a mid-season, tall cultivar with red, round, mid-sized fruit. They are suitable for growing in the unheated greenhouses.

Cultivars of sweet pepper "Alanta" and "Reda" are mid-season with a plant height of 0.75–0.80 m, and are self-pollinating [31]. The cylindrical fruit of "Alanta" are distinguished with massive, orange, tasty fruit. This variety is suitable for growing in unheated greenhouses. The shape "Reda" is cut-cylindrical, with external attractiveness, and dark red colour. It grows well in unheated greenhouses, under covers of plastic film, and in the open field.

3.2.3. Onions

In *Allium* L. breeding, the new cultivars "Žiemiai" and "Vasariai" of garlic were created. A flat-round bulb shape is typical for "Žiemiai", the outer skin of the bulb is white, and the cloves are big with 6–9 pieces in a bulb. Bulbils are small, about 150–170 in an inflorescence. The cultivar is disease and pest resistant. "Vasariai" is a mid-season cultivar suitable for planting in spring. The outer skin of the bulb is white, and the bulb consists of 10–16 cloves that are arranged and concentrated into two rounds. The bulb size is 21–27 g.

Homozygotic lines of onions were created by applying a gynogenesis method [32,33]. The cultivar "Babtų didieji" is a mid-season type, with bulbs suitable for fresh use, storage, and processing. The shape of the bulb is round, it weighs 95–116 g, the skin is yellow-orange, and the fresh is white and mildly sweet.

The chive "Aliai" is a perennial plant grown as a bunch. Leaves of "Aliai" start growth early in spring, and the first yield can be harvested 20 days later. Leaves are suitable for fresh use until flowering. The leaves of this cultivar are thin, small and tasty.

3.2.4. Cabbage

The white cabbage "Bagočiai" is a mid- to late-season cultivar. The shape of the head is oval or flat-oval. The colour of the outer leaves is grey-green. "Bagočiai" is distinguished with a good biochemical composition. They are suitable for pickling. Fresh heads can be stored until March.

3.2.5. Legumes

The bean "Baltija" belongs to the bush pod group. The cultivar has a mid-early season. The pod length may reach 11–13 cm. Pods are straight or straight-curved. Seeds are white and oblong. The weight of 1000 seeds is 300–350 g.

4. Conclusions

Cultivars deemed best suited for organic production include: blackcurrant ("Gagatai", "Pilėnai","Tauriai", "Blizgiai", "Vakariai", "Dainiai", "Kriviai", "Kupoliniai"), strawberry ("Dangė" and "Saulenė"), apple ("Aldas", "Skaistis", "Rudenis"), pear ("Lukna", "Gaisra", "Liepona"), sweet cherry ("Jurgita", "Mindaugė", "Jurga", "Meda"), sour cherry ("Vytėnų žvaigždė", "Notė"), plum ("Gynė", "Jūrė", "Rausvė", "Kauno vengrinė", "Aleksona"), carrot ("Ieva" H, "Rokita" H, "Svalia" H, "Garduolės", "Šatrija BS"), red beet ("Joniai", "Rikiai", "Ilgiai"), radish ("Babtų žara", "Liliai"), cucumber ("Gintai", "Krukiai BS", "Daugiai"), tomato ("Svara", "Viltis", "Balčiai", "Skariai"), sweet pepper ("Alanta", "Reda"), garlic ("Žiemiai", "Vasariai"), white cabbage ("Bagočiai"), onion ("Babtų didieji"), chive ("Aliai"), and bean ("Baltija"). In addition, the new vegetable cultivars were included in the EU Common catalogue of varieties of vegetable species in 2015.

Acknowledgments: This work was carried out within the framework of the long-term research program "Horticulture: agro-biological basics and technologies" implemented by the Lithuanian Research Centre for Agriculture and Forestry.

Author Contributions: This work was a product of the joint effort of all of the authors, and the authors equally contributed to the manuscript writing and revisions.

Conflicts of Interest: The authors declare no conflict of interest.

References

1. Liekis, A. Augalų Selekcija (Mokslo Straipsnių Rinkinys). Available online: http://elibrary.lt/inf_res4.phtml?id=69103 (accessed on 10 March 2016). (In Lithuanian)
2. Armolaitienė, J. Morkų veislė "Garduolės 2". *Sod. ir darž.* **1997**, *16*, 47–51. (In Lithuanian)
3. Gaučienė, O. Morkų hibridas "Svalia" F$_1$. *Sod. ir darž.* **1997**, *16*, 57–62. (In Lithuanian)
4. Petronienė, D. "Ilgiai"—Nauja raudonųjų burokėlių veislė. *Sod. ir darž.* **2001**, *20*, 42–47. (In Lithuanian)
5. Petronienė, D.; Viškelis, P. Biochemical composition and preservation of various red beet cultivars. *Sod. ir darž.* **2004**, *23*, 89–97.
6. Karklelienė, R. Morkų ir burokėlių lietuviškų veislių bei hibridų ypatumai ekologinėje ir intensyvioje daržininkystėje. *Sod. ir darž.* **2006**, *25*, 193–200. (In Lithuanian)
7. Baniulis, D.; Gelvonauskienė, D.; Rugienius, R.; Sasnauskas, A.; Stanienė, G.; Stepulaitienė, I.; Frercks, B.; Sikorskaitė, S.; Lukoševičiūtė, V.; Mažeikienė, I.; et al. Orchard plant breeding, genetics, and biotechnology research at the Institute of Horticulture. LRCAF. *Sod. ir darž.* **2013**, *32*, 21–48.
8. Karklelienė, R.; Radzevičius, A.; Juškevičienė, D.; Maročkienė, N.; Bobinas, Č. Daržo augalų selekcijos tyrimų apžvalga. *Sod. ir darž.* **2013**, *32*, 69–86. (In Lithuanian)
9. Gelvonauskiene, D.; Stanys, V. Screening of scab resistant apple seedlings using isolated cotyledons. *Fruit Sci.* **2000**, *207*, 56–60.
10. Sikorskaite, S.; Gelvonauskiene, D.; Stanys, V.; Baniulis, D. Characterization of microsatellite loci in apple (*Malus* × *domestica* Borkh.) cultivars. *Zemd. Agric.* **2012**, *99*, 131–138.
11. Sikorskaite, S.; Gelvonauskiene, D.; Bendokas, V.; Stanys, V.; Baniulis, D. *Malus* sp.-*V. inaequalis* interaction characteristics among local apple cultivars in Lithuania. *Acta Hortic.* **2013**, *976*, 567–572. [CrossRef]
12. Lukoševičius, A. *Kriaušių veislės (katalogas)*; Lietuvos Sodininkystės ir Daržininkystės Institutas: Babtai, Lithuania, 2003. (In Lithuanian)
13. Stanys, V.; Frercks, B.; Šikšnianienė, J.B.; Stepulaitienė, I.; Gelvonauskienė, D.; Stanienė, G.; Bobinas, Č. Identification of sweet cherry (*Prunus avium* L.) cultivars using AFLP and SSR markers. *Zemd. Agric.* **2012**, *99*, 437–444.
14. Stepulaitienė, I.; Žebrauskienė, A.; Stanys, V. Frost resistance is associated with development of sour cherry (*Prunus cerasus* L.) generative buds. *Zemdirb. Agric.* **2013**, *100*, 175–178. [CrossRef]
15. Lukoševičius, A. *Slyvų Veislės (Katalogas)*; Lietuvos Sodininkystės ir Daržininkystės Institutas: Babtai, Lithuania, 2002; p. 110. (In Lithuanian)
16. Stanys, V.; Staniene, G.; Shikshnianas, T.; Bobinas, C. Interspecific hybridization in *Ribes* genus. *Acta Hortic.* **2004**, *663*, 861–864. [CrossRef]
17. Sasnauskas, A.; Siksnianas, T.; Stanys, V.; Bobinas, C. Evaluation of agronomical characters of blackcurrant cultivars and selections in Lithuania. *Acta Hortic.* **2012**, *946*, 189–194. [CrossRef]
18. Bendokas, V.; Mazeikiene, I.; Baniulis, D.; Stanys, V.; Siksnianas, T. Application of *P* gene donors in breeding of black currant resistant to gall mite. *Acta Hortic.* **2013**, *976*, 523–527. [CrossRef]
19. Siksnianas, T.; Sasnauskas, A. Black currant cultivars Gagatai, Kriviai, Kupoliniai. *Sod. ir darž.* **1998**, *17*, 23–33.
20. Siksnianas, T. Development of productive, resistant to fungal diseases and gall mite black currant cultivars. *Sod. ir darž.* **2005**, *24*, 16–24.

21. Mazeikiene, I.; Bendokas, V.; Stanys, V.; Siksnianas, T. Molecular markers linked to resistance to the gall mite in blackcurrant. *Plant Breed.* **2012**, *131*, 762–766. [CrossRef]

22. Rugienius, R.; Sasnauskas, A.; Shikshnianas, T. "Saulene" and "Dange"—Two recent Lithuanian strawberry cultivars. *Acta Hortic.* **2004**, *649*, 73–76. [CrossRef]

23. Karkleliené, R.; Radzevičius, A.; Bobinas, Č. Productivity and root-crop quality of Lithuanian carrot (*Daucus Sativus* Röhl.) breeder lines. *Proc. Latv. Acad. Sci.* **2009**, *63*, 63–65. [CrossRef]

24. Karkleliené, R.; Radzevičius, A.; Dambrauskiené, E.; Surviliené, E.; Bobinas, Č.; Duchovskiené, L.; Kavaliauskaité, D.; Bundiniené, O. Root yield, quality and plant resistance to diseases of organically grown carrot hybrids and cultivars. *Zemdirb. Agric.* **2012**, *99*, 393–398.

25. Petroniené, O.D.; Viškelis, P. Įvairių veislių tipų ir grupių raudonųjų burokėlių (*Beta vulgaris* L.) biocheminė sudėtis. *Maisto Chemija ir Technologija* **2005**, *38*, 42–47. (In Lithuanian)

26. Karkleliené, R.; Viškelis, P.; Radzevičius, A.; Duchovskiené, L. Evaluation of productivity and biochemical composition of perspective red beet breeding number. *Proc. Fourth Balkan Symp. Vegetables Potatoes.* **2009**, *1*, 255–260. [CrossRef]

27. Petroniené, D. "Joniai"—Nauja burokėlių veislė. *Sod. ir darž.* **2000**, *19*, 81–86. (In Lithuanian)

28. Karkleliené, R.; Radzevičius, A.; Maročkiené, N.; Juškevičiené, D.; Dambrauskiené, E. *Raudonojo burokėlio* (*Beta vulgaris* L. subsp. *vulgaris* convar. *vulgaris* var. *vulgaris*) veislė "Rikiai". *Sod. ir darž.* **2013**, *32*, 49–54. (In Lithuanian)

29. Dambrauskas, E. *Agurkų hibridai* "Krukiai" ir "Žalsviai". *Sod. ir darž.* **2001**, *20*, 76–82. (In Lithuanian)

30. Dambrauskas, E. *Trumpavaisiai partenokarpiniai* agurkai "Pūkiai" F_1, "Troliai" F_1 ir "Ulonai" F_1. *Sod. ir darž.* **2001**, *20*, 32–41. (In Lithuanian)

31. Maročkiené, N.; Karkleliené, R.; Bobinas, Č. *Saldžiosios paprikos* veislės "Alanta" biologinių ūkinių savybių įvertinimas. *Sod. ir darž.* **2009**, *28*, 127–134. (In Lithuanian)

32. Juškevičiené, D.; Stanys, V.; Bobinas, Č. *Gynogenesis pecularities* of *Allium* L. vegetables grown in Lithuania. *Biologija* **2005**, *3*, 6–9.

33. Juškevičiené, D.; Stanys, V. *Valgomojo svogūno* (*Allium cepa* L.) ginogenezė ir homozigotinių linijų kūrimas. *Sod. ir darž.* **2007**, *20*, 180–187. (In Lithuanian)

9

The Application of Mycorrhizal Fungi and Organic Fertilisers in Horticultural Potting Soils to Improve Water Use Efficiency of Crops

Frank Eulenstein [1,*], Marion Tauschke [1], Axel Behrendt [1], Jana Monk [2], Uwe Schindler [1], Marcos A. Lana [1] and Shaun Monk [3]

[1]   Leibniz Centre for Agricultural Landscape Research (ZALF), Müncheberg 15374, Germany; mtauschke@zalf.de (M.T.); abehrendt@zalf.de (A.B.); uschindler@zalf.de (U.S.); Marcos.Lana@zalf.de (M.A.L.)
[2]   AgResearch Ltd., Christchurch 8140, New Zealand; jana.monk@outlook.com
[3]   Grasslanz Technology Ltd., Palmerston North 4442, New Zealand; Shaun.Monk@grasslanz.com
*   Correspondence: feulenstein@zalf.de

Abstract: In recent years, the addition of microorganisms such as Plant Growth-Promoting Bacteria (PGPB) and mycorrhiza are becoming more popular, both in research as well as in practical use. While inoculants are usually not necessary for plants cultivated outdoors on biologically active soil, they can be useful on sterile substrates, newly created artificial landscapes, and also in soils that have been managed using non-selective sterilization methods, such as fumigation. In a multi-year lysimeter experiment, we investigated the influence of a commercial mycorrhizal inoculum on water use efficiency and biomass production of maize (*Zea mays*), sunflower (*Helianthus annuus*), sweet clover (*Melilotus officinalis*), sweet sorghum (*Sorghum bicolor*), cup-plant (*Silphium perfoliatum*) and tall wheatgrass (*Elymus elongatus* subsp. *ponticus* cv. Szarvasi-1) when exposed to high or low ground-water levels. Results showed that all plants benefited from the mycorrhizal association. Mycorrhizal-inoculated plants were more successful in terms of dry matter production and water use than the non-mycorrhizal plants. The source of the mycorrhiza—autochthonous or introduced—made no significant difference. The results indicate that inoculation with mycorrhiza and promotion of the naturally abundant mycorrhiza in agricultural production systems can significantly contribute to a sustainable production of crops. Effects depended on plant species, cultivar, soil type, ground-water level and the mycotrophy of the individual crop species.

Keywords: water use efficiency; mycorrhizal fungi; mycotrophy; multi-year lysimeter experiment; sustainable and resource conserving management; horticultural substrates; growing media

1. Introduction

Since 1992, there are horticultural potting soils available that contain *Bacillus subtilis* and commercial endomycorrhiza inoculum (predominantly *Funneliformis mosseae* and *Rhizophagus irregularis*). Sustainable and resource conserving management of horticultural production systems includes efficient management of soil microorganisms such as mycorrhizas [1–4]. El Husseini et al. [5] and Chowdhury et al. [6] demonstrated the beneficial impact of organic fertilizer containing *Bacillus subtilis* (FZB24®) or *Bacillus amyloliquefaciens* (RhizoVital®42), respectively on crop production systems. The addition of *Enterobacter radicincitans* and commercially produced mycorrhizal fungi also resulted in a promotion of plant growth [7,8].

Usually, a commercial endomycorrhiza inoculum is produced in greenhouses on expanded clay, with proliferation of mycorrhizal fungi on suitable host plants such as *Zea mays* cv. Badischer Landmais, a cultivar with a high level of mycotrophy and *Tagetes* spp. At the end of the production cycle, the expanded clay contains between 50 and 150 spores per mL of clay volume. Twenty kilogram (60 L) of the inoculated expanded clay is mixed with one ton of organic fertilizer. This fertilizer called "MYKO-AKTIV" (Cuxin DCM, Telgte, Germany) is used for the preparation of different horticultural potting soil. One cubic metre of horticultural potting soil and 4 to 6 kg of MYKO-AKTIV fertilizer are mixed. Currently there are about 60 different nursery substrates and horticultural potting soils marketed in Europe.

Arbuscular mycorrhiza (AM), the symbiotic association between soil fungi and plant roots, are known to protect host plants from the harmful effects of drought [7,9–12] and can improve the nutrient uptake and growth of plants under water stress conditions. Various experiments under controlled and field conditions have shown that mycorrhizal colonisation of roots increased drought tolerance of different crops such as maize [13,14] wheat [15], soybean [16], onion [17], lettuce [18–21], and red clover [22]. One of the mechanisms of the mycorrhizal symbiosis on host plant water balance is increased root biomass and, subsequently, plant size. In particular, the mobilization and uptake of phosphorus is often related to an increase in plant size [23].

This study focused on the influence of a commercial mycorrhizal inoculum on water use efficiency and biomass production of maize, sunflower, sweet clover, sweet sorghum, cup-plant (*Silphium perfoliatum*), and tall wheatgrass (*Elymus elongatus* subsp. *ponticus*) when exposed to high or low ground-water levels in lysimeter experiments.

2. Experimental Section

Site Characteristics and Experimental Setup

The experiment was carried out in 24 lysimeters at the Lysimeter Station Paulinenaue in northeast Germany with 515 mm precipitation/year (30 year average). The average precipitation during the crop season is 318 mm, and the average annual temperature is 8.9 °C.

Stainless steel lysimeter vessels were filled in 1968 with undisturbed hydromorphic mineral soil monoliths of low-level moors, half-bogs, humus gley, and sand gley, as well as loamy substrates. The vessels have a surface area of 1 m^2 and are 1.5 m deep with fully adjustable ground water levels at 40, 70, 100, and 120 cm below the surface (Figure 1). The latter three groundwater levels simulate drought conditions.

The maize cultivar Nolween (at 30 kg seeds/ha), sunflower cultivar Aloa (at 7 seeds/m^2), sweet clover (at 22 kg/ha), grain sorghum cultivar Lussi (at 30 kg/ha), sudangrass cultivar Nutri Honey (at 30 kg/ha), cup-plant (at 5 plants/m^2), and tall wheatgrass cultivar Szarvazi-1 (at 20 kg/ha) were planted at two different ground water levels (40 or 100 cm), and at more in some instances.

A commercial mycorrhizal product with two different species of mycorrhizal fungi (*Rhizophagus interadices* and *Claroideoglomus etunicatum*) was applied to each treatment according to the producer's recommendation (75–100 g/m^2) incorporated into the upper 20 cm of the soil. Lysimeters without mycorrhiza application served as control. The planted lysimeters are shown in Figure 2.

Root colonisation was monitored by staining fresh roots with a lacto glycerol-trypan blue solution (0.05% *w/v*) to determine colonization progress [24]. The mean percentage of root colonization was counted by the gridline intersection method (Brundrett et al.) [25]. A total of 100 root segments were observed for each plant. In the 4 year experiment (2010–2014), 163 measurements were performed for all species, for maize 43. The "deeper water level" and the mycorrhizal inoculation were compared to control, with multifactorial ANOVA. The averages were determined by post hoc comparison using Fisher's least significant difference test ($p = 0.95$).

(a)

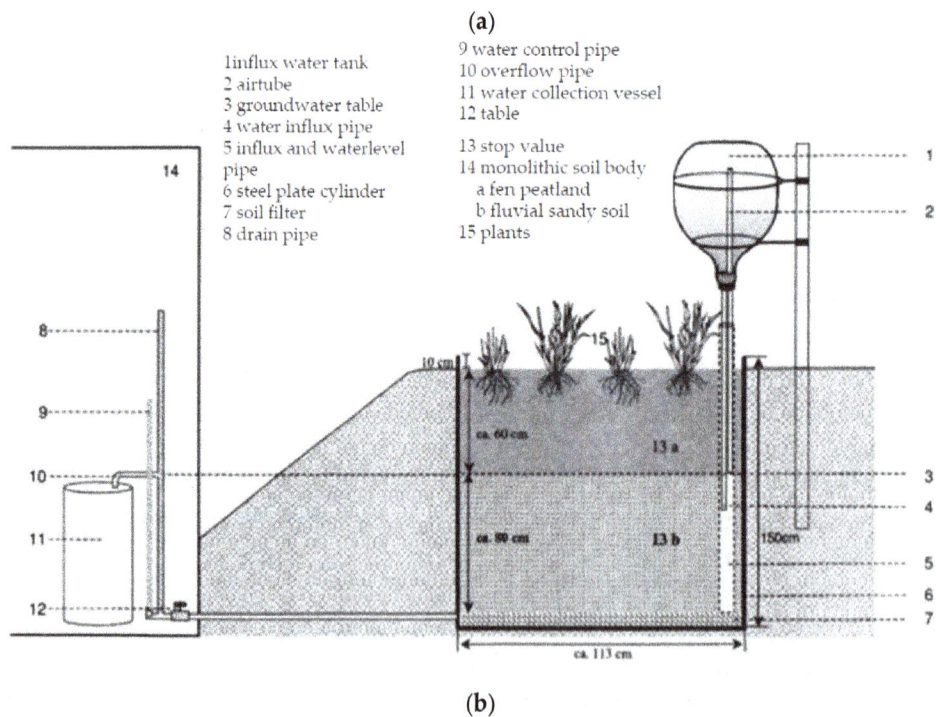

1 influx water tank
2 airtube
3 groundwater table
4 water influx pipe
5 influx and waterlevel pipe
6 steel plate cylinder
7 soil filter
8 drain pipe

9 water control pipe
10 overflow pipe
11 water collection vessel
12 table

13 stop value
14 monolithic soil body
  a fen peatland
  b fluvial sandy soil
15 plants

(b)

Figure 1. Map of Germany with the ZALF experimental station in Paulinenaue in the state Brandenburg (**a**) and lysimeter design (**b**).

Figure 2. Experimental setup of the crop lysimeter experiment (Photo: Marion Tauschke 2014).

3. Results and Discussion

3.1. Inoculation Success

Root staining revealed that inoculated as well as control (non-inoculated) plants were colonised with mycorrhizal fungi. The mycorrhizal colonisation of inoculated plants varied between 34% and 70% and was not significantly different from that of control plants over the entire monitoring period of four years (data not shown). Because of the colonisation potential of the naturally abundant mycorrhizal fungi population present in the lysimeters, the mycorrhizal colonisation of control plant roots was as high as 68%, e.g., in maize in 2011 and cup-plant in 2013.

The percentage of colonised root was not affected by the addition of the commercial inoculum (Figure 3). However, in maize the colonization of roots was significantly increased ($r = 0.53$; $p < 0.05$) after inoculation with the commercial inoculum in almost all treatments (Figure 4).

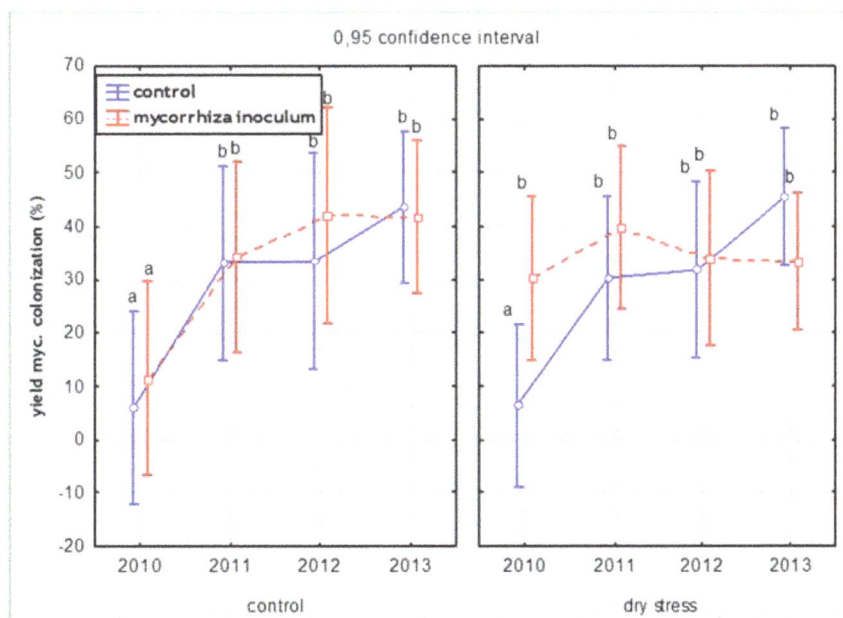

Figure 3. Mycorrhizal colonisation of control and inoculated plants across all species.

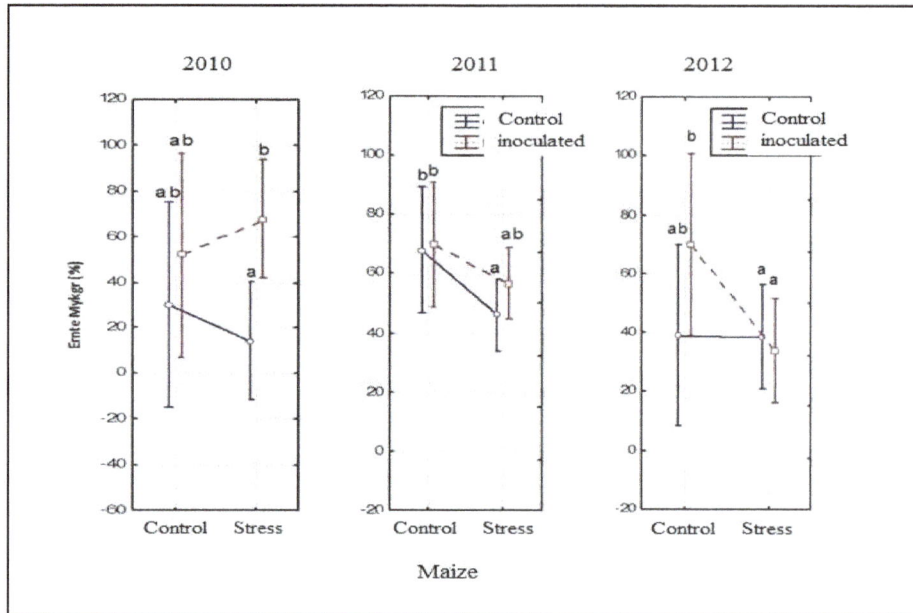

Figure 4. Mycorrhizal colonisation of maize plants.

3.2. Dry Matter

Although for most plant species there was no additional colonisation after application of the commercial inoculum, the effect of colonisation by autochthonous populations and applied mycorrhiza on plant development could be seen. The degree of mycorrhizal colonisation of the different plant species was directly correlated ($r = 0.71$; $p < 0.05$) to dry matter production/m^2. Plants with mycorrhizal root colonisation of 60% formed 30% more dry matter than plants with only 30% mycorrhizal colonisation (Figure 5). Moreover, the P and N content was significantly correlated ($p < 0.05$) with the mycorrhizal colonisation of the roots with $r = 0.68$, and $r = 0.66$, respectively.

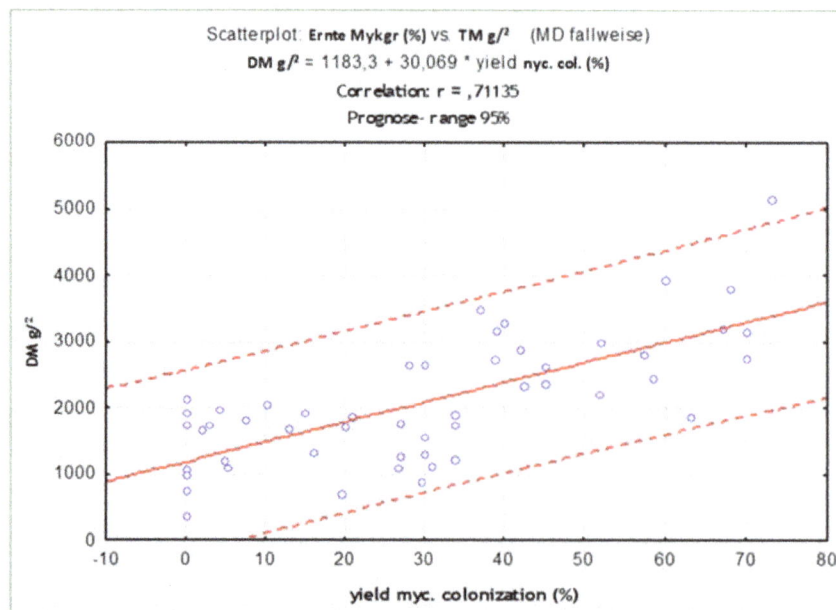

Figure 5. Correlation of mycorrhizal colonisation with plant dry matter across all species and years.

3.3. Water Use Efficiency

Except for grain sorghum, the specific water use efficiency was improved for all plants with a high percentage of mycorrhizal colonisation (r = 0.46). Mycorrhizal plants required less water than non-colonised plants to produce 1 kg of dry matter. Plants with 60% root colonisation required about 25% less water to produce 30% more dry matter than plants with 30% colonisation (Figure 6), showing a higher water use efficiency. Regardless of the degree of mycotrophy in the analysed plant species this correlation was not affected.

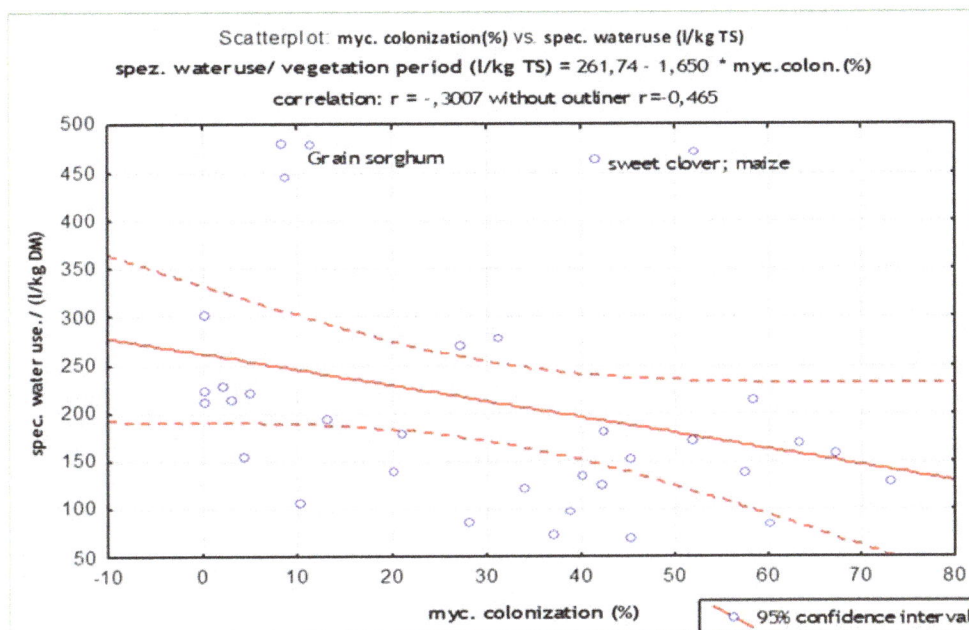

Figure 6. Correlation of root mycorrhizal colonisation and water use efficiency.

4. Conclusions

In our experiments all plant species benefited from mycorrhizal association. Colonised plants of all species had greater dry matter production and less water use than non-mycorrhizal plants, independent of the fungal origin-applied or autochthonous. The different species varied in the degree of their reaction to the mycorrhiza. The effect was more pronounced the lower the ground water level. Results showed that mycorrhiza can not only improve the growth of high performance crops, but also promote their adaptation to different environmental conditions such as deep groundwater levels. The results also indicated that inoculation with mycorrhiza in biologically active soils is not necessary. However, they may be useful in sterile substrates, newly-created artificial landscapes, and also in soil that has been managed using non-selective sterilization methods, such as fumigation.

Acknowledgments: This work was supported by the Federal Ministry of Food and Agriculture of Germany (BMEL) and by the Royal Society of New Zealand.

Author Contributions: Frank Eulenstein is the producer of mycorrhiza inoculum, prepared the manuscript and was the Initiator of the project. Marion Tauschke designed the mycorrhiza experiments detected mycorrhiza in soil and roots and assisted in statistical analysis and preparation of the manuscript. Axel Behrendt implemented the lysimeter experiments, the soil- and plant-sampling and the groundwater analysis. Jana Monk stained the roots, evaluated the mycorrhizal colonization, counted the spores and provided assistance with the manuscript. Shaun Monk produced the mycorrhiza inoculum, advised the soil- plant sampling and the groundwater analysis. Marcos A. Lana calculated the water balances and assisted in the preparation of the manuscript. Uwe Schindler designed the lysimeter experiments and calculated the water balances.

Conflicts of Interest: The authors declare no conflict of interest.

References

1. Jeffries, P.; Gianinazzi, S.; Perotto, S.; Turnau, K.; Barea, J.M. The contribution of arbuscular mycorrhizal fungi in sustainable maintenance of plant health and soil fertility. *Biol. Fertil. Soils* **2003**, *37*, 1–16.

2. Bunemann, E.K.; Schwenke, G.D.; Van Zwieten, L. Impact of agricultural inputs on soil organism—A review. *Aust. J. Soil Res.* **2006**, *44*, 379–406. [CrossRef]

3. Vosátka, M.; Albrechtová, J. Microbial strategies for crop improvement. In *Benefits of Arbuscular Mycorrhizal Fungi to Sustainable Crop Production*; Khan, M.S., Zaidi, A., Musarrat, J., Eds.; Springer: Dordrecht, The Netherlands, 2009; pp. 205–225.

4. Gianinazzi, S.; Gollotte, A.; Binet, M.N.; van Tuinen, D.; Redecker, D.; Wipf, D. Agroecology: The key role of arbuscular mycorrhizas in ecosystem services. *Mycorrhiza* **2010**, *20*, 519–530. [CrossRef] [PubMed]

5. El Husseini, M.M.; Bochow, H.; Junge, H. The biofertilising effect of seed dressing with PGPR *Bacillus amyloliquefaciens* FZB 42 combined with two levels of mineral fertilising in African cotton production. *Arch. Phytopathol. Plant Prot.* **2012**, *45*, 2261–2271. [CrossRef]

6. Chowdhury, S.P.; Dietel, K.; Rändler, M.; Schmid, M.; Junge, H.; Borriss, R.; Hartmann, A.; Grosch, R. Effects of *Bacillus amyloliquefaciens* FZB42 on lettuce growth and health under pathogen pressure and its impact on the rhizosphere bacterial community. *PLoS ONE* **2013**. [CrossRef] [PubMed]

7. Brock, A.K.; Berger, B.; Mewis, I.; Ruppel, S. Impact of the PGPB *Enterobacter radicincitans* DSM 16656 on growth, glucosinolate profile, and immune responses of *Arabidopsis thaliana*. *Microb. Ecol.* **2013**, *65*, 661–670. [CrossRef] [PubMed]

8. Nadeem, S.M.; Ahmad, M.; Zahir, Z.A.; Javaid, A.; Ashraf, M. The role of mycorrhizae and plant growth promoting rhizobacteria (PGPR) in improving crop productivity under stressful environments. *Biotechnol. Adv.* **2014**, *32*, 429–448. [CrossRef] [PubMed]

9. Auge, R.M. Water relations, drought and vesicular-arbuscular mycorrhizal symbiosis. *Mycorrhiza* **2001**, *11*, 3–42. [CrossRef]

10. Ruiz-Lozano, J.M. Arbuscular mycorrhizal symbiosis and alleviation of osmotic stress. New perspectives for molecular studies. *Mycorrhiza* **2003**, *13*, 309–317. [CrossRef] [PubMed]

11. Boomsma, C.R.; Vyn, T.J. Maize drought tolerance: Potential improvements through arbuscular mycorrhizal symbiosis? *Field Crops Res.* **2008**, *108*, 14–31. [CrossRef]

12. Tauschke, M.; Behrendt, A.; Monk, J.; Lentzsch, P.; Eulenstein, F.; Monk, S. Improving the water use efficiency of crop plants by application of mycorrhizal fungi. In *Moving Farm Systems to Improved Nutrient Attenuation*; Currie, L., Burkitt, K.L., Eds.; Fertilizer and Lime Research Centre, Massey University: Palmerston North, New Zealand, 2015; pp. 1–8.

13. Sylvia, D.E.; Hammond, L.C.; Bennet, J.M.; Hass, J.H.; Linda, S.B. Field response of maize to a VAM fungus and water management. *Agron. J.* **1993**, *85*, 193–198. [CrossRef]

14. Subramanian, K.S.; Charest, C. Influence of arbuscular mycorrhizae on the metabolism of maize under drought stress. *Mycorrhiza* **1995**, *5*, 273–278. [CrossRef]

15. Bryla, D.R.; Eissenstat, D.M. Respiratory costs of mycorrhizal associations. In *Plant Respiration: From Cell to Ecosystem. Advances in Photosynthesis and Respiration*; Lambers, H., Ribas-Carbo, M., Eds.; Springer: Dordrecht, The Netherlands, 2005; pp. 207–224.

16. Bethlenfalvay, G.J.; Schuepp, H. Arbuscular mycorrhizas and agroecosystem stability. In *Impact of Arbuscular Mycorrhizas on Sustainable Agriculture and Natural Ecosystems*; Gianinazzi, S., Schuepp, H., Eds.; Birkhäuser Verlag: Basel, Switzerland, 1994; pp. 117–131.

17. Azcon, R.; Tobar, R.M. Activity of nitrate reductase and glutamine synthetase in shoot and root of mycorrhizal *Allium cepa*. Effects of drought stress. *Plant Sci.* **1998**, *133*, 1–8. [CrossRef]

18. Tobar, R.M.; Azcon, R.; Barea, J.M. Improved nitrogen uptake and transport from 15N- labeled nitrate by external hyphae of arbuscular mycorrhizae under water-stressed conditions. *New Phytol.* **1994**, *126*, 119–122. [CrossRef]

19. Tobar, R.M.; Azcon, R.; Barea, J.M. The improvement of plant N acquisition from an ammonium-treated, drought-stressed soil by the fungal symbiont in arbuscular mycorrhizae. *Mycorrhiza* **1994**, *4*, 105–108. [CrossRef]

20. Azcon, R.; Gomez, M.; Tobar, R. Physiological and nutritional responses by *Lactuca sativa* L. to nitrogen sources and mycorrhizal fungi under drought conditions. *Biol. Fertil. Soils* **1996**, *22*, 156–161. [CrossRef]

21. Ruiz-Lozano, J.M.; Azcon, R. Mycorrhizal colonization and drought stress exposition as factors affecting nitrate reductase activity in lettuce plants. *Agric. Ecosyst. Environ.* **1996**, *60*, 175–181. [CrossRef]

22. Fitter, A.H. Water relations of red clover *Trifolium pratense* L. as affected by VA mycorrhizal infection and phosphorus supply before and during drought. *J. Exp. Bot.* **1988**, *39*, 595–604. [CrossRef]

23. Subramanian, K.S.; Santhanakrishnan, P.; Balasubramanian, P. Responses of field grown tomato plants to arbuscular mycorrhizal fungal colonization under varying intensities of drought stress. *Sci. Hortic.* **2006**, *107*, 245–253. [CrossRef]

24. Phillips, I.M.; Hayman, D.S. Improved procedures for clearing roots and staining parasitic and vesicular-arbuscular mycorrhizal fungi for rapid assessment of infection. *Trans. Br. Mycol. Soc.* **1970**. [CrossRef]

25. Brundrett, M.; Bougher, N.; Dell, B.; Grove, T.; Malajczuk, N. *Working with Mycorrhizas in Forestry and Agriculture*; Australian Centre for International Agricultural Research (ACIAR): Canberra, Australia, 1996; p. 374.

10

Evaluating the Efficiency of Wicking Bed Irrigation Systems for Small-Scale Urban Agriculture

Niranjani P. K. Semananda *, James D. Ward and Baden R. Myers

School of Natural and Built Environments, University of South Australia, Mawson Lakes Campus, Adelaide 5095, Australia; James.Ward@unisa.edu.au (J.D.W.); Baden.Myers@unisa.edu.au (B.R.M.)
* Correspondence: semnp001@mymail.unisa.edu.au

Abstract: A wicking bed (WB) is a plant driven system where plants receive water through capillary rise from a self-contained coarse material-filled subsoil reservoir. WBs have been widely promoted as a water-efficient irrigation solution for small-scale and urban food gardens. However, little published research exists to support popular claims about their effectiveness. In this study, the performance of WBs was compared with best-practice, precision surface irrigation in terms of water use efficiency (WUE), fruit yield, fruit quality and labour input, using tomato (*Solanum lycopersicum*) as the experimental crop. The influence of WB design variables (reservoir depths and soil bed depths) was tested. Results showed that WBs performed as well or better than precision surface irrigated pots, showing statistically significant improvement in WUE, yield and fruit quality. The results also suggest an optimum design exists for soil depth (where 300 mm outperformed 600 mm) but not reservoir depth (no difference between 150 and 300 mm). The WBs were more labour efficient, requiring significantly less frequent watering to achieve the same or better WUE. WBs are inherently low-tech and scalable and appear well-suited to a variety of urban agriculture settings.

Keywords: capillary; irrigation; tomato; urban agriculture; water use efficiency; wicking

1. Introduction

More than a half of the world's population lives in urban and peri-urban areas, and this proportion can be up to 90% in developed countries [1,2]. However, land available for crop production within cities is considerable [3]. In the UK, about 87% of households have access to a private garden and these contribute between 22% and 36% of the total urban area [4]. In Australia, more than half of households are growing some of their own food [5]. A wide acceptance and expansion of urban agriculture can be seen even in densely populated cities like New York, USA, as a measure towards sustainable use of land, and in the pursuit of economic and food security [6,7]. There is an increasing trend towards growing crops (especially organic products) in household and community gardens in urban areas. This is due to many factors, including concerns over the side effects of herbicides, insecticides and fertilizers in commercial agriculture [8,9]. Urban gardens also provide a significant contribution to urban green infrastructure, providing benefits to the urban ecosystem. Such benefits include reduced urban heat island impacts, improved human health and well-being, and improved aesthetics [2,10,11]. Urban agriculture can also provide a source of food or additional income for households in developing countries [12].

The growing interest in urban food production adds pressure to municipal water supplies. Municipal treated water is the main water supply used to irrigate urban gardens, and in Australia, about 34% of household water consumption is used to irrigate residential gardens [13]. In addition to supply constraints, the water used for irrigation in urban settings is extremely expensive compared to rural water supplies for commercial horticulture [14,15]. There is therefore a great interest in the development of efficient irrigation systems which can secure urban water resources and provide urban gardeners with an economically efficient means of irrigation that is not technologically complex [16–18].

Although improved practices exist for efficient crop irrigation in commercial farm management, there has been limited adoption of these techniques in urban areas due to the barriers of cost and knowledge to maintain them [18]. In urban gardens, the issues principally revolve around determining when and how much water needs to be applied to crops. For example, precision irrigation scheduling has been widely applied for efficient irrigation of large-scale monoculture crops. At the domestic and community garden scale, however, plants tend to be mixed and irrigation cannot be optimally scheduled to meet the needs of multiple crop water use patterns [19,20]. As a result, homeowners frequently apply more water than necessary, which leads to water loss as subsoil drainage [21]. Another major problem for urban growers is the labour invested in watering which may conflict with other activities [22]. In some regions, those interested in growing crops are not able to do so due to the presence of legacy contamination [23] or a shallow water table [24].

Capillary wick irrigation has been investigated for some years as a method to deliver water to plants in container gardens [25]. Existing approaches include wicking mats, self-watering pots, and sub-irrigated planters [24,26]. The more recent "wicking bed" (WB) [27] has been characterized by a saturated media-filled reservoir beneath an unsaturated soil (root zone). Water is delivered by capillary action to the root zone in response to the water requirement of the plant by allowing individual plants to uptake water according to their demand.

In principle, in a WB, each plant should get precisely the right amount of water at a particular time, making the system a possible solution to the urban irrigation issues identified above. Indeed, WBs are gaining popularity among the gardening community because they are relatively simple and scalable, and informal recommendations abound as to their construction details. A review of grey literature, including garden magazines and websites, reveals claims that WBs can save up to 90% of water compared to traditional surface irrigation [28]. Despite increasing enthusiasm for, and claims about, WB performance, there is no comprehensive study to investigate the performance of these systems in terms of crop growth, water use efficiency (WUE) or labour needs. Moreover, while recent research [24] on a sub-irrigated planter design (akin to, though not labelled, a WB) indicated that the system may have higher yield and lower maintenance than other container gardening approaches, it did not explicitly investigate the water use or labour efficiency of the irrigation system, and recommended further research into these areas.

A substantial knowledge gap exists relating to the engineering design of WBs, specifically the depth of both soil and saturated reservoir. We were unable to find any published research supporting design recommendations for these depths. It has long been understood that the capillary rise of water in soil reaches a maximum height, with moisture content decreasing with distance above the saturated water level [29,30]. It follows that in a WB reservoir, especially when the media used is coarse (e.g., gravel), there is a critical depth beyond which wicking into the soil will not occur, rendering the additional depth of storage useless. Meanwhile, insufficient soil may lead to evaporation from the surface, and excess soil may limit the rate of water delivery to the root zone. Thus, it is reasonable to expect an optimum soil bed height for a given plant rooting habit, which would balance water delivery and crop yield against minimal soil evaporation.

The objective of this study was therefore to evaluate, for the first time, the performance of WB systems relative to equivalent best-practice conventional surface irrigation techniques, with a focus on WUE, fruit yield, fruit quality, dry matter production and irrigation labour (indicated by the number of irrigation events required). The study also compared the effect of different reservoir and soil bed depths to better inform the design of WBs for maximum WUE. The outcomes of this research are of high significance in the context of water consumption and efficiency in urban agriculture, and should be of specific interest to decision-makers, practitioners and proponents of WBs as a water-efficient irrigation technology.

2. Materials and Methods

2.1. Experimental Conditions

The experiment was conducted in a glasshouse at the University of South Australia, Mawson Lakes campus in Adelaide, Australia (34.81°S, 138.61°E). The glasshouse provided a protected environment in which to compare WB and surface irrigation efficiency without the influence of rainfall. The growing period was during spring and summer 2014–2015. Soil temperature, indoor air temperature and relative humidity were recorded at hourly intervals during the crop growing period. Soil temperature was measured in a single replicate of each treatment by installing TR 10 temperature data loggers at two depths—10 and 15 cm from the soil surface. Temperature was assumed not to vary significantly across the plots within the greenhouse (area 5 m × 3 m). Indoor air temperature (±1 °C) and relative humidity (±3%) were recorded using a USB temperature/humidity data logger (LCD—QP6014, Jaycar electronics, Sydney, Australia).

One of the most common varieties of tomato (*Solanum lycopersicum*), cv. "Mighty Red", was obtained from a local plant nursery in Virginia, South Australia. The experimental soil was purchased from a local supplier, and was a type especially developed for gardening, so was taken as representative of urban garden soils. The soil was a poorly graded sandy soil, according to the unified soil classification system (USCS) (90% sand and <5% fines). It had a field capacity of 30%, porosity of 49%, dry bulk density of 1.33 g/cm^3, organic matter of 4.5%, and pH of 7.7.

2.2. Experimental Design and Treatments

The main features of a WB are the reservoir, separation material, vertical pipe and overflow pipe. All beds were prepared using opaque plastic containers with an internal diameter of 56 cm and 100 cm initial height, which were cut down to provide the required soil plus reservoir depth. Different pots were built to compare the WUE of standard surface irrigation, conventional wicking beds, and wicking beds with a soil column, each with two soil bed heights (300, 600 mm) and two WB reservoir depths (150, 300 mm). A coarse-grained material (quartzite gravel) was used to fill the reservoir. The use of gravel adds strength to the WB to support the soil weight, conforms to the size and shape of the container, and creates a medium that is both porous and permeable so that it both holds a substantial volume of water and is fast to fill. A soil column within the reservoir and underneath the main soil body (Figure 1) to improve wicking consisted of a perforated polyvinyl chloride (PVC) pipe of 30 cm diameter, sealed at the base, and wrapped with a geotextile fabric. To measure evaporation solely from the soil surface, two pots with different soil bed heights (300 and 600 mm) were used without plants. Accordingly, there were eight treatments in total and each treatment was in triplicate. Treatments were labelled as treatment number (T1 to T8), followed by the irrigation method including depth of reservoir and soil bed in mm (Figure 1).

A transparent, flexible tube was fixed to the side of each wicking treatment as a manometer to view the water level in the reservoir. A vertical tube was placed in the reservoir through which water was added to the WB. A PVC access tube sealed at the base was installed into each pot to facilitate access by the soil moisture probe as shown in Figure 1 (this is not a part of typical WB). Aggregate (10 mm Quartzite) was then placed as the reservoir material and a geotextile fabric was laid over the aggregate. Soil was placed into pots in 10 cm layers, each layer being lightly compacted. Granulated slow-release fertilizer (20% N, 6.1% P, 11% K as *w/w*), was added in equal amounts to all pots (30 g/plant), and mixed into the soil layer from 10 to 20 cm below surface. The same plastic containers were used for the surface irrigated pots but with several holes drilled in the base to drain any surplus water. All pots were arranged inside the glasshouse with treatments in a randomized complete block design. The reservoirs were filled with water three days before the seedlings were transplanted. One tomato seedling, 4-weeks old, was transplanted into the centre of each bed.

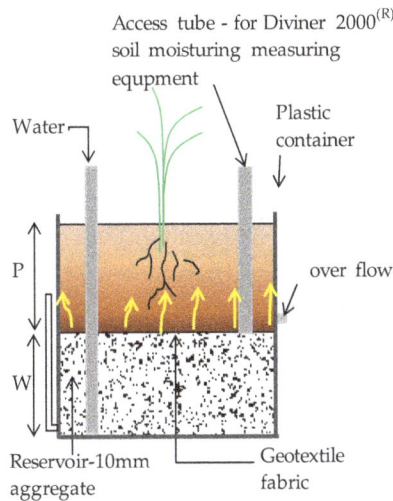

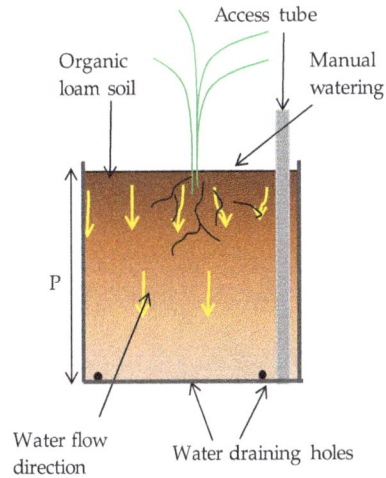

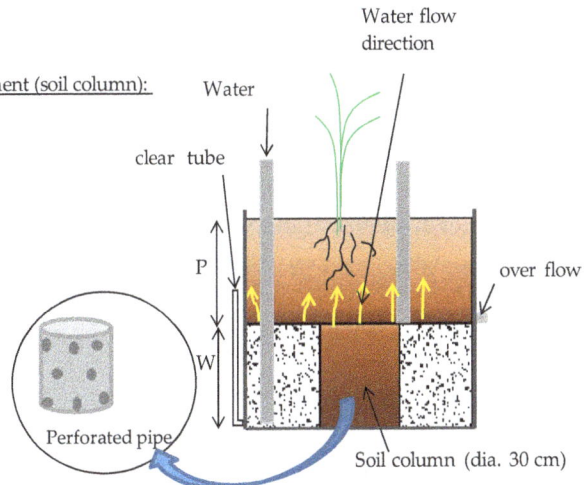

Treatment notation:
*W-wicking, WS-wicking with soil column, WN- wicking without plant.

Figure 1. Treatment types and identification labels. The numbers following a letter for each irrigation method treatment indicate the depth of soil (P) or reservoir (W) in mm. S indicates surface irrigation.

2.3. Application of Irrigation

Pots were manually irrigated with small irrigation volumes immediately after each seedling was transplanted, and thereafter every fourth day during the crop establishment period to support early root growth (10.25 L for shallow beds and 26.25 L for deep beds in total). This period lasted about 14 days after transplanting (DAT) for wicking beds with a 300 mm soil bed depth and 21 DAT for 600 mm deep beds. The volume of water applied to each pot was manually recorded. Following the crop establishment period, irrigation scheduling and volume were determined based on the soil-water balance method outlined below.

The volumetric soil moisture content (VMC, $v/v\%$) in each pot was monitored every day at 10 cm intervals using a portable Diviner 2000® soil moisture monitoring probe (Sentek Sensor Technol., Stepney, SA, Australia) [31]. The Diviner 2000® probe was calibrated gravimetrically for the experimental soil during the progress of the experiment. The summation of volumetric moisture content (VMC) in each layer was calculated and a decision on whether to irrigate the crops was made based when the total VMC in each treatment had decreased to 75% of the moisture content field capacity (*FC*) (Equation (1)):

$$If\ Total\ VMC = \sum_{i=1}^{n}(\theta_i) > \sum_{i}^{n}[0.75(\theta_{FC})],\ then\ no\ watering \tag{1}$$

where θ = volumetric moisture content (%), θ_i = volumetric soil moisture content (%) in layer i, n = number of soil layers, and *FC* = volumetric moisture content at field capacity.

The surface irrigation volume was determined according to Equation (2), which aimed to reach the *FC* in the upper layers and 75% of *FC* in the bottom layer, as:

$$Irrigation\ water\ (mm^3) = \sum_{i=1}^{n-1}(\theta_{FC} - \theta_i){\cdot}A + \sum_{i=n}(0.75\theta_{FC} - \theta_n){\cdot}A \tag{2}$$

where θ_i = soil moisture content in layer i, n = number of soil layers, *FC* = volumetric moisture content at field capacity, and A = pot area (mm^2).

The soil moisture data were collected until the end of the cropping period. A summation of daily moisture content depletion was used to determine the ET in all treatments. Evapotranspiration over the total growing period (ET$_c$) was calculated by summing up the ET values for each data recording period using Equation (3) below.

$$ET = I + \sum_{i=1}^{n}\frac{(\theta_{i,\,t-1} - \theta_{i,\,t})}{100}{\cdot}(D_i) \tag{3}$$

where ET = evapotranspiration (mm), I = irrigation volume (surface treatment) or water added to the reservoir (wicking treatment) (mm), n = number of soil layers in each pot, $\theta_{i,t-1}$ = volumetric moisture content (%) in the ith layer on the previous day, $\theta_{i,t}$ = volumetric moisture content (%) in the ith layer on the current day, and D_i = depth of the ith layer (mm).

2.4. Plant Performance and Water Use Efficiency

The total crop growing period was divided into four stages according to experimental observations for each treatment. These stages were the seedling stage (Stage I, from transplant to first flowering), flowering stage (Stage II, from first flowering to first fruit set), fruit development stage (Stage III, from first fruit set to the first harvest) and fruit maturity and harvesting stage (Stage IV, from first harvest to end of the experiment). The average length of each stage was 14.7 ± 4.8, 18.2 ± 0.8, 39.9 ± 3.3 and 67.2 ± 2.8 days, respectively. First fruit was harvested about 72.8 ± 2.8 DAT. All plants were trained to one stem by pruning all side shoots and by topping plants at the end of the crop, taken as the point when each plant had formed 5 fruit bearing clusters [32]. Tomato fruits were harvested when more than 90% of the fruit were red in colour. The experiment was carried out for a period of 140 DAT.

The size of the tomato fruit was measured using two parameters: fruit weight and diameter. The fruit were weighed using a digital balance with an accuracy of ± 0.01 g, and individual fruit length and crosswise diameter were recorded using a Vernier calliper. All harvested fruit were graded into various marketable and non-marketable grades according to Ontario commercial standard [33]. Marketable tomatoes had three grades based on diameter: small (40 to 50 mm), large (55 to 75 mm), and extra large (>75 mm). Unmarketable fruit were graded as culls which included the fruit that did not fit into the marketable grades above, and damaged or diseased fruit (blossom end rot or other visible marks or injury). Each fruit was cut into thin slices (~5 mm) and dried in a dehydrator at 55 °C to obtain the dry fruit weight.

Plant growth and performance were evaluated by measuring plant height and stem diameter and counting the number of leaves at 14 day intervals. The plant height was recorded from the soil surface to the apex, while the stem diameter was measured 5 cm above the soil surface. At the end of the experiment, the stem and foliage weight of each plant were weighed separately, and then were dried in a forced air oven at 55 °C for subsequent dry weight determination. The total biomass accumulation was calculated as the dry weight of each vegetative part plus the dry mass of fruit.

Water use efficiency was calculated as the ratio of total marketable fruit yield to either the amount of water used by the plants (WUE) or that was applied through irrigation (iWUE). WUE and iWUE were calculated using Equations (4) and (5) below [34]:

$$WUE = Y/ET_c \qquad (4)$$

$$iWUE = Y/I \qquad (5)$$

where Y = total marketable fruit yield (g), I = irrigation water volume (L), and ET_c = total crop evapotranspiration (L). ET_c values were obtained by multiplying Equation (3) above with the area of the pot (A).

2.5. Root Sampling and Analysis

The root distribution pattern of each treatment was investigated at the end of the experiment. Root samples were collected at the end of the cropping period from one representative plant per treatment. Soil cores were extracted using a 6 cm diameter soil auger at three different depths (0–10, 10–20, and 20–30 cm) for 30 cm soil beds and five depths (0 to 10, 10 to 20, 20 to 30, 30 to 45, and 45 to 60 cm) for the 60 cm soil beds. Cores were removed at eight different surface locations, along two perpendicular transects through the middle of the circular pot (Figure 2). Roots were separated from the soil by manually rinsing and sieving using the method of Smit et al. [35]. The root mass contained in each 10 cm soil core was measured with a precision of 10^{-5} g and used to calculate the root mass density (RMD) (g of root per L of soil). The average RMD in each layer was compared along the soil depth. RMD contour maps were plotted using the online "EZplot®" Excel contour plotting tool [36] to observe the root distribution pattern within the soil bed.

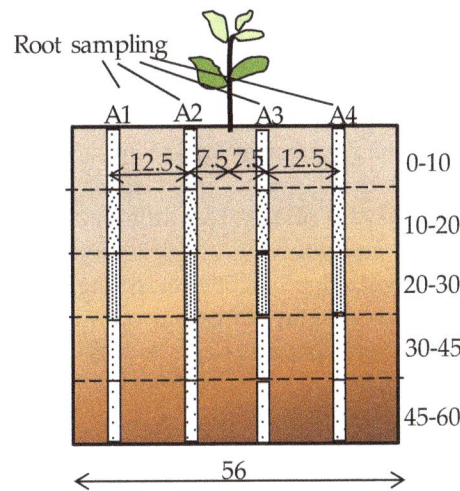

Figure 2. Root sampling diagram (all dimensions in cm).

2.6. Statistical Analysis

Statistical analyses were performed using the one-way analysis of variance (One-Way ANOVA) procedure, computed using SPSS version 22 (IBM Corp., New York, NY, USA). The procedure was used to compare treatments with respect to total plant biomass accumulation, fruit yield, ET_c and WUE. Dixon's Q-test was used to decide whether suspected outliers between replicates could be legitimately rejected or not, with a confidence level of 95%. The test of homogeneity of variances was carried out to evaluate the confidence level of accepting the null hypothesis (i.e., variances are equal) and whether the homogeneity of variance assumption had been met. When the F value was statistically significant, Post-hoc multiple comparisons were performed using Fisher's Least Significant Difference (LSD) test at $P \leq 0.05$ to determine if treatment means were significantly different.

3. Results

3.1. Relative Humidity, Air and Soil Temperature

Soil temperature is an important parameter contributing to healthy plant growth. Figure 3 shows the variation in relative humidity (RH), indoor air temperature and soil temperature at 50 and 150 mm depths for 5 consecutive days during the fruit development stage (Stage III). The soil temperatures shown are only for treatments T2 S0/P600 and T5 W300/P300. The soil temperature of each soil depth followed the same trend as indoor air temperature but with a lag of about 4 to 6 h.

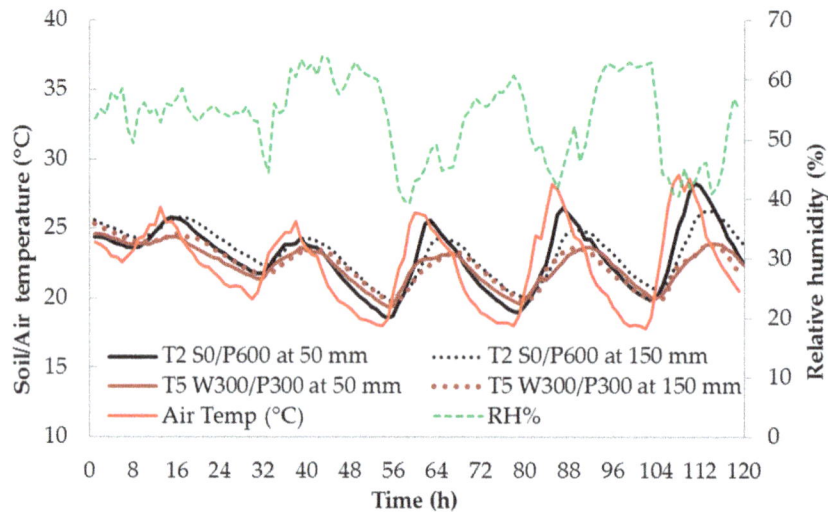

Figure 3. Daily relative humidity (%), indoor air temperature (°C) and soil temperature at 50 mm and 150 mm depth for five consecutive days in the fruit development stage (Stage III). See Figure 1 for treatment descriptions.

The difference in soil temperature between the two treatments was large when the air temperature was high—the wicking treatment T5 had a consistently lower maximum temperature than the surface treatment T2. This is most likely due to the greater thermal mass in the volume of water contained in the reservoir and saturated soil layers of T5. The difference decreased at lower temperatures. In both treatments, the highest soil temperature was typically recorded at 1500 h and the lowest was around 0800 h for the shallow soil bed at a depth of 50 mm. At the 150 mm depth, temperature maxima and minima lagged behind the shallower temperatures by about 2–3 h (Figure 4). The RH varied in the opposite direction to the air temperature, i.e., RH was high at low temperature, and vice versa (Figure 3).

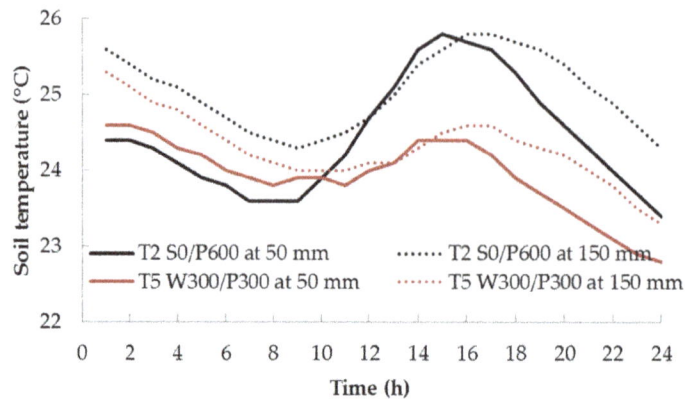

Figure 4. Diurnal variation of soil temperature 48 days after transplanting for surface (T2) and wicking (T5) treatments. See Figure 1 for treatment descriptions.

3.2. Effect of Irrigation Treatment on Plant Growth, Yield and Fruit Quality

At 70 DAT, there was no significant difference among treatments for crop height, stem diameter and number of leaves. The crop height ranged from 137.7 ± 16.6 to 146.7 ± 14.6 cm. At the completion of the study period, the total and marketable fruit yields were higher for the WB treatments T4, T5, and T6 than for the surface-irrigated treatment T7, and marketable yield of T6 was greater than T1 and T7, with all other treatments at intermediate values. The highest number of marketable fruits was recorded from the WB treatment with no soil column (T5 W300/P300, 28.3 ± 1.5 fruit) and the lowest was with a surface treatment (T7 S0/P300, 22 ± 3.1 fruit).

The diameter of red, ripe fruit ranged from 22 to 83 mm, and fruit weight ranged from 18 to 249 g across treatments. All WB treatments with the 300 mm soil bed produced more large-sized fruit than the surface treatment with the same soil depth (Figure 5a). Plants in the deep soil pots (600 mm) with surface or wick irrigation did not differ from the other treatments. More small fruit were produced by both surface irrigated treatments than the treatment with wick irrigation and a soil column (T6 WS300/P300). While yields of small and extra-large fruit were considerably lower than yields of large fruit, WB treatments T4, T5, and T6 produced greater marketable yields of large fruit than surface irrigated treatment T7 (Figure 5b). The wicking treatment with 150 mm reservoir depth (T4 W150/P300) produced more extra large fruit and greater marketable yield than wicking treatments T1 W300/P600 and T5 W300/P300 as well as the surface irrigated T7 S0/P300 treatment.

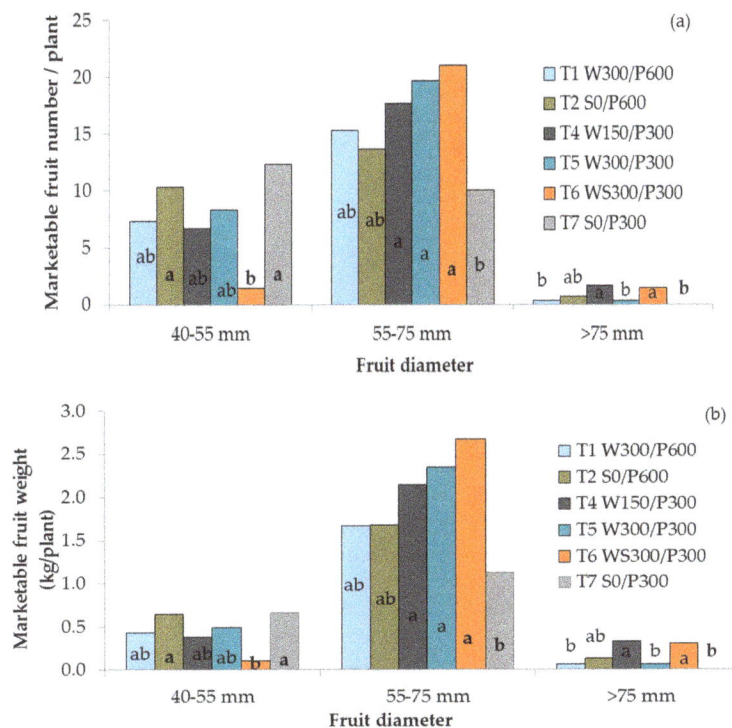

Figure 5. Number (**a**) and yield (**b**) of small, large and extra-large marketable tomato fruit as influenced by irrigation treatments. Columns within a group with different lower case letters are significantly different ($P \leq 0.05$, LSD). W = wicking bed; S = surface irrigation; WS = wicking bed with soil column; WN = wicking bed without plant; P = soil bed. The numbers following a letter for each treatment indicate the depth of reservoir (W) and soil (P), respectively, in mm.

The percentage of unmarketable yield of each treatment was calculated as a proportion of the total harvested yield using the data in Table 1. It varied widely across the treatments, with no discernible pattern. The lowest values were 0.60% ± 0.55% and 1.1% ± 1.9% unmarketable fruit in WBs with a 150 mm reservoir (T4 W150/P300) and wicking with a soil column (T6 WS300/P300), respectively. The wicking treatment with a 300 mm reservoir had 5.2% ± 3.2% rejected fruits, while the deep soil bed (both wicking and surface) had 7.6% ± 6.5 and 9.8% ± 3.8% rejected, respectively. The greatest value was 16.8% ± 7.2% unmarketable fruit for the shallow soil surface treatments (T7 S0/P300).

3.3. Effect of Irrigation Treatment on Water Use Efficiency

Neither WUE or iWUE were significantly different between the WB and surface treatments with 600 mm soil bed depths, T1 and T2, respectively (Table 1). However, there was a significant difference within the 300 mm soil bed depths where both WUE and iWUE were higher in WBs than the surface treatment, T4, T5, and T6 versus T7. The maximum ET_c value was by T2 S0/P600, a significantly higher value than all other treatments. The number of irrigation events over the growing period included water added at the surface (to all treatments) during the crop establishment period. Surface treatments needed 40 to 50 irrigation events (watering once every 2 to 3 days), compared with less than 26 for WB treatments (Figure 6). Table 2 compares the plausible water saving by WBs with respect to the corresponding surface treatment. A yield increase of 62 to 73% was obtained for shallow soil WBs, with water use varying from a 6% loss to a 9% savings relative to surface irrigated plots. In contrast, the deeper WBs had a higher water saving (22%), but with an 11% loss of yield; hence, the lack of significant difference in WUE for these treatments.

Table 1. ANOVA summary for total yield, marketable yield, total water, crop evapotranspiration (ET_c), irrigation water use efficiency (iWUE) and water use efficiency (WUE) as affected by each irrigation treatment at the completion of the study.

Treatment	Yield (kg/Plant)		Total Water (L)	Crop Water Use (L/kg)	ET_c (L)	iWUE (g/L)	WUE (g/L)
	Total	Marketable					
T1 W300/P600 [z]	2.36 ab [y]	2.18 b	123.1 c	0.06 ab	123.9 bc	17.8 bc	17.6 bc
T2 S0/P600	2.72 ab	2.46 abc	158.1 a	0.06 ab	160.6 a	15.5 c	15.3 c
T3 WN300/P600	* [x]	*	32.2 d	*	18.5 d	*	*
T4 W150/P300	2.86 a	2.85 ab	121.7 c	0.04 b	118.8 c	23.5 a	24.0 a
T5 W300/P300	3.06 a	2.89 ab	125.2 c	0.04 b	123.0 bc	23.2 a	23.6 a
T6 WS300/P300	3.08 a	3.08 a	143.0 ab	0.05 b	136.2 bc	21.5 ab	22.6 ab
T7 S0/P300	2.11 b	1.78 c	134.8 bc	0.08 a	137.4 b	13.2 c	13.0 c
T8 WN300/300	*	*	30.2 d	*	23.6 d	*	*

[z] W = wicking bed; S = surface irrigation; WS = wicking bed with soil column; WN = wicking bed without plant; P = soil bed. The numbers following a letter for each treatment indicate the depth of reservoir (W) and soil (P), respectively, in mm; [y] Means within columns followed by different letters are significantly different for irrigation treatments ($P \leq 0.05$, LSD); [x] An asterisk (*) indicates no data, or calculation was not possible as they relate to a treatment with no plant.

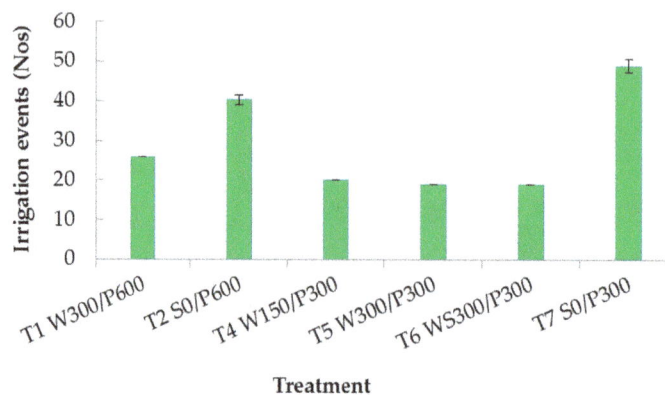

Figure 6. Histogram of average total number of irrigation events throughout the experimental period. Vertical bars indicate standard error. W = wicking bed; S = surface irrigation; WS = wicking bed with soil column; WN = wicking bed without plant; P = soil bed. The numbers following a letter for each treatment indicate the depth of reservoir (W) and soil (P), respectively, in mm. Error bars indicate standard deviation.

Table 2. Comparison of irrigation water reduction and tomato yield increase between wicking and surface treatments.

Treatment	Irrigation Water (L)	Marketable Yield (kg)	Water Reduction (S-W)/S %	Yield Increase (W-S)/S %
T1 W300/P600 ^{z}	123.1 ^{y}	2.18	22.1	−11.2
T2 S0/P600	158.1	2.46	-	-
T4 W150/P300	121.7	2.85	9.7	59.6
T5 W300/P300	125.2	2.89	7.1	62.5
T6 WS300/P300	143.0	3.08	−6.1	72.8
T7 S0/P300	134.8	1.78	-	-

^{z} Treatment types are labelled with treatment number followed by the irrigation method, reservoir depth and the soil bed height. W = wicking bed; S = surface irrigation; WS = wicking bed with soil column; WN = wicking bed without plant; P = soil bed. The numbers following a letter for each treatment indicate the depth of reservoir (W) and soil (P), respectively, in mm; ^{y} Statistical differences for Irrigation Water and Marketable Yield are shown in Table 1. Percentage Water Reduction and Yield Increase for W versus S were calculated separately for each soil bed height.

3.4. Effect of Irrigation Treatments on Fruit and Shoot Dry Matter

The dry weight of fruit, stems, and foliage are presented in Table 3. There was no significant difference in dry stem weight between the treatments. The dry foliage weight of surface treatments (both 300 and 600 mm soil bed depths) was significantly higher than that of WB treatments. However, overall dry biomass accumulation (stem, foliage and fruit) of T2 S0/P600 was significantly greater than T1 W300/P600 and T5W 300/P300. Generally, the higher the water consumption, the higher the biomass and foliage weight. Supporting this, results showed that the total dry biomass weight was linearly related to the amount of water applied in all other treatments ($R^2 > 0.99, P < 0.05$), excluding treatments T1 W300/P600 and T5 W300/P300. The lower foliage or biomass dry weight (in wicking treatments T1W300/600 and T5 W300/P300) may be due to less water availability and uptake by the plant in the former case, and lateral root growth and elongation observed in subsurface layers in the latter case (Figures 5 and 6, discussed later).

Table 3. Above-ground dry biomass as affected by irrigation treatment at the completion of the study.

Treatment	Above-ground Average Dry Biomass Weight (g/Plant)			
	Fruits	Stem	Foliage	Total
T1 W300/P600	154.7 b	50.3 a	69.3 c	274.4 b
T2 S0/P600	203.6 ab	56.0 a	93.1 ab	352.7 a
T4 W150/P300	182.6 ab	58.8 a	75.9 bc	317.2 ab
T5 W300/P300	188.8 ab	45.2 a	61.0 c	294.9 b
T6 WS300/P300	204.6 a	55.3 a	77.6 bc	337.5 ab
T7 S0/P300	177.8 ab	54.9 a	99.2 a	331.9 ab

W = wicking bed; S = surface irrigation; WS = wicking bed with soil column; WN = wicking bed without plant; P = soil bed. The numbers following a letter for each treatment indicate the depth of reservoir (W) and soil (P), respectively, in mm; Means within columns followed by different letters are significantly different for irrigation treatments ($P \leq 0.05$, LSD).

The distribution of soil-water content in different layers during the fruit maturity and development stage (Stage IV, 74 to 134 days) is presented in Figure 7. In surface treatments (T2 S0/P600 and T7 S0/P300), a frequent fluctuation of moisture content was observed at the top (0 to 10 cm) layer, as would be expected due to the frequent irrigation events. In WB treatments, the soil water content was more stable and was driest at the top (0 to 10 cm). However, there was no clear pattern among the different WB treatments.

In contrast to the surface treatments, the moisture content increased substantially with depth in the wicking treatments, as may be reasonably expected given that the source of water was at the bottom of the soil. In the second layer (10 to 20 cm), the difference in moisture content among different

WB treatments was greater than in the first layer. A high moisture content was recorded for treatment T6 WS300/P300 followed by T5 W300/P300, T4 W150/P300 and T1 W300/P600 (for treatments with plants). The moisture content gradually decreased over time in this layer of the wicking treatments until the reservoir was refilled. For both surface treatments (deep and shallow), the moisture content in this layer corresponded to the lower end of the range of moisture content in the WBs. In WBs, the soil moisture content at the bottom layer was close to saturation, increasing when water was added and drying in the interim.

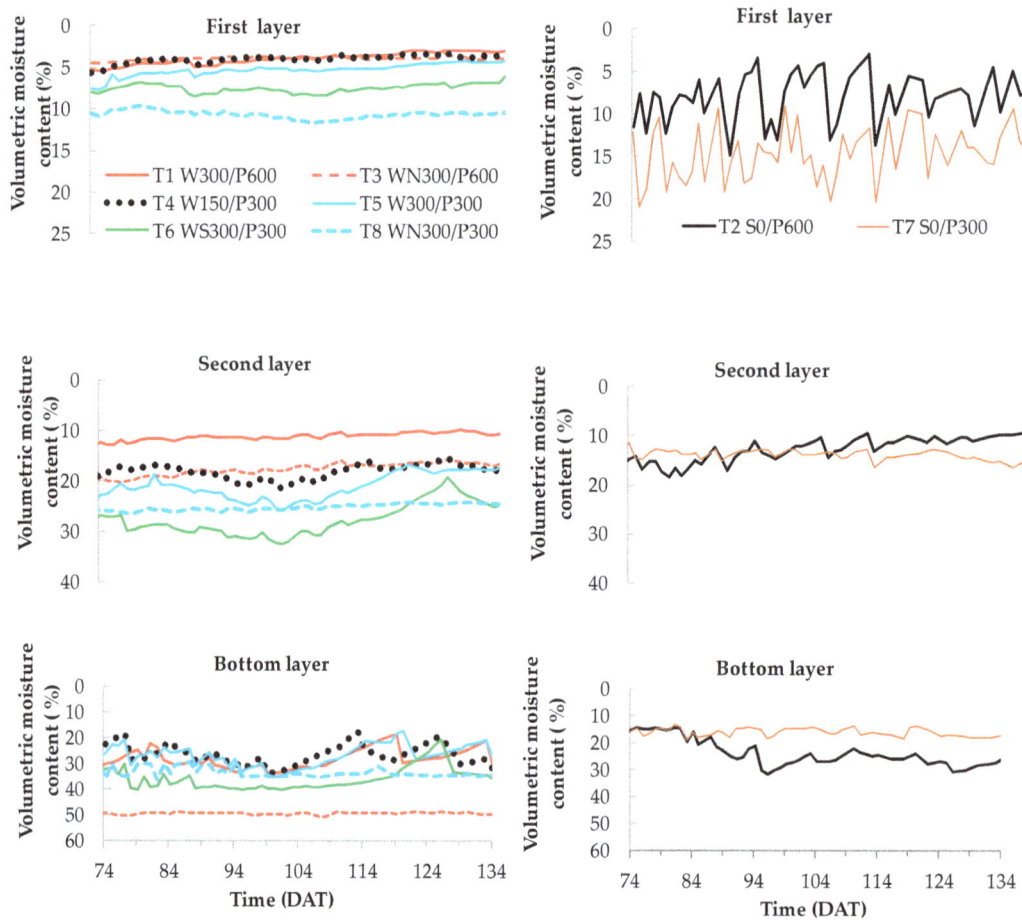

Figure 7. Average soil water content distribution (as % volumetric moisture content) at different soil depths during fruit maturity and harvest stage for wicking bed (**left**) and surface (**right**) treatments. W = wicking bed; S = surface irrigation; WS = wicking bed with soil column; WN = wicking bed without plant; P = soil bed. The numbers following a letter for each treatment indicate the depth of the reservoir (W) and soil (P), respectively, in mm. First layer—0 to 10 cm, second layer—10 to 20 cm, bottom layer—20 to 30 (300 mm bed) and 45 to 60 cm (600 mm bed).

A vertical distribution of moisture content at the end of the experiment is shown in Figure 8. All treatments had a similar distribution pattern (moisture content increasing with depth). WBs of reservoir depths 150 and 300 mm and the WB with soil column had the same distribution pattern, with the soil column having a marginally (about 2%) higher moisture content. A higher moisture content at the surface level was recorded in the shallow wicking treatment without a plant, possibly due to the lack of interception and uptake of capillary water by roots, allowing more water to reach the surface.

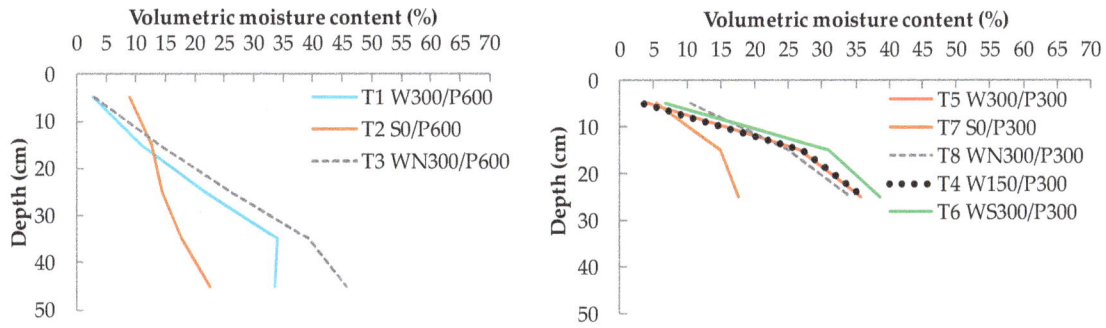

Figure 8. Average soil-water content distribution (as % volumetric moisture content) at the end of experiment for 600 mm soil bed depths (**left**) and 300 mm soil bed depths (**right**) treatments. W = wicking bed; S = surface irrigation; WS = wicking bed with soil column; WN = wicking bed without plant; P = soil bed. The numbers following a letter for each treatment indicate the depth of the reservoir (W) and soil (P), respectively, in mm.

3.5. Root Mass Density

The root mass density (RMD) for each treatment revealed a possible difference in rooting patterns among irrigation treatments (Figure 9). A large concentration of roots was generally found in the top layer (0 to 10 cm), except in treatment T6 WS300/P300. A higher RMD was observed near the surface for surface irrigated compared with WB treatments, with roots decreasing approximately exponentially with the soil depth. A contour plot of RMD across the bed showed that the RMD was preferentially concentrated near the plant and within the top 0 to 20 cm layer (Figure 10). Interestingly, no discernible differences or trends could be observed in RMD between the surface and wicking treatments, except that the wicking bed with soil column exhibited a more uniform, and generally lower, RMD compared to all other treatments. Although these results are therefore inconclusive, previous studies have indicated that a lower RMD was produced when there was more water available [37] near the water table, which may have been the case for the WB with soil column which had a higher water demand.

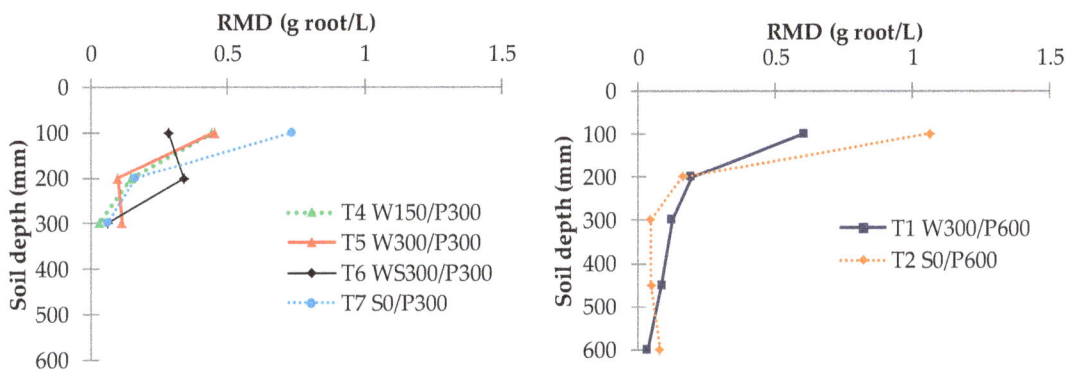

Figure 9. Root mass density (RMD) distribution for different treatments. W = wicking bed; S = surface irrigation; WS = wicking bed with soil column; WN = wicking bed without plant; P = soil bed. The numbers following a letter for each treatment indicate the depth of reservoir (W) and soil (P), respectively, in mm.

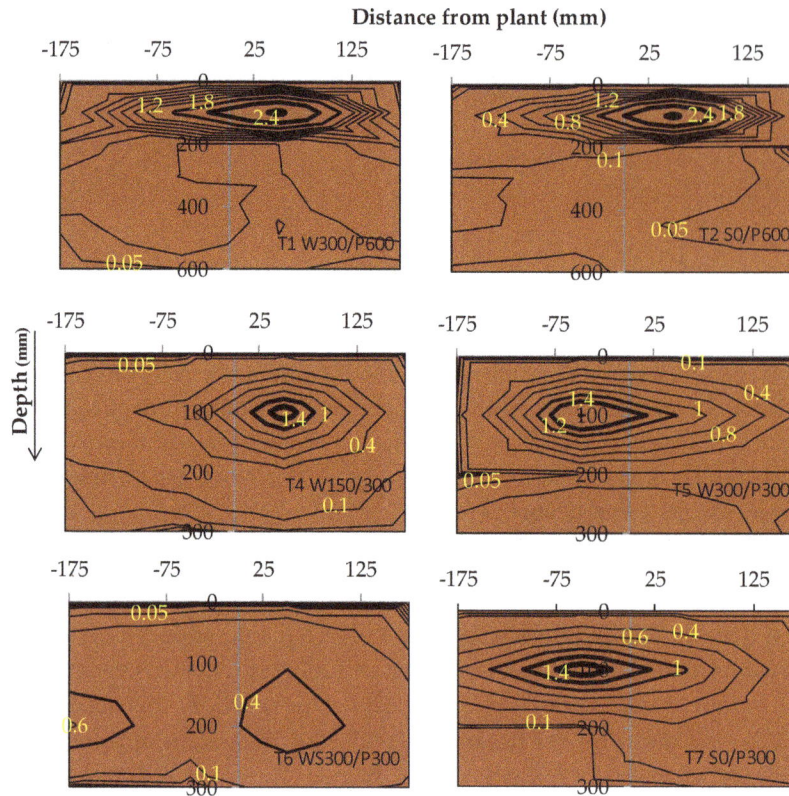

Figure 10. Tomato root mass density (RMD) distribution (mg/cm^3) as affected by soil depth and distance from the plant at the end of the experiment. Bolder contours indicated higher RMD. W = wicking bed; S = surface irrigation; WS = wicking bed with soil column; WN = wicking bed without plant; P = soil bed. The numbers following a letter for each treatment indicate the depth of reservoir (W) and soil (P), respectively, in mm.

4. Discussion

The results showed that there was a significant difference between the wicking and surface treatments for total biomass, yield and water use efficiency. The total water use was significantly different between surface and wicking treatments in the two deep soil bed depths (600 mm); however, the difference was not significant between the treatments for the 300 mm soil bed depth or between the two reservoir depths. An average irrigation amount of approximately 122 L/plant was the optimum amount of water for maximizing the WUE for tomatoes in these study conditions. This optimum WUE occurred in the 150 mm reservoir wicking bed (T4 W150/P300). The average amount of water needed per plant was about 0.9 L/plant/day in this WB treatment, depending on the growth stage and the season. In comparison, Snyder [38] recommended about 50 mL/day for newly transplanted crops and 2.7 L/plant/day at maturity on sunny days for glasshouse-grown tomatoes. Water requirements for drip irrigated tomatoes grown in greenhouse conditions in the tropics were reported to range from 0.3 to 0.4 L/plant/day, a figure derived from a pot area of 706.50 cm^2, similar to this study [39]. For context, the water requirement for an average field-grown tomato cultivar is 100 to 150 L/plant (a 400 to 600 mm irrigation rate) for a 90 to 120 day growth period (FAO, 2015).

Despite the application of larger irrigation volumes in surface treatments, yield and WUE were significantly greater in wicking treatments. The highest marketable yield of 3.1 kg/plant was obtained in the soil column WB (T6 WS300/300), while the lowest (1.8 kg/plant) occurred in the shallow soil (300 mm) surface irrigated treatment (T7 S0/P300). To place the yields in this experiment in some context, the FAO (2015) stated that a good commercial yield under irrigation was 1.1 to 1.6 kg/plant for field-grown tomatoes, while well-managed greenhouse-grown tomatoes in a soil-free culture should yield 3.3 to 6 kg/plant [40,41].

The deep soil bed wicking treatment (T1 W300/P600) had a lower yield than its corresponding surface treatment, and was also lower than the shallower wicking treatments. The reason for the low yield and WUE in T1 W300/P600 was initially considered to be insufficient supply of water to the root zone, due to the greater height and the resultant difficulty in maintaining water supply to the roots by capillary rise (noting that the amount of capillary water present in soil decreased with distance above the water table). In support of this theory, soil evaporation observed in the unplanted deep wicking treatment (T3 WN300/P600) was low when compared with the 300 mm WB T8 WN300/P300 treatment (Table 1), suggesting that less water was able to reach the surface by capillary rise. However, the ET_c in the deep soil WBs was not significantly different from the ET_c in the shallower wicking treatments, suggesting that a similar amount of water must have been delivered to the plant, despite the greater soil depth. Although there was about a 22% water reduction in T1 W300/P600 relative to the surface treatment (T2 S0/P600) which had an equivalent soil depth (Table 2), the wicking treatment delivered lower yields and thus failed to produce a significant improvement in WUE. It should be noted that the percentage of unmarketable fruits was lower in WBs compared to surface treatments for the shallow soil plots, but not for the deep soil treatments. On this basis, it is inferred that the use of wicking irrigation to grow tomatoes with a 600 mm soil depth was not effective. This depth of wicking bed may, however, be suitable for deeper rooted plants. It should be noted that the main reason for discarding fruit was blossom end rot. Previous studies have mentioned that insufficient calcium, or either too much or too little water, can cause blossom end rot in tomatoes [38,42]. It is therefore possible that the soil chemistry, the irrigation method, and/or the irrigation volume may have affected the reliability of fruit production in the experiment, and inferences drawn from these results (especially for the shallow 300 mm soil treatments) should be moderated accordingly.

The wicking bed with a soil column had a higher ET_c compared to other shallow wicking treatments. However, to compensate, it also had the highest yield, plant height and vegetative mass. Studies have shown that water moves upward slowly and over long distances in fine textured soils, but not so for coarse grained soils [43]. This may provide a plausible explanation for the improved performance as the soil column in the reservoir potentially carried more capillary water to the soil surface by maintaining a connection of saturated soil, thus allowing plants to consume more water than other treatments. This may be the reason for higher yield, plant vegetative mass, and water consumption (Tables 1 and 2). The yield was highest in T6 WS300/P300 but, perhaps due to the increased vegetative mass, the WUE was not as high as the other wicking treatments. However, there was no significant difference in the crop water use with respect to yield among shallow WBs. As such, a wicking bed with a soil column would appear to be a better option for areas where there is little concern about water availability (Table 1). Nevertheless, we may expect a better result by optimizing this system design further. For example, changes could be made to the size of the soil column, the frequency of irrigation events or by the application of mulch to the soil surface.

There was no significant difference in WUE between the 150 mm and 300 mm reservoir depths. This finding has implications for design of cost-effective wicking beds, i.e., there was little gain from a deeper reservoir. Based on having a smaller reservoir depth and achieving an equal maximum WUE, treatment T4 W150/P300 may be considered the most economically efficient wicking bed system tested in this study. Overall, shallow wicking treatments were the more favourable treatments in terms of WUE and yield. This may be related to the relatively shallow (medium) rooting depth of tomatoes.

The labour efficiency of each treatment was evaluated by counting number of irrigation events over the growing period of each replicate and averaged. Surface treatments needed frequent watering at approximately 3 day intervals. In contrast, the wicking treatment pots lasted up to 4 weeks without watering during the developing stage (after the initial two weeks of surface watering to establish the plant), and from 1 to 2 weeks during the maturity stage. These results indicated that WBs reduced manual watering requirements. Most importantly, the WBs eliminated the problem of deciding when and how much water to irrigate. Efficient surface irrigation depended on precise soil moisture measurements and water balance calculations which may not be available to non-commercial growers. In practice, of course, the extent of the labour-savings will depend on the size of the pot, soil bed height and reservoir volume, and the type of crop or root behaviour of the plant selected.

The low soil moisture content in the top layer of all replicates was likely due to increased water evaporation from the soil surface along with transpiration, as the root density was highest at the upper soil layer (Figures 9 and 10). Moreover, the moisture content in the upper layer was higher in the treatment with no plants than that of corresponding plant treatments (Figure 7). This suggests that transpiration was responsible for a significant amount of moisture depletion in the soil, and that the combined effect of evaporation and transpiration was responsible for lowering the moisture content in the soil.

A regular pattern of soil moisture content distribution was observed in wicking treatments without plants (T3 WN300/P600 and T8 WN300/P300) (Figure 7). The total water consumption of these pots may be assumed to be due to soil water evaporation. Soil moisture content of the first and second layer of treatment T3 WN300/P600 was always lower than that of treatment T8 WN300/P300. This was likely due to a decrease in matric potential (i.e., becoming more negative) when the soil bed depth increased (600 mm as opposed to 300 mm). Interestingly, this also led to reduced surface water evaporation (Table 1), as the evaporating water was likely not be replaced as quickly through the taller soil beds. As such, the bottom layer of treatment T3 WN300/P600 held a maximum moisture content, which was similar to the saturated moisture content of the soil.

The surface layer accumulates heat more during the day due to solar radiation than the underlying layer, resulting in a temperature gradient in the soil [44]. The soil temperature was always greater in T2 S0/600 than in the T5 W300/300 at high temperatures. The frequent application of water in the surface treatment increased the soil moisture content in the upper layer compared to the wicking treatment. Counterintuitively, the higher moisture content in the top layer of surface treatments may be the reason for the higher temperature, due to an increase in the thermal conductivity and diffusivity of wet soil [44–46]. Roots survive and grow where adequate water, temperature and oxygen are present [47]. The observed maximum RMD in the second layer of treatment T6 WS300/P300 may have been due to sufficient moisture content and other suitable growing conditions (Figure 9). The moisture content was very high (approaching saturation) at the bottom layer of wicking treatments, and as such, high moisture content may have lowered the amount of oxygen trapped in the soil and restricted root growth. As such, the unfavourable environment may have lead the root system to develop horizontally rather than vertically [48].

It is understood that container-grown plants have different root growth patterns compared to field-grown crops. For instance, where a container restricts roots, this may result in fewer primary roots and an increase in lateral roots [49,50]. The large diameter of the pot may have produced a high concentration of lateral rather than vertical roots, in which case a different root pattern may occur if smaller diameter pots were used. It is also noteworthy that there is a potential for salt accumulation at the surface of wicking beds, a phenomenon which has been previously noted in sub-irrigated systems. Follow-up research is being undertaken to determine if this occurs in WBs.

5. Conclusions

Wicking beds (WBs) have been widely promoted and adopted as an efficient irrigation system for urban agriculture. However, there has been little published research to support popular claims about the effectiveness of WBs. This study rigorously tested the performance of WBs relative to best-practice, precision surface irrigation systems in terms of total water use, marketable yield, fruit quality, water use efficiency (WUE) and irrigation frequency. The suitability of WBs for growing tomatoes was compared using two soil bed depths (300 vs. 600 mm), two reservoir depths (150 vs. 300 mm), and introduced a soil column in the reservoir to improve wicking. The WB with a 150 mm reservoir and 300 mm soil bed was the most effective of all treatments tested, as it was cost-effective (requiring minimal material) with a high WUE.

Overall, the results of this study indicated that WBs matched (or exceeded) WUE and yield achieved with best-practice surface irrigation, and offered a potentially substantial labour saving for gardeners. Moreover, given that surface irrigation in urban agriculture is likely to fall well short of the precision irrigation method used for comparison in this study, the relative WUE improvement of WBs in practice is probably greater. As WBs can be made out of recycled (and recyclable) material, they may be a low-tech and low-cost system. This study therefore provides a much-needed scientific basis for the widespread adoption of WBs in urban agriculture.

There are opportunities to further improve the system to provide more benefit to cities, not only for saving water in gardens, but ultimately to help facilitate the other benefits of urban food production. For example, they may also provide an interesting glasshouse model for smaller-scale commercial producers. There is a possibility to connect multiple WB pots in series and parallel for glasshouse settings. The reservoir can be filled manually or automatically, making it even more practical. However, sustainable irrigation practices will always be site specific, which depends on local climate, soil, topography, and the quality of water. One limitation of this trial is that this is a single site and a single season. Further studies are intended to improve our understanding of the effectiveness of the WB irrigation system for different crop types and environmental settings, as well as soil salinity and sodicity with the use of recycled wastewater.

Acknowledgments: The authors would like to acknowledge that this research was conducted with the assistance of an Australian Government Endeavour Awards Scholarship. Funding was also provided by the Goyder Institute for Water Research, a partnership between the South Australian Government through the Department of Environment, Water and Natural Resources, CSIRO, Flinders University, the University of Adelaide, and the University of South Australia.

Author Contributions: N.P.K.S., J.D.W. and B.R.M. conceived and designed the experiments; N.P.K.S. performed the experiments and analysed the data; J.D.W. and B.R.M. contributed to the analysis process; N.P.K.S. wrote the paper; J.D.W. and B.R.M. contributed to the manuscript.

Conflicts of Interest: The authors declare no conflict of interest.

References

1. Kadenyeka, M.V.; Omutimba, D.; Harriet, N. Urban agriculture livlihoods and household food security: A case of Eldoret, Kenya. *J. Agric. Biol. Sci.* **2013**, *8*, 90–96.
2. Cameron, R.W.F.; Blanuša, T.; Taylor, J.E.; Salisbury, A.; Halstead, A.J.; Henricot, B.; Thompson, K. The domestic garden—Its contribution to urban green infrastructure. *Urban For. Urban Green.* **2012**, *11*, 129–137. [CrossRef]
3. Grewal, S.S.; Grewal, P.S. Can cities become self-reliant in food? *Cities* **2012**, *29*, 1–11. [CrossRef]
4. Gibbons, S.; Mourato, S.; Resende, G.M. The amenity value of English nature: A hedonic price approach. *Environ. Resour. Econ.* **2014**, *57*, 175–196. [CrossRef]
5. The Australia Institute. *Grow Your Own: The Potential Value and Impacts of Residential and Community Food Gardening*; The Australia Institute: Canberra, Australia, 2014.
6. Goldstein, M.; Bellis, J.; Morse, S.; Myers, A.; Ura, E. *Urban Agriculture: A Sixteen City Survey of Urban Agriculture Practices across the Country*; Turner Environmental Law Clinic: Atlanta, GA, USA, 2011.
7. Lovell, S.T. Multifunctional urban agriculture for sustainable land use planning in the United States. *Sustainability* **2010**, *2*, 2499–2522. [CrossRef]
8. Forget, G. Balancing the need for pesticides with the risk to human health. In *Impact of Pesticide Use on Health in Developing Countries*; Forget, G., Goodman, T., de Villiers, A., Eds.; International Development Resesrch Centre: Ottawa, ON, Canada, 1990.
9. Pimentel, D.; Culliney, T.W.; Bashore, T. Public Health Risks Associated with Pesticides and Natural Toxins in Foods. Available online: http://ipmworld.umn.edu/pimentel-public-health (accessed on 25 June 2016).
10. Breuste, J.H.; Artmann, M. Allotment gardens contribute to urban ecosystem service: Case study Salzburg, Austria. *J. Urban Plan. Dev.* **2014**. [CrossRef]
11. Taylor, J.R.; Lovell, S.T. Urban home gardens in the global north: A mixed methods study of ethnic and migrant home gardens in Chicago, IL. *Renew. Agric. Food Syst.* **2015**, *30*, 22–32. [CrossRef]
12. Poulsen, M.N.; McNab, P.R.; Clayton, M.L.; Neff, R.A. A systematic review of urban agriculture and food security impacts in low-income countries. *Food Policy* **2015**, *55*, 131–146. [CrossRef]
13. Prime Minister's Science, Engineering and Innovation Council (PMSEIC). *Recycling Water for Our Cities*; Prime Minister's Science, Engineering and Innovation Council (PMSEIC): Canberra, Australia, 2003; p. 44.
14. Dima, S.J.; Ogunmokun, A.A.; Nantanga, T. *The Status of Urban and Peri-Urban Agriculture in Windhoek and Oshakati*; Food and Agriculture Organisation (FAO): Windhoek, Nambia, 2002.

15. Ward, J.D.; Ward, P.J.; Saint, C.P.; Mantzioris, E. The urban agriculture revolution: Implications for water use in cities. *J. Aust. Water Assoc.* **2014**, *41*, 69–74.

16. Levidow, L.; Zaccaria, D.; Maia, R.; Vivas, E.; Todorovic, M.; Scardigno, A. Improving water-efficient irrigation: Prospects and difficulties of innovative practices. *Agric. Water Manag.* **2014**, *146*, 84–94. [CrossRef]

17. Liu, F. Irrigation strategies for sustainable environmental and influence on human health. In *Encyclopedia of Environmental Health*; Nriagu, J.O., Ed.; Elsevier: Burlington, NY, USA, 2011; pp. 297–303.

18. Russo, T.; Alfredo, K.; Fisher, J. Sustainable water management in urban, agricultural, and natural systems. *Water* **2014**, *6*, 3934–3956. [CrossRef]

19. Jones, K.S.; Costello, L.R. *Irrigating the Home Landscape*; University of California Cooperative: Half Moon Bay, CA, USA, 2003.

20. Peters, T. *Drip Irrigation for the Yard and Garden*; Washington State University: Pullman, WA, USA, 2011; p. 7.

21. Hayden, L.; Cadenasso, M.L.; Haver, D.; Oki, L.R. Residential landscape aesthetics and water conservation best management practices: Homeowner perceptions and preferences. *Landsc. Urban Plan.* **2015**, *144*, 1–9. [CrossRef]

22. Langellotto, G.A. What Are the Economic Costs and Benefits of Home Vegetable gardens? Available online: http://www.joe.org/joe/2014april/rb5.php (accessed on 2 June 2016).

23. Kessler, R. Urban gardening: Managing the risks of contaminated soil. *Environ. Health Perspect.* **2013**. [CrossRef] [PubMed]

24. Sullivan, C.; Hallaran, T.; Sogorka, G.; Weinkle, K. An evaluation of conventional and subirrigated planters for urban agriculture: Supporting evidence. *Renew. Agric. Food Syst.* **2015**, *30*, 55–63. [CrossRef]

25. Kirkham, M.; Gabriels, D. Water and nutrient uptake of wick-grown plants. *Hortic. Res.* **1979**, *19*, 3–13.

26. Schuch, U.K.; Kelly, J.J. Capillary mats for irrigating plants in the retail nursery—And saving water. *Southwest Hortic.* **2006**, *23*, 24–25.

27. Austin, C. Wicking Bed—A New Technology for Adapting to Climate Change. Available online: http://www.waterright.com.au/ (accessed on 20 October 2015).

28. ModBOX © Modbox Raised Garden Beds. Available online: http://www.modbox.com.au/wicking-garden-beds/ (accessed on 10 June 2016).

29. Lane, K.S.; Washburn, D.E. Capillarity tests by capillarimeter and by soil filled tubes. In *Highway Research Board Proceedings*; Highway Research Board: Washington, DC, USA, 1947; Volume 26, pp. 460–473.

30. Li, X.; Zhang, L.M.; Fredlund, D.G. Wetting front advancing column test for measuring unsaturated hydraulic conductivity. *Can. Geotech. J.* **2009**, *46*, 1431–1445. [CrossRef]

31. Sentek Pty Ltd. *User Guide: Diviner 2000*; version 1.2; Sentek Pty Ltd.: Stepney, South Australia, Australia, 2003; p. 66.

32. Peet, M.; Welles, G. Greenhouse tomato production. In *Tomatoes (Crop production science in Horticulture)*; Heuvelink, E., Ed.; CAB International: Oxfordshire, UK, 2005; pp. 257–304.

33. Saha, U.K.; Papadopoulos, A.P.; Hao, X.; Khosla, S. Irrigation strategies for greenhouse tomato production on rockwool. *HortScience* **2008**, *43*, 484–493.

34. Abuarab, M.; Mostafa, E.; Ibrahim, M. Effect of air injection under subsurface drip irrigation on yield and water use efficiency of corn in a sandy clay loam soil. *J. Adv. Res.* **2013**, *4*, 493–499. [CrossRef] [PubMed]

35. Smit, A.L.; Bengough, A.G.; Engels, C.; Noordwijk, M.; Pellerin, S.; Geijn, S.C. *Root Methods: A Handbook*; Springer: Berlin/Heidelberg, Germany, 2000; pp. 177–182.

36. Office Expander EZplot for Excel. Available online: http://www.officeexpander.com/ (accessed on 12 July 2015).

37. Sorenson, S.K.; Dileanis, P.D.; Branson, F.A. *Soil Water and Vegetation Responses to Precipitation and Changes in Depth to Ground Water in Owens Valley, California*; United States Geological Survey Water-Supply (USGPO): Washington, DC, USA, 1991.

38. Snyder, R.G. *Greenhouse Tomato Handbook*; Mississippi State University: Starkville, MS, USA, 2010.

39. Salokhe, V.; Babel, M.; Tantau, H. Water requirement of drip irrigated tomatoes grown in greenhouse in tropical environment. *Agric. Water Manag.* **2005**, *71*, 225–242.

40. Maboko, M.; Du Plooy, C.; Bertling, I. Comparative performance of tomato cultivars cultivated in two hydroponic production systems. *S. Afr. J. Plant Soil* **2011**, *28*, 97–102. [CrossRef]

41. Rodriguez, J.C.; Cantliffe, D.J.; Shaw, N. Performance of greenhouse tomato cultivars grown in soilless culture in North Central Florida. *Proc. Fla. State Hortic. Soc.* **2001**, *114*, 303–306.

42. Whiting, D.; O'Meara, C.; Wilson, C. *Water Conservation in the Vegetable Garden*; Colorado State University: Fort Collins, CO, USA, 2015.

43. Kuo, J. *Practical Design Calculations for Groundwater and Soil Remediation*, 2nd ed.; Taylor and Francis: Hoboken, NJ, USA, 2014.

44. Chacko, T.P.; Renuka, G. Temperature mapping, thermal diffusivity and subsoil heat flux at Kariavattom of Kerala. *J. Earth Syst. Sci.* **2002**, *111*, 79–85. [CrossRef]

45. Ismail, S.M.; Ozawa, K.; Khondaker, N.A. Influence of single and multiple water application timings on yield and water use efficiency in tomato (var. First Power). *Agric. Water Manag.* **2008**, *95*, 116–122. [CrossRef]

46. Roxy, M.S.; Sumithranand, V.B.; Renuka, G. Estimation of soil moisture and its effect on soil thermal characteristics at Astronomical Observatory, Thiruvananthapuram, South Kerala. *J. Earth Syst. Sci.* **2014**, *123*, 1793–1807. [CrossRef]

47. Phillips, L. Read about Root Development. Available online: http://gibneyce.com/1-read-about-root-development.html (accessed on 15 July 2015).

48. Stokes, A. *The Supporting Roots of Trees and Woody Plants: Form, Function and Physiology*; Springer: Dordrecht, The Netherlands, 2000.

49. Nesmith, D.S.; Duval, J.R. The effect of container size. *HortTechnology* **1998**, *8*, 495–498.

50. Peterson, T.A.; Reinsel, M.D.; Krizek, D.T. Tomato (Lycopersicon esculentum mill, cv. "Better Bush") plant response to root restriction. *J. Exp. Bot.* **1991**, *42*, 1233–1240. [CrossRef]

11

Agroecology: A Global Paradigm to Challenge Mainstream Industrial Agriculture

Hector Valenzuela

Department of Plant and Environmental Protection Sciences, University of Hawaii at Manoa, 3190 Maile Way No. 307, Honolulu, HI 96822, USA; hector@hawaii.edu

Abstract: Considerable controversy continues to exist in scientific and policy circles about how to tackle issues of global hunger, malnutrition, and rural economic decline, as well as environmental issues, such as biodiversity loss and climate change adaptation. On the one hand, powerful vested interests, with close ties to government, media, and academic institutions, propose high-input technology-based solutions, speculative and neoliberal "market-based" solutions, and export-oriented agricultural models. On the other hand, an international scientific and grassroots Food Movement has emerged, calling for a redesign of the Global Food System in support of small-scale agroecological farming systems. A call to re-evaluate our current Food Systems was made in 2008 by the International Assessment of Agricultural Knowledge, Science and Technology for Development (IAASTD). Here, using the IAASTD study as a backdrop, we review the recent literature to outline key contentious points in the controversy between the need for high-input and "techno-based" *versus* agroecological farming models. A critical assessment is made of proposed strategies to protect soil resources, improve nutrient and energy cycles, protect agrobiodiversity, and promote social well-being in rural communities. With an increase in the number of affluent consumers (*i.e.*, the middle class) in the developing world, and with the continued problem of extreme and chronic poverty with other larger sectors of society, Organic Farming and Agroecology models are put forward as a sound social, scientific, and rural development strategy.

Keywords: agroecology; agrobiodiversity; biodiversity; farming systems; organic farming; ecological farming; soil management

1. Introduction

Considerable controversy continues to exist about how to tackle issues of global hunger, malnutrition, and rural economic decline, as well as on strategies to address environmental decline and climate change adaptation. Because global agricultural activities are strong determinants of human well-being and environmental quality, the judicious design and management of agricultural systems are central to discussions about food security and climate change, and about the conservation of natural resources such as land, water, energy, and biodiversity. A call for a re-evaluation of our current Food Systems was made in 2008 by the International Assessment of Agricultural Knowledge, Science and Technology for Development (IAASTD) [1]. Here, we review the recent literature to outline key contentious points in the controversy between the merits of "techno-based" industrial methods of farming *versus* alternative agroecological farming models.

2. Reviews about the Future of Agriculture

A review of agricultural systems in the early 1990s queried "Is the environmental price of America's unprecedented agricultural productivity too high?" and warned that "farms have become more specialized and more dependent on off-farm inputs" [2]. National scientific panel reports from the 1980s determined that in the USA alternative farming practices were practical, as productive, often more profitable, and that their adoption "would result in even greater benefits to farmers and environmental gains for the nation" [2]. Today, international agricultural development experts agree that current agricultural systems "must be transformed" [3,4]. An agricultural transformation is required for health and environmental reasons because "today's farming systems undermine the well-being of communities in many ways", including the destruction of "huge regions of natural habitat" and "an untold loss of ecosystem services" [3]. Similarly, it is recognized that agricultural systems need to shift towards alternative methods of farming that move away from "today's brand of resource-intensive, environmentally destructive agriculture" and that conventional methods of farming represent "a poor option" [5]. Along the same lines, a recent international panel report called for "the fundamental transformation of agriculture" following a dual approach that "drastically reduces the environmental impact of conventional agriculture" and that opens the door for "agro-ecological production methods" [6].

Despite the call for alternative methods of production over the years, the paradigm of industrial or conventional agriculture still dominates and permeates most mainstream academic and policy discussions about the future of agriculture [4]. An evaluation of agricultural systems from 1991 to 2003 determined that the conventional systems of industrial agriculture continued to predominate [7]. According to the authors, "the paradigm of industrial agriculture (High External Input Agriculture) has been simply amplified, by doing more of the same, with only minor adjustments in special countries. For those looking for a major transition toward a different pattern of production more focused on rural development, ecological compatibility and quality food, this is a reason for concern" [7]. Furthermore government policy and subsidies [2,8] and even the University Research and Extension system have helped to validate, strengthen and to perpetuate the expansion of current models of industrial agriculture [2,9].

Nevertheless, considerable field research has demonstrated the potential benefit to rural communities and to the agricultural sector of adopting alternative production systems [2,10]. For instance, a recent global analysis determined that organic systems are more profitable than conventional systems, even when external costs of production are not counted. While organic systems had an overall 10%–15% greater labor demand, this was considered to be less of a constraint in regions where labor is cheaper or where there is a surplus of labor [11]. As about 80% of global food demands are met by small-scale farms, agricultural development programs need to re-focus their programmatic activities to improve the productivity of small farms in the tropics [12,13]. However, for development and research programs to be meaningful, it is critical that socioeconomic conditions be considered by following participatory approaches [14]. Specific guidelines and steps required to "democratize" and establish participatory research and to promote food sovereignty for small farms in West Africa were outlined by Pimbert *et al.* [14].

3. Challenges for the Future of Agriculture

The paradigm and future prospect of modern industrialized agricultural systems is being challenged on several fronts because of its dependence on capital, external energy and agrochemical inputs, and for its adverse impact on biodiversity and on human health [2,15–20]. According to the American Medical Association, in reference to industrial agriculture, "these methods have contributed to the development of antibiotic resistance; air and water pollution; contamination of food and water with animal waste, pesticides, hormones, and other toxins; increased dependence on nonrenewable fossil fuels (including fertilizers); and a food system that is increasingly vulnerable to accidental or intentional contamination" [4,21]. Warnings have thus been raised and documented about the adverse health and environmental impacts from the intensification of both crop and livestock production systems [4,22–28].

Concerns have also been raised about the increased homogeneity of the food supply at a regional and global scale, resulting in a general decline in global food security with 85% of countries showing marginal or low food self-sufficiency indices [29–31]. Calls have also been made to revisit issues of agricultural sustainability concerning the impending environmental impacts of climate change and its effect on agriculture [3,32–34]. With potential global crop yield losses of over 50%, calls have thus been made to develop more "resilient" production systems to better withstand the impending impacts of climate change [16,35–39].

Concerns have also been raised that industrial agricultural practices are exacerbating the anthropogenic causes of climate change by contributing about 25% of global greenhouse gas emissions, about 60% of nitrous oxide emissions from the use of synthetic chemical nitrogen fertilizers and pesticides, and from its adverse impact on biodiversity [4]. Furthermore, new data indicate that previous estimates of nitrous oxide emissions from industrial agricultural systems may have been grossly underestimated, and that when "riverine" watershed emissions are considered, the levels of nitrous oxide emissions from areas such as the MidWestern USA may be up to 40% greater than earlier estimated [40]. The overall global environmental impact of these increased emissions on climate change could be significant as other similar regions of the world where intensive industrial farming practices are followed represent, globally, an area of over 230 million ha.

4. Industry and Industry-Funded Academics Support Techno-Based Solutions

Despite the many calls for a transformation of the food and conventional agricultural system, both the agrochemical industry and industry-funded scientists continue to espouse the value of industrial and "techno-based" solutions [5,33,41]. A paper authored 25 years ago by scientists from the agrochemical industry made a call for "patriotic" Sustainable Agriculture through a reliance on agrochemicals and on genetically modified organisms, emphasizing that "Sustainable agriculture is possible only with biotechnology and imaginative chemistry" [42]. A more recent article on Nature Magazine reiterated this earlier line of thought, indicating that "Feeding the world is going to require the scientific and financial muscle of agricultural biotechnology companies" [5,43]. Thus, according to this report, the world is placing its hopes about meeting future food global demands on the agrochemical industry, despite the fact that "the majority of companies' R&D spending and effort still goes towards blockbuster crops with traits, such as pest control, that benefit agribusiness, leaving neglected many crops that are important in the developing world" and despite the fact that the lack of progress to achieve these goals to date "is in large part a consequence of the hold that the private sector has on intellectual property rights to crucial technology, such as genetic markers, and the sequences of key genes and 'promoters' that drive gene expression" [43].

Thus, the mainstream thought in high level industry, media, the academic community, and many policy circles continues to espouse the idea that high capital and high-input dependent agricultural systems are the key to feed the world [4,33,36,41]. A recent United Nations Conference on Trade and Development (UNCTAD) panel report confirms that globally today the "priority remains heavily focused on increasing industrial agricultural production" [6]. For instance, according to a prime minister from Uganda attending a Global Food Summit "modern agriculture requires capital and technology", and as such the need for "both local and foreign investors" [44]. It is well-recognized that the agrochemical industry in general promotes "the benefits of chemical crop protection and biotechnology products, their importance to sustainable agriculture and food production" [16,33,41]. Thus, a central feature of agricultural assistance programs in Africa organized by the World Food Program, with support from the Bill and Melinda Gates Foundation, was to provide financial assistance to small farmers through the introduction of a regional speculative market for grains, to allow them to "acquire better seeds, fertilizers, and pesticides, more advanced irrigation systems". The access to cash, as explained by Bill Gates, would thus "impel small farmers to purchase more loans, more pesticides, more seeds" [44]. However, when implemented, the strategy to assist small farmers went belly-up, as the speculative markets led to a 13% increase in the price of millet in local markets and to a 7% price increase at a national level, a program which "would eventually send more people into poverty and starvation. The monetary gift triggered all manner of unforeseen consequences" [44].

Models of agricultural "intensification" based on a continued reliance on external inputs, chemicals and proprietary seed have been espoused for Africa by several academic and policy groups such as the Montpellier Panel [38] and by the Africa Progress Panel [45]. Similar calls for agricultural "intensification" have been made for Asia and other regions by leading academics, based on the use of "genetically modified organisms and transgenic animals" and on "high-yield plant varieties, irrigation, adding fertilizer and pest control measures" [46,47]. In echoes of the Green Revolution of the 1970s, many of these industrial agriculture models have been sponsored and implemented through subsidized agricultural programs and through neoliberal "public-private" partnerships [5,13].

5. Does Conventional Agriculture Meet Basic Sustainability Criteria?

Several studies have challenged the claims of sustainability made by proponents of modern industrial or conventional agriculture farming systems. A life-cycle analysis of conventional agriculture found that central features of the model failed to meet key sustainability criteria, including its dependency on high fossil-fuels inputs, a trend towards food industry consolidation, adverse human health impacts, a loss of agrobiodiversity, and soil degradation, among others [48]. The unsustainability of current conventional production practices was also outlined more recently by an international panel of 63 scientists [6] and as described from a pest management perspective by Ramon Seidler [49] a former Senior Scientist at the USA Environmental Protection Agency.

The overarching trend over the past twenty years for the "single tactic" pest management approach used on most of the major grain crops grown globally, and predominantly in the USA, was also deemed to be unsustainable by a team of weed scientists [4,50]. According to this review, the adoption of single-tactic approaches in conventional agriculture, such as herbicide resistant traits, has "potential negative consequences for environmental quality" plus "the short-term fix provided by the new traits will encourage continued neglect of public research and extension in integrated weed management [50]. Such a general structural decline over the past 15 years in the physical infrastructure and human capital required to establish and support Integrated Pest Management Programs has already been documented in expansive agricultural areas such as Texas [51]. An analytical comparison of mainstream agronomic systems in the USA that are based on the "single tactic" pest management approach, as compared to the more diversified systems in Europe, found a generally lower sustainability in the USA systems [39]. The lower sustainability of the USA vs. European cropping systems was observed in the form of relatively lower yields, increased pesticide use, increased consolidation of the supply industry, and a general narrowing of the germplasm diversity in the Midwestern USA, as compared to the European cropping systems [39].

6. Calls for an Agroecological Approach

Over 35 years ago, the International Commission on International Development, and later others, called for a shift towards alternative agroecological models of agriculture that were cognizant of local socioeconomic conditions and that protected ecological balance [16,52,53]. A similar call for the adoption of agroecology in Asia was made at a Conference organized by the Asian Productivity Organization [54]. Referring to some of the "green revolution" technologies that prompted the call for these alternative models of agriculture, such as the "ill-advised application of agricultural chemicals", Ekstrom and Ekbom [16] indicated that these concerns 30 years later "are as much a reality today as then".

Table 1. Some strategies for the establishment and implementation of agroecological farming systems.

Follow a Participatory Approach, Based on Indigenous or Local Knowledge

- Reliance on Indigenous knowledge to maintain agrobiodiversity such as for the preservation and use of herbs and medicinal plants [55].
- Use of experiential knowledge in the preparation of research and outreach programs [56].
- Bottom-up approaches for the design of research and outreach programs [13,54].
- Follow a Farming Systems Research/Extension and Development approach [57,58].

Regeneration and Maintenance of Soil Quality

- A healthy soil is needed to strengthen system resiliency [59].
- The value of cover crops and organic matter to soil quality [60].
- The value of soil quality to manage pests on the farm [61].

Resource Conservation and Establishment of Eco-Efficient and Integrated Systems

- Conservation of basic farm natural resources [60].
- Improved nutrient use efficiency and integration of farm activities [62,63].
- Integrated crop-livestock systems [17,62].
- Improved nutrient cycles [31,63].
- Improved ecosystem services [64].
- Management of physical and biological resources to manage pests on the farm [61].

Agrobiodiversity

- Germplasm conservation by local communities [65].
- Value of organic farms to support biodiversity [17].
- Promoting vegetational diversity to improve below-ground biodiversity [20].
- The importance of small farms to maintain biodiversity [66].
- The importance of public seed sources to maintain agrobiodiversity [39,67].
- The importance of biodiversity and habitat management for pest control [61].

Landscape-Wide Management Programs

- Value of landscape approach to maintain biodiversity [17].
- Landscape approach to maintain environmental "stability" [68].
- Landscape approach to facilitate community management of natural resources [13].

Socio-Economics or Social Considerations

- Need to consider socio-economical conditions [34,69].
- Value of promoting multifunctional agriculture [1,70].
- Need to include ethical considerations [69].
- Need to promote and maintain socially equitable systems [62].
- Need to incorporate gender considerations [13,71].

Research Considerations

- Need to create new research protocols, to study agroecosystems from a holistic and landscape perspective [72].
- Bioindicators need to be implemented for assessment of soil and environmental quality [73].
- Need to elucidate the ecosystem services provided by soil biota to restore ecological balance on the farm [74].

More recent calls have been made towards a wider adoption of agroecology, as outlined in Table 1, given the many documented benefits such as improved resource utilization, reduced externality costs, and less adverse impacts on the environment or human health [1,4,6,17,37,62]. Among the benefits reported from the adoption of agroecological practices include increased profitability [11,75]; comparable yields and pest controls [76–79]; improved water use efficiency in horticultural crops [80], as well as crop performance during drought years [78]; improved soil quality and organic matter content [77,78]; improved and more uniform Nitrogen mineralization, increased organic matter content and soil microbial activity in rotations with tomato [81]; improved biodiversity, ecosystem services, and resilience at both the farm and landscape levels [20,36,75,82,83] and as observed in Kiwi fruit orchards [84]; an improved sustainability index as observed with cacao in Mexico [39,85]; improved nutritional profiles, as observed on long-term trials with tomato [75,86,87]; as well as a reduction of pesticide residues in the body [75,86,88,89], including in children [90]. In turn, the greater ecological balance obtained in agroecological systems through crop diversification and increased soil health often results in a greater activity of above- and below-ground beneficial organisms, resulting in enhanced internal biocontrol mechanisms on the farm [76,79]. Another concern of the industrial agriculture model is a narrowing of the germplasm base, and a decline in agrobiodiversity within the farm and at a national level, as observed in the USA over the past 35 years [39,91]. Crop diversification and a wide germplasm base are considered integral to establish and maintain sustainable and resilient agricultural systems, because "In a future of climate change, public breeding and *in situ* conservation are likely to be fundamental to the survival of billions of people" [29,35,39].

The most prominent recommendation in support of agroecology was made by the international IAASTD assessment, which consisted of a panel of over 400 scientists from over 60 countries, as part of a multi-year global-scale study sponsored by the United Nations and the World Bank [1,4,16,70]. It is recognized that alternative production systems are available to maintain productivity and protect valuable natural resources, but that for their wider implementation, "it might be required of our society that it changes some of its paradigms and 'values' in order to preserve our support system, the soil and its health, for the future generations" [36].

Thus "the conclusion that emerges is that a radical rethink is needed in the orientation of agriculture" and that "The solutions will not be narrow sectoral or technical innovations but nested sets of innovations at the scale of the plant, the agronomic system, the landscape, and the institutional environment" [72]. This paradigm shift, according to an international panel of scientists, would consist of a "significant shift from conventional, monoculture-based and high external-input dependent industrial production towards mosaics of sustainable, regenerative production systems that also considerably improve the productivity of small-scale farmers", as well as a "shift from a linear to a holistic approach in agricultural management" [6].

7. Conclusions

Because of increased global population pressures, of the impending impacts of climate change on food production, and of the increased trend toward the market price volatility of the major global staple crops, there have been increased calls for a transformation of modern agricultural systems. The debate and the narrative about the future of agriculture is permeated by the narrative of powerful vested interests with close ties to government and academic institutions that make a call for a continuation of capital and input-intensive technological based solutions for agriculture. However, scientific surveys and reviews have documented a range of human health and environmental externality costs from industrial or conventional production systems, and these surveys have questioned the sustainability of such systems because of their potential adverse impacts on the long-term quality of the soil, natural resources, and on future generations.

As a result of the concerns about the lack of sustainability and lack of resiliency observed in modern industrial agricultural production practices, calls have been made for a paradigm shift in the design of agricultural systems. Agroecological approaches have been put forward as viable solutions to increase agricultural productivity, to increase economic well-being as well as the social and gender equity in rural communities, and to increase agricultural productivity while minimizing reliance on external proprietary technology, capital and synthetic chemical inputs. Key features of an agroecological approach include the decentralization of the production and marketing process, the need to follow a holistic and integrated participatory approach, an emphasis on minimizing erosion and enhancing soil quality, the conservation of natural resources, the promotion of agrobiodiversity and of ecosystem services both at the farm and landscape or watershed level, and the need to fully integrate socioeconomic, social and gender equity considerations in all phases of the agricultural research, extension, and developmental process.

Conflicts of Interest: The author declares no conflict of interest.

References

1. McIntyre, B.D.; Herren, H.R.; Wakhungu, J.; Watson, R.T. *International Assessment of Agricultural Knowledge, Science and Technology for Development (IAASTD): Global Report*; Island Press: Washington, DC, USA, 1978; p. 606.

2. Carpenter, S.J. Farm chemicals, soil erosion, and sustainable agriculture. *Stanf. Environ. Law J.* **1994**, *13*, 190.

3. Sachs, J.; Remans, R.; Smukler, S.; Winowiecki, L.; Andelman, S.J.; Cassman, K.G.; Sanchez, P.A. Monitoring the world's agriculture. *Nature* **2010**, *466*, 558–560. [CrossRef] [PubMed]

4. Barker, D. Genetically Engineered (GE) crops: A misguided strategy for the twenty-first century? *Development* **2014**, *57*, 192–200. [CrossRef]

5. Anonymous. Editorial: How to feed a hungry world. *Nature* **2010**, *466*, 531–532.

6. United Nations Conference on Trade and Development. *Wake Up before It Is Too Late: Make Agriculture Truly Sustainable Now for Food Security and Changing Climate*; United Nations: Geneva, Switzerland, 2013; p. 341.

7. Arizpe, N.A.; Giampietro, M.; Ramos-Martin, J. Food security and fossil energy dependence: An international comparison of the use of fossil energy in agriculture (1991–2003). *Crit. Rev. Plant Sci.* **2011**, *30*, 45–63. [CrossRef]

8. Fausti, S.W. The causes and unintended consequences of a paradigm shift in corn production practices. *Environ. Sci. Policy* **2015**, *52*, 41–50. [CrossRef]

9. National Research Council. *Publicly Funded Agricultural Research and the Changing Structure of U.S. Agriculture*; Committee to Review the Role of Publicly Funded Agricultural Research on the Structure of U.S. Agriculture, National Academy Press: Washington, DC, USA, 2001; p. 158.

10. Pretty, J.N.; Noble, A.D.; Bossio, D.; Dixon, J.; Hine, R.E.; Penning de Vries, F.W.; Morison, J.I. Resource-conserving agriculture increases yields in developing countries. *Environ. Sci. Technol.* **2006**, *40*, 1114–1119. [CrossRef] [PubMed]

11. Crowder, D.W.; Reganold, J.P. Financial competitiveness of organic agriculture on a global scale. *Proc. Natl. Acad. Sci. USA* **2015**, *112*, 7611–7616. [CrossRef] [PubMed]

12. Fan, S.; Rosegrant, M. Investing in agriculture to overcome the world food crisis and reduce poverty and hunger. In *IFPRI Policy Brief 3*; International Food Policy Research Institute: Washington, DC, USA, 2008.

13. Vargas-Lundius, R. Sustainable smallholder agriculture: Feeding the world, protecting the planet. In Proceedings of the Thirty-fifth Session of IFAD's Governing Council, Rome, Italy, 22–23 February 2012.

14. Pimbert, M.; Barry, B.; Berson, A.; Tran-Thanh, K. *Democratising Agricultural Research for Food Sovereignty in West Africa*; IIED, CNOP, Centre Djoliba, IRPAD, Kene Conseils, URTEL: Bamako, Mali; London, UK, 2010; p. 70.

15. Pimentel, D. Environmental and economic costs of the application of pesticides primarily in the United States. *Environ. Dev. Sustain.* **2005**, *7*, 229–252. [CrossRef]

16. Ekström, G.; Ekbom, B. Pest control in agro-ecosystems: An ecological approach. *Crit. Rev. Plant Sci.* **2011**, *30*, 74–94. [CrossRef]

17. Gomiero, T.; Pimentel, D.; Paoletti, M.G. Environmental impact of different agricultural management practices: Conventional *vs.* organic agriculture. *Crit. Rev. Plant Sci.* **2011**, *30*, 95–124. [CrossRef]

18. Beketov, M.A.; Kefford, B.J.; Schäfer, R.B.; Liess, M. Pesticides reduce regional biodiversity of stream invertebrates. *Proc. Natl. Acad. Sci. USA* **2013**, *110*, 11039–11043. [CrossRef] [PubMed]

19. Stehle, S.; Schulz, R. Agricultural insecticides threaten surface waters at the global scale. *Proc. Natl. Acad. Sci. USA* **2015**, *112*, 5750–5755. [CrossRef] [PubMed]

20. Tiemann, L.K.; Grandy, A.S.; Atkinson, E.E.; Marin-Spiotta, E.; McDaniel, M.D. Crop rotational diversity enhances belowground communities and functions in an agroecosystem. *Ecol. Lett.* **2015**, *18*, 761–771. [CrossRef] [PubMed]

21. Report 8 of The Council on Science and Public Health (A-09). Available online: http://www.ama-assn.org/ama1/pub/upload/mm/475/refcomd.pdf (accessed on 28 March 2010).

22. Pelletier, N.; Tyedmers, P. Forecasting potential global environmental costs of livestock production 2000–2050. *Proc. Natl. Acad. Sci. USA* **2010**, *107*, 18371–18374. [CrossRef] [PubMed]

23. American Academy of Pediatrics (AAP). Policy statement: Pesticide exposure in children. *Pediatrics* **2012**, *130*, e1757–e1763.

24. American College of Obstetricians and Gynecologists. Exposure to toxic environmental agents. Committee Opinion No. 575. *Obstet. Gynecol.* **2013**, *122*, 931–935.

25. Jones, B.A.; Grace, D.; Kock, R.; Alonso, S.; Rushton, J.; Said, M.Y.; McKeever, D.; Mutua, F.; Young, J.; McDermott, J.; *et al.* Zoonosis emergence linked to agricultural intensification and environmental change. *Proc. Natl. Acad. Sci. USA* **2013**, *110*, 8399–8404. [PubMed]

26. Lopez, S.L.; Aiassa, D.; Benitez-Leite, S.; Lajmanovich, R.; Manas, F.; Poletta, G.; Sanchez, N.; Simoniello, M.F.; Carrasco, A.E. Pesticides used in south American GMO-based agriculture: A review of their effects on humans and animal models. *Adv. Mol. Toxicol.* **2012**, *6*, 41–75.

27. Mesnage, R.; Defarge, N.; Spiroux de Vendômois, J.; Séralini, G.E. Potential toxic effects of glyphosate and its commercial formulations below regulatory limits. *Food Chem. Toxicol.* **2015**, *84*, 133–153. [CrossRef] [PubMed]

28. Guyton, K.Z.; Loomis, D.; Grosse, Y.; El Ghissassi, F.; Benbrahim-Tallaa, L.; Guha, N.; Scoccianti, C.; Mattock, H.; Straif, K. Carcinogenicity of tetrachlorvinphos, parathion, malathion, diazinon, and glyphosate. *Lancet Oncol.* **2015**, *5*, 490–491.

29. Khoury, C.K.; Bjorkman, A.D.; Dempewolf, H.; Ramirez-Villegas, J.; Guarino, L.; Jarvis, A.; Struik, P.C. Increasing homogeneity in global food supplies and the implications for food security. *Proc. Natl. Acad. Sci. USA* **2014**, *111*, 4001–4006. [CrossRef] [PubMed]

30. Khoury, C.K.; Jarvis, A. *The Changing Composition of the Global Diet: Implications for CGIAR Research*; CIAT Policy Brief No. 18.; Centro Internacional de Agricultura Tropical: Cali, Colombia, 2014; p. 6.

31. Puma, M.J.; Bose, S.; Chon, S.Y.; Cook, B.I. Assessing the evolving fragility of the global food system. *Environ. Res. Lett.* **2015**, *10*, 024007. [CrossRef]

32. Minami, K. The global nitrogen/carbon cycles and agricultural practice for environmental sustainability. In *Asian Productivity Organization (APO)*; Workshop on Environment-Friendly Agriculture: Tokyo, Japan, 2003; pp. 8–15.

33. Fedoroff, N.V.; Battisti, D.S.; Beachy, R.N.; Cooper, P.J.M.; Fischhoff, D.A.; Hodges, C.N.; Zhu, J.K. Radically rethinking agriculture for the 21st century. *Science* **2010**, *327*, 833–834. [CrossRef] [PubMed]

34. Vermeulen, S.J.; Challinor, A.J.; Thornton, P.K.; Campbell, B.M.; Eriyagama, N.; Vervoort, J.M.; Smith, D.R. Addressing uncertainty in adaptation planning for agriculture. *Proc. Natl. Acad. Sci. USA* **2013**, *110*, 8357–8362. [CrossRef] [PubMed]

35. Lin, B.B. Resilience in agriculture through crop diversification: Adaptive management for environmental change. *BioScience* **2011**, *61*, 183–193. [CrossRef]

36. Gomiero, T.; Pimentel, D.; Paoletti, M.G. Is there a need for a more sustainable agriculture? *Crit. Rev. Plant Sci.* **2011**, *30*, 6–23. [CrossRef]

37. Wijeratna, W. *Fed Up, Now's the Time to Invest in Agro-Ecology*; Action Aid: Washington, DC, USA, 2012; p. 42.

38. The Montpellier Panel. *Sustainable Intensification: A New Paradigm for African Agriculture*; Agriculture for Impact; Imperial College: London, UK, 2013.

39. Heinemann, J.A.; Massaro, M.; Coray, D.S.; Agapito-Tenfen, S.Z.; Wen, J.D. Sustainability and innovation in staple crop production in the U.S. midwest. *Int. J. Agric. Sustain.* **2014**, *12*, 71–88. [CrossRef]

40. Turner, P.A.; Griffis, T.J.; Lee, X.; Baker, J.M.; Venterea, R.T.; Wood, J.D. Indirect nitrous oxide emissions from streams within the U.S. Corn Belt scale with stream order. *Proc. Natl. Acad. Sci. USA* **2015**, *112*, 9839–9843.

41. Soetan, K.O. The role of biotechnology towards attainment of a sustainable and safe global agriculture and environment–A review. *Biotechnol. Mol. Biol. Rev.* **2011**, *6*, 109–117.

42. Schneiderman, H.A.; Carpenter, W.D. Planetary patriotism: Sustainable agriculture for the future. *Environ. Sci. Technol.* **1990**, *24*, 466–473. [CrossRef]

43. Gilbert, N. Food: Inside the hothouses of industry. *Nature* **2010**, *466*, 548–551. [CrossRef] [PubMed]

44. Kaufman, F. *Let Them Eat Cash! Can Bill Gates Turn Hunger into Profit*; Harper's Magazine: New York, NY, USA, 2009; pp. 51–59.

45. Watkins, K. *Grain Fish. Money Financing Africa's Green and Blue Revolutions*; Africa Progress Report; Africa Progress Panel: Geneva, Switzerland, 2014; p. 180.

46. Tso, T.C. Agriculture of the future. Commentary. *Nature* **2004**, *428*, 215–217. [CrossRef] [PubMed]

47. Tollefson, J. The global farm. *Nature* **2010**, *466*, 554–556. [CrossRef] [PubMed]

48. Heller, M.C.; Keoleian, G.A. Assessing the sustainability of the U.S. food system: A life cycle perspective. *Agric. Syst.* **2003**, *76*, 1007–1041.

49. Pesticide Use on Genetically Engineered Crops. Available online: http://static.ewg.org/agmag/pdfs/pesticide_use_on_genetically_engineered_crops.pdf (accessed on 15 May 2015).

50. Mortensen, D.A.; Egan, J.F.; Maxwell, B.D.; Ryan, M.R.; Smith, R.G. Navigating a critical juncture for sustainable weed management. *BioScience* **2012**, *62*, 75–84. [CrossRef]

51. Allen, C. History of pest management in Texas and the southern United States and how recent grower adoption of preventative pest management technologies have diminished the capability for IPM delivery. *Outlooks Pest Manag.* **2015**, *26*, 52–55. [CrossRef]

52. Altieri, M.A. Agroecology: A new research and development paradigm for world agriculture. *Agric. Ecosyst. Environ.* **1989**, *27*, 37–46. [CrossRef]

53. Edwards, C.A.; Grove, T.L.; Harwood, R.R.; Colfer, C.P. The role of agroecology and integrated farming systems in agricultural sustainability. *Agric. Ecosyst. Environ.* **1993**, *46*, 99–121. [CrossRef]

54. Palmer, J.J. Synthesis of experiences on better agricultural practices for environmental sustainability. In *Workshop on Environment-Friendly Agriculture*; Asian Productivity Organization (APO): Tokyo, Japan, 2003; pp. 1–8.

55. Turner, N.J.; Jakub Łuczaj, L.; Migliorini, P.; Pieroni, A.; Dreon, A.L.; Sacchetti, L.E.; Paoletti, M.G. Edible and tended wild plants, traditional ecological knowledge and agroecology. *Crit. Rev. Plant Sci.* **2011**, *30*, 198–225. [CrossRef]

56. Francis, C.A.; Jordan, N.; Porter, P.; Breland, T.A.; Lieblein, G.; Salomonsson, L.; Sriskandarajah, N.; Wiedenhoeft, M.; DeHaan, R.; Braden, I.; *et al.* Innovative education in agroecology: Experiential learning for a sustainable agriculture. *Crit. Rev. Plant Sci.* **2011**, *30*, 226–237.

57. Shaner, W.W.; Philipp, P.F.; Schmehl, W.R. *Farming Systems Research and Development: Guidelines for Developing Countries*; Westview Press: Boulder, CO, USA, 1982.

58. Farming and Rural Systems Research: A Constellation of Systemic and Interdisciplinary Perspectives. NSS Dialogues 2010. Available online: http://www.nss-dialogues.fr/IMG/pdf/Farming_and_Rural_Systems_Research-2.pdf (accessed on 1 August 2014).

59. Borenstein, S. *Overlooked in the Global Food Crisis: A Problem with Dirt*; Associated Press: Washington, DC, USA, 2008; Available online: http://usatoday30.usatoday.com/tech/science/2008-05-08-3388434369_x.htm (accessed on 13 March 2016).

60. Pimentel, D. Food for thought: A review of the role of energy in current and evolving agriculture. *Crit. Rev. Plant Sci.* **2011**, *30*, 1–44. [CrossRef]

61. Valenzuela, H. Pest and disease control strategies for sustainable pacific agroecosystem. In *Agroforestry Landscapes for Pacific Islands: Creating Abundant and Resilient Food Systems*; Elevitch, C.R., Ed.; Permanent Agriculture Resources (PAR): Kona, Hawaii Island, 2015; p. 332. Available online: http://agroforestry.org/images/pdfs/Sustainable_Pest_and_Disease_Control_Valenzuela.pdf (accessed on 13 March 2016).

62. Francis, C.A.; Porter, P. Ecology in sustainable agriculture practices and systems. *Crit. Rev. Plant Sci.* **2011**, *30*, 64–73. [CrossRef]

63. Carberry, P.S.; Liang, W.L.; Twomlow, S.; Holzworth, D.P.; Dimes, J.P.; McClelland, T.; Keating, B.A. Scope for improved eco-efficiency varies among diverse cropping systems. *Proc. Natl. Acad. Sci. USA* **2013**, *110*, 8381–8386. [CrossRef] [PubMed]

64. Boreux, V.; Kushalappa, C.G.; Vaast, P.; Ghazoul, J. Interactive effects among ecosystem services and management practices on crop production: Pollination in coffee agroforestry systems. *Proc. Natl. Acad. Sci. USA* **2013**, *110*, 8387–8392. [CrossRef] [PubMed]

65. Jarvis, D.I.; Hodgkin, T.; Sthapit, B.R.; Fadda, C.; Lopez-Noriega, I. An heuristic framework for identifying multiple ways of supporting the conservation and use of traditional crop varieties within the agricultural production system. *Crit. Rev. Plant Sci.* **2011**, *30*, 125–176. [CrossRef]

66. Jarvis, D.I.; Brown, A.H.; Cuong, P.H.; Collado-Panduro, L.; Latournerie-Moreno, L.; Gyawali, S.; Hodgkin, T. A global perspective of the richness and evenness of traditional crop-variety diversity maintained by farming communities. *Proc. Natl. Acad. Sci. USA* **2008**, *105*, 5326–5331. [CrossRef] [PubMed]

67. Howe, L.; Redfeather, N.; Valenzuela, H. *The Hawaii Public Seed Initiative*; Hanai' Ai/The Food Provider: Honolulu, HI, USA, 2012; p. 6. Available online: http://www.ctahr.hawaii.edu/sustainag/news/articles/V11-Valenzuela-seedinitiative.pdf (accessed on 13 March 2016).

68. Sayer, J.; Cassman, K.G. Agricultural innovation to protect the environment. *Proc. Natl. Acad. Sci. USA* **2013**, *110*, 8345–8348. [CrossRef] [PubMed]

69. Hall, S.J.; Hilborn, R.; Andrew, N.L.; Allison, E.H. Innovations in capture fisheries are an imperative for nutrition security in the developing world. *Proc. Natl. Acad. Sci. USA* **2013**, *110*, 8393–8398. [CrossRef] [PubMed]

70. Rivera-Ferre, M.G. The future of agriculture. *EMBO Rep.* **2008**, *9*, 1061–1066. [CrossRef] [PubMed]

71. Wakhungu, J.W. *Gender Dimensions of Science and Technology: African Women in Agriculture*; African Centre for Technology Studies: Nairobi, Kenya, 2010; p. 8.

72. Sayer, J.; Sunderland, T.; Ghazoul, J.; Pfund, J.L.; Sheil, D.; Meijaard, E.; Buck, L.E. Ten principles for a landscape approach to reconciling agriculture, conservation, and other competing land uses. *Proc. Natl. Acad. Sci. USA* **2013**, *110*, 8349–8356. [CrossRef] [PubMed]

73. Paoletti, M.G. *Invertebrate Biodiversity as Bioindicators of Sustainable Landscapes. Practical Use of Invertebrates to Assess Sustainable Land Use*; Elsevier: Amsterdam, The Netherlands, 2012; p. 446.

74. Paoletti, M.G. *Ecological Implications of Minilivestock. Insects, Rodents, Frogs and Snails*; Science Publishers Inc.: Enfield, NH, USA, 2005; p. 648.

75. Reganold, J.P.; Wachter, J.M. Organic agriculture in the twenty-first century. *Nat. Plants* **2016**. [CrossRef]

76. Crowder, D.W.; Northfield, T.D.; Strand, M.R.; Snyder, W.E. Organic agriculture promotes evenness and natural pest control. *Nature* **2010**, *466*, 109–112. [CrossRef] [PubMed]

77. Delate, K.; Cambardella, C.; Chase, C.; Johanns, A.; Turnbull, R. The long-term agroecological research (LTAR) experiment supports organic yields, soil quality, and economic performance in Iowa. *Crop Manag.* **2013**, *12*. [CrossRef]

78. Pimentel, D.; Hepperly, P.; Hanson, J.; Douds, D.; Seidel, R. Environmental, energetic, and economic comparisons of organic and conventional farming systems. *BioScience* **2005**, *55*, 573–582. [CrossRef]

79. Zehnder, G.; Gurr, G.M.; Kühne, S.; Wade, M.R.; Wratten, S.D.; Wyss, E. Arthropod pest management in organic crops. *Annu. Rev. Entomol.* **2007**, *52*, 57–80. [CrossRef] [PubMed]

80. Wheeler, S.A.; Zuo, A.; Loch, A. Watering the farm: Comparing organic and conventional irrigation water use in the Murray-Darling Basin, Australia. *Ecol. Econ.* **2015**, *112*, 78–85. [CrossRef]

81. Marinari, S.; Lagomarsino, A.; Moscatelli, M.C.; Di Tizio, A.; Campiglia, E. Soil carbon and nitrogen mineralization kinetics in organic and conventional three-year cropping systems. *Soil Tillage Res.* **2010**, *109*, 161–168. [CrossRef]

82. Petit, S.; Munier-Jolain, N.; Bretagnolle, V.; Bockstaller, C.; Gaba, S.; Cordeau, S.; Lechenet, M.; Mézière, D.; Colbach, N. Ecological intensification through pesticide reduction: Weed control, weed biodiversity and sustainability in arable farming. *Environ. Manag.* **2015**, *56*, 1078–1090.

83. Tuck, S.L.; Winqvist, C.; Mota, F.; Ahnström, J.; Turnbull, L.A.; Bengtsson, J. Land-use intensity and the effects of organic farming on biodiversity: A hierarchical meta-analysis. *J. Appl. Ecol.* **2014**, *51*, 746–755. [CrossRef] [PubMed]

84. Todd, J.H.; Malone, L.A.; McArdle, B.H.; Benge, J.; Poulton, J.; Thorpe, S.; Beggs, J.R. Invertebrate community richness in New Zealand kiwifruit orchards under organic or integrated pest management. *Agric. Ecosyst. Environ.* **2011**, *141*, 32–38. [CrossRef]

85. Priego-Castillo, G.A.; Galmiche-Tejeda, A.; Castelán-Estrada, M.; Ruiz-Rosado, O.; Ortiz-Ceballos, A. Evaluación de la sustentabilidad de dos sistemas de producción de cacao: Estudios de caso de unidades de producción rural en Comalcalco, Tabasco. *Univ. Cienc.* **2009**, *25*, 39–57. (In Spanish)

86. Barański, M.; Średnicka-Tober, D.; Volakakis, N.; Seal, C.; Sanderson, R.; Stewart, G.B.; Benbrook, C.; Biavati, B.; Markellou, E.; Giotis, C.; *et al.* Higher antioxidant and lower cadmium concentrations and lower incidence of pesticide residues in organically grown crops: A systematic literature review and meta-analyses. *Br. J. Nutr.* **2014**, *112*, 794–811. [PubMed]

87. Mitchell, A.E.; Hong, Y.J.; Koh, E.; Barrett, D.M.; Bryant, D.E.; Denison, R.F.; Kaffka, S. Ten-year comparison of the influence of organic and conventional crop management practices on the content of flavonoids in tomatoes. *J. Agric. Food Chem.* **2007**, *55*, 6154–6159. [CrossRef] [PubMed]

88. Benbrook, C.M.; Baker, B.P. Perspective on dietary risk assessment of pesticide residues in organic food. *Sustainability* **2014**, *6*, 3552–3570. [CrossRef]

89. Magnér, J.; Wallberg, P.; Sandberg, J.; Cousins, A.P. *Human Exposure to Pesticides from Food*; Report No. U 5080; Swedish Environmental Research Institute: Stockholm, Sweden, 2015.

90. Bradman, A.; Quirós-Alcalá, L.; Castorina, R.; Schall, R.A.; Camacho, J.; Holland, N.T.; Barr, D.B.; Eskenazi, B. Effect of organic diet intervention on pesticide exposures in young children living in low-income urban and agricultural communities. *Environ. Health Perspect.* **2015**, *123*, 1086–1093. [CrossRef] [PubMed]

91. Aguilar, J.; Gramig, G.G.; Hendrickson, J.R.; Archer, D.W.; Forcella, F.; Liebig, M.A. Crop species diversity changes in the United States: 1978–2012. *PLoS ONE* **2015**, *10*, e0136580. [CrossRef] [PubMed]

12

Trends of Soybean Yields under Climate Change Scenarios

Frank Eulenstein [1,2,*], Marcos Lana [1], Sandro Schlindwein [3], Askhad Sheudzhen [2],
Marion Tauschke [1], Axel Behrend [1], Edgardo Guevara [4] and Santiago Meira [4]

[1] Leibniz-Zentrum für Agrarlandschaftsforschung (ZALF) Müncheberg, Eberswalder Straße 84,
Müncheber 15374, Germany; Marcos.Lana@zalf.de (M.L.); mtauschke@zalf.de (M.T.);
abehrendt@zalf.de (A.B.)

[2] Department Agro-Chemistry, Kuban State Agrarian University, Krasnodar 350044, Russia;
rgpzkrs@mail.kuban.ru

[3] Departamento de Engenharia Rural, Universidade Federal de Santa Catarina, Florianópolis 88034-000,
Brazil; sandro.schlindwein@ufsc.br

[4] Department of Crop Production INTA—Instituto Nacional de Tecnologia Agropecuaria, Pergamino 2700,
Argentina; guevara.edgardo@inta.gob.ar (E.G.); meira.santiago@inta.gobv.ar (S.M.)

* Correspondence: feulenstein@zalf.de

Abstract: Soybean is a very important crop, cultivated mainly as stock feed for animal production, but also for other uses such as biodiesel. Brazil is the second largest producer of soybeans, and the main exporter. About 10% of total Brazilian production is aimed for biodiesel production. The aim of this work was to assess the impact of climate change scenarios on soybean yield and evaluate two simple adaptation strategies: cultivar and planting date. Tests were done for soil profiles from two important producing regions: Chapecó-Red Oxisol, and Passo Fundo-Rodic Hapludox. Two commercial soybean cultivars (CD202 and CD204) and seven regional circulation models (RCM) were used. All simulations were done with DSSAT. After model calibration, eleven planting dates were run for two periods (2011–2040 and 2071–2100) using the RCM's. There were no differences between cultivars. For Chapecó, the majority of RCM's projected yield reductions, with few RCM's projecting increments, and for only few planting dates (November). The response pattern for both time periods were identical, although the end-of-century period presented a further yield reduction. The main reason was reduced water holding capacity in soil, high temperatures, and changes in rainfall distribution along the cropping season. For Passo Fundo, 2011–2040 yields were distinct, depending on the RCM. Simulated yields tended to follow the actual yield pattern along the different planting dates, besides discrepancies. For 2071–2100, all but one RCM indicated yields equal or lower to actual levels. Regarding planting dates, no significant changes were identified, although reductions were observed for the early planting dates (August–September). The scenarios suggest that soybean yields will be reduced, jeopardizing the viability of this crop and biodiesel production in the studied regions.

Keywords: climate change; crop model; efficiency use

---

1. Introduction

Climate change is a worldwide concern, and regardless of all the scientific improvements to comprehend and forecast changes, the determination of future climate is still a very hard task. The high level of complexity and the nature of climatic interactions is a challenge to forecasting, although there are scenarios that point to possible directions of change.

The impact of climate change on agricultural production is actually the core issue of several investigations. Rising seasonal temperatures are expected to increase more than the annual averages, with reduced precipitation expected to accompany higher temperatures in some regions. Additionally, heat waves are expected to increase in frequency, intensity, and duration [1]. End-of-century growing season temperatures in the tropics and subtropics may exceed even the most extreme seasonal temperatures measured to date [2]. Not considering all the inherent variability of crop production factors, all climate changes described above can lead to modifications of crop yields, posing a threat to agricultural systems that will affect the whole production and consumption chain, impacting especially agroecosystems and populations with low availability of or access to financial and natural resources. The global food and financial crises of 2007 and 2008, which have pushed an additional 115 million into hunger, highlight the severity of the hunger and poverty crisis that has challenged the world for decades [3]. Price volatility remains a concern, with weather-related yield variability the main threat as long as stocks remain low [4]. This risky situation will be worsened by the present effects of drought on soybean yields of USA [5], which will impact the whole world's food supply.

Increasing the prediction capacity of climate change impacts for stakeholders has become a major challenge in Southern Brazil, where economic wealth strongly depends on agriculture [6,7]. In this region, the agricultural landscape has faced major changes during the last 30 years due to new technologies for crops, to a strong increase in cereal and oil crop world demands, and also to favorable climate conditions with increases of about 20%–30% in annual precipitation over large parts of the region [8].

Crop models can be a useful tool to assess the influence of climatic and other environmental or management factors on crop development and yield [9]. The Decision Support System for Agrotechnology Transfer—DSSAT v. 4.5 contains the CROPGRO-Soybean model [10], and is used to (a) determine best planting dates [11], (b) for fertilization timing [12], (c) in precision agriculture [13], and (d) also for detecting/investigating potential impacts of climate change on agriculture [14]. In the embedded model the development and growth of the crop is simulated on a daily basis from the planting until physiological maturity. The model calculations are based on environmental and physiological processes that control the phenology and dry matter accumulation in the different organs of the plant. The DSSAT also has other embedded models that can simulate the flow of nutrients and water balance in the soil. The minimum data set necessary to run DSSAT [15] consists of daily weather data of maximum and minimum temperature, rainfall and solar radiation, soil chemical and physical parameters for each layer, genetic coefficients for each cultivar with information about development and biomass accumulation, and management information, such as soil preparation, planting dates, plant density, fertilization amounts and timing or other agricultural practices. Experimental data like soil available water, plant phenology, biomass partitioning and other morphological components like leaf area index are necessary to calibrate the genetic coefficients and check the accuracy of the model [16].

2. Experimental Section

In order to run simulations for soybeans, data from field experiments and literature were used. For simulation in the Brazilian sites, data from literature was obtained from Dallacort et al. [10], who performed experiments in Parana State evaluating four soybean cultivars. The cultivars were characterized, calibrated and validated for the CROPGRO-Soybean. The four cultivars, namely CD 202, CD 204, CD 206 and CD 210, were tested for both Brazilian sites using census data and generic agronomic management. The two cultivars with lowest RMSE for yield were selected to run further analyses.

After calibrating and validating the genetic parameters and the model itself, scenarios provided by CLARIS LPB Project WP5 (2011–2040 and 2071–2100 periods) were downloaded and formatted for the DSSAT standard using Weatherman Software [17]. From the CLARIS-LPB Project Data Archive Center, seven weather series of RCM's (and matching the same location of the study sites weather stations) were downloaded, converted and adjusted to be used as weather input for DSSAT using Weatherman software. The RCM's are RCA1, RCA2 and RCA3, from the Rossby Centre Regional Climate model [18]; PROMES, from Universidad de Castilla-La Mancha [19]; LMDZ version 4 Configuration South America with IPSLA1B and EC5OM-R3 boundaries, from Laboratoire de Meteorologie Dynamique [20]; and ETA, from Instituto Nacional de Pesquisas Espaciais [21].

The crop model was run with each one of the seven RCM's for the target periods (2011–2040 and 2071–2100). Simulations of the impact of RCM's scenarios on soybean cultivars (CD202 and CD204) planted on eleven different dates, in two locations (Chapecó and Passo Fundo), and two time periods (2011–2040 and 2071–2100, summing 29 cropping seasons each). Black lines represent median yields simulated with RCM's while black bars represent the standard error of each planting date for the 29 seasons; the grey lines represent actual yields with respective planting dates and standard error.

3. Results and Discussion

The analyses (Figure 1) showed the impact of seven RCM's on the yield of the soybean cultivars CD202 and CD204 in two locations and two time periods (2011–2040 and 2071–2100). It is important to mention that both soybean cultivars, besides having differences in genetic coefficients, presented very similar results.

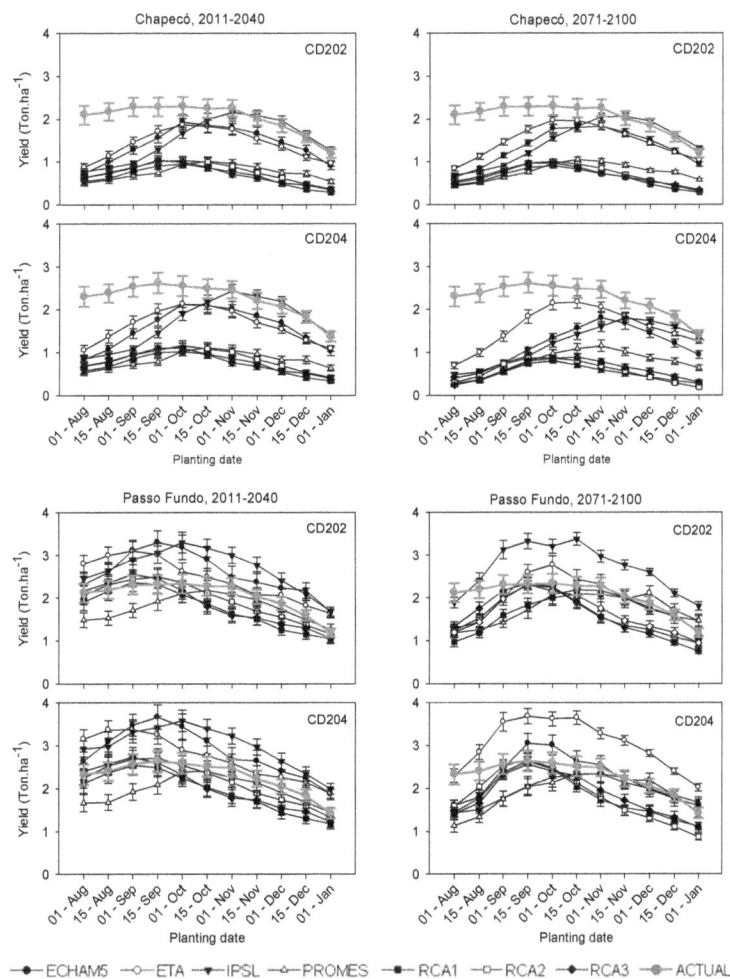

Figure 1. Simulations of the impact of RCM's scenarios on soybean cultivars (CD202 and CD204) planted on eleven different dates, in two locations (Chapecó and Passo Fundo), and two time periods (2011–2040 and 2071–2100): black lines represent yields simulated with RCM's and black bars represent the standard error of each planting date; the grey lines represent actual yields with respective planting dates and standard error.

For the Chapecó 2011–2040 period, the majority of RCM's projected very low yields when compared with actual yields. Only ETA, IPSL and ECHAM5 presented a trend of increase in yields, and after the 1 October planting date. Even so, only IPSL could mimic the actual yields for the late planting dates. This assessment is also applicable for the 2071–2100 period, but with a further reduction of projections of all RCMs. An integrated analysis indicates with high level of agreement that early planting dates—prior to 1 October—will generate lower yields; planting after 1 October shows that three out of seven RCM's (namely, ETA, ECHAM5 and IPSL) have a tendency to follow the actual yields, while the others remain with very low yields, jeopardizing the viability of this crop in the region.

The results presented for Passo Fundo showed significant difference from the ones of Chapecó, with RCM yields following the trend of actual yield. It also presents a situation where RCMs project even significant increments in yield for the 2011–2040 period. This can be observed especially in the early planting dates, where all but one RCM are equal or significantly higher than the actual yield. For the end-of-century period a generalized reduction of yield was calculated, with exception of IPSL, which showed significant increases. Though a trend of yield reduction, all RCMs presented at least one planting date that did not differ significantly from the actual best yields.

4. Conclusions

Both genotypes tested (CD202 and CD204) did not present remarkable differences among them when in the same region. Unfortunately, no other suitable soybean data sets are available to calibrate and validate the crop model in the study region, undermining the assessment of the role of cultivar as adaptation strategy. The impact of climate scenarios on soybean yield was directly influenced by location: in Chapecó region yields tend to decrease, while for Passo Fundo region yields could even increase.

Acknowledgments: This work in the CLARIS LPB Project was supported the Environmental Risks Unit, Directorate Environment, DG-Research, European Commission (http://www.claris-eu.org).

Author Contributions: Marcos Lana, Edgardo Guevara and Santiago Meira organized the data, run the calibration and generated the results. Frank Eulenstein, Sandro Schlindwein, Askhad Sheudzhen, Marion Tauschke and Axel Behrend interpreted the results and wrote this text. All authors contributed equally to this paper.

Conflicts of Interest: The authors declare no conflict of interest.

References

1. Tebaldi, C.; Hayhoe, K.; Arblaster, J.M.; Meehl, G.A. Going to the extremes. *Clim. Chang.* **2006**, *79*, 185–211. [CrossRef]
2. Battisti, D.S.; Naylor, R.L. Historical warnings of future food insecurity with unprecedented seasonal heat. *Science* **2009**, *323*, 240–244. [CrossRef] [PubMed]
3. Viatte, G.; De Graaf, J.; Demeke, M.; Takahatake, T.; Rey de Arce, M. *Responding to the Food Crisis: Synthesis of Medium-Term Measures Proposed in Inter-Agency Assessments*; FAO: Rome, Italy, 2009.
4. *OECD-FAO Agricultural Outlook 2012*; OECD: Paris, France, 2012.
5. University of Nebraska-Lincoln Crop Watch 2012. Available online: http://cropwatch.unl.edu/2012archive (accessed on 18 May 2015).
6. FAO Global Information System of Water and Agriculture. Available online: http://www.fao.org (accessed on 13 May 2015).
7. Eulenstein, F.; Lana, M.; Schlindwein, S.L.; Sheudzhen, A.K.; Tauschke, M.; Guevara, E.; Meira, S. Impact of climate scenarios on soybean yields in southern Brazil. In Proceedings of the Moving Farm Systems to Improved Nutrient Attenuation: 28th Annual Fertilizer and Lime Research Centre Workshop, Palmerston North, New Zealand, 10–12 February 2015; p. 89.
8. Magrin, G.O.; Travasso, M.I.; Rodríguez, G.R. Changes in climate and crop production during the 20th century in Argentina. *Clim. Chang.* **2005**, *72*, 229–249. [CrossRef]
9. Reidsma, P.; Ewert, F.; Lansink, A.O.; Leemans, R. Adaptation to climate change and climate variability in European agriculture: The importance of farm level responses. *Eur. J. Agron.* **2010**, *32*, 91–102. [CrossRef]

10. Dallacort, R.; Freitas, P.S.L.; Faria, R.T.; Goncalves, A.C.A.; Rezende, R.; Guimaraes, R.M.L. Simulation of bean crop growth, evapotranspiration and yield in Paraná State by the CROPGRO-Drybean model. *Acta Sci. Agric.* **2011**, *33*, 429–436.

11. Banterng, P.; Hoogenboom, G.; Patanothai, A.; Singh, P.; Wani, S.P.; Pathak, P.; Tongpoonpol, S.; Atichart, S.; Srihaban, P.; Buranaviriyakul, S.; et al. Application of the Cropping System Model (CSM)—CROPGRO—Soybean for determining optimum management strategies for soybean in tropical environments. *J. Agron. Crop Sci.* **2010**, *196*, 231–242. [CrossRef]

12. D'orgeval, T.; Boulanger, J.P.; Capalbo, M.J.; Guevara, E.; Penalba, O.; Meira, S. Yield estimation and sowing date optimization based on seasonal climate information in the three CLARIS sites. *Clim. Chang.* **2010**, *98*, 565–580. [CrossRef]

13. Asadi, M.E.; Clemente, R.S. Simulation of maize yield and N uptake under tropical conditions with the CERES-Maize model. *Trop. Agric.* **2001**, *78*, 211–217.

14. Thorp, K.R.; DeJonge, K.C.; Kaleita, A.L.; Batchelor, W.D.; Paz, J.O. Methodology for the use of DSSAT models for precision agriculture decision support. *Comput. Electron. Agric.* **2008**, *64*, 276–285. [CrossRef]

15. Fischer, G.; Shah, M.; Tubiello, F.N.; Van Velhuizen, H. Socio-economic and climate change impacts on agriculture: An integrated assessment, 1990–2080. *Philos. Trans. R. Soc. B Biol. Sci.* **2005**, *360*, 2067–2083. [CrossRef] [PubMed]

16. Jones, J.W.; Hoogenboom, G.; Porter, C.H.; Boote, K.J.; Batchelor, W.D.; Hunt, L.A.; Wilkens, P.W.; Singh, U.; Gijsman, A.J.; Ritchie, J.T. The DSSAT cropping system model. *Eur. J. Agron.* **2003**, *18*, 235–265. [CrossRef]

17. Wilkens, P.W. DSSAT v4 weather data editing program (Weatherman). In *Data Management and Analysis Tools—Decision Support System for Agrotechnology Transfer Version 4.0: DSSAT v4: Data Management and Analysis Tools*; University of Hawaii: Honolulu, HI, USA, 2004; Volume 2, pp. 92–151.

18. Samuelsson, P.; Jones, C.G.; Willen, U.; Ullrstig, A.; Gollvik, S.; Hansson, U.; Jansoon, C.; Kjellström, E.; Nikulin, G.; Wyser, K. The rossby centre regional climate model RCA3: Model description and performance. *Tellus A* **2011**, *63*, 4–23. [CrossRef]

19. Domínguez, M.; Gaertner, M.; de Rosnay, P.; Losada, T. A regional climate model simulation over West Africa: Parameterization tests and analysis of land-surface fields. *Clim. Dyn.* **2010**, *35*, 249–265. [CrossRef]

20. Hourdin, F.; Musat, I.; Bony, S.; Braconnot, P.; Codron, F.; Dufresne, J.L.; Fairhead, L.; Filiberti, M.A.; Friedlingstein, P.; Grandpeix, J.Y. The LMDZ4 general circulation model: Climate performance and sensitivity to parametrized physics with emphasis on tropical convection. *Clim. Dyn.* **2006**, *27*, 787–813. [CrossRef]

21. Marengo, J.; Chou, S.; Kay, G.; Alves, L.; Pesquero, J.; Soares, W.; Santos, D.; Lyra, A.; Sueiro, G.; Betts, R.; et al. Development of regional future climate change scenarios in South America using the Eta CPTEC/HadCM3 climate change projections: Climatology and regional analyses for the Amazon, São Francisco and the Paraná River basins. *Clim. Dyn.* **2012**, *38*, 1829–1848. [CrossRef]

13

Apple Tree Responses to Deficit Irrigation Combined with Periodic Applications of Particle Film or Abscisic Acid

Khalid M. Al-Absi [1] and Douglas D. Archbold [2,*

[1]   Department of Plant Production, Faculty of Agriculture, Mu'tah University, Al-Karak 61710, Jordan; absikhal@mutah.edu.jos
[2]   Department of Horticulture, University of Kentucky, Lexington, KY 40546, USA
*   Correspondence: darchbol@uky.edu

Abstract: The objective of this study was to determine if the application of two antitranspirant compounds would moderate water deficit stress effects on physiological responses of "Granny Smith", "Royal Gala" and "Golden Delicious" apple (*Malus domestica* Borkh.) trees on MM106 rootstock that occur during deficit irrigation. Uniform trees were grown in pots under water supply regimes of 30%, 60%, and 80% depletion of available water (DAW) before irrigation to runoff and received applications of kaolin particle film (PF) or abscisic acid (ABA) at 0, 30 and 60 days. At 120 days, genotype and deficit irrigation affected nearly all leaf traits, but antitranspirant treatment had no significant effects. As the % DAW increased, the net photosynthetic rate (Pn), transpiration rate (T), stomatal conductance, leaf water and pressure potential, variable-to-maximal chlorophyll fluorescence, leaf number, and leaf N, P and K contents were reduced. A significant genotype by deficit irrigation interaction was evident on T, water use efficiency (WUE), and leaf osmotic pressure potential. A significant deficit irrigation by antitranspirant interaction was evident on only leaf Pn, with PF and ABA reducing it at 30% DAW and only PF reducing it at 80% DAW. However, the periodic use of PF or ABA during deficit irrigation did not alleviate most physiological effects of water deficit stress due to deficit irrigation.

Keywords: water deficit; kaolin; abscisic acid; leaf gas exchange; leaf water relations; chlorophyll fluorescence

1. Introduction

Drought is a critical environmental stress that disrupts crop growth and performance that will be exacerbated as the climate changes [1]. To avoid water deficits that reduce crop production, irrigation is applied, but water availability is the main limiting factor for productive agriculture in most arid and semi-arid regions. In Jordan, it has been estimated that crop irrigation utilizes about 70% of the total demand for water [2]. As the population grows and demand increases, the water available to agriculture will decline. Irrigation below the evapotranspiration (ET) requirements of a crop is considered deficit irrigation. If the water available for irrigation declines, deficit irrigation may become more common [3].

Plant species respond to water deficits through molecular changes, biochemical and physiological modifications, and ultimately morphological adaptations [4]. Plants have developed three major drought-resistance strategies in their adaptation to drought stress, including drought tolerance, drought avoidance, and drought escape [5]. Under water deficit conditions, most plant species close their stomata and modulate their leaf area (drought avoidance), adjusting the loss of water from the canopy, and may osmotically adjust (drought tolerance) [6]. Drought escape, completing the life cycle prior to severe stress, is most common in annual plant species.

Apple (*Malus domestica* Borkh.) is one of most widely grown and valuable deciduous fruit crops worldwide, and is extensively grown in the arid southern regions of Jordan. Several selection criteria have been proposed for identifying drought-tolerant genotypes based on their performance in stress and non-stress environments. Differences in canopy structure and water use efficiency (WUE) [7], as well as in leaf water potential (Ψ_w), osmotic adjustment, net photosynthetic rate (Pn), and stomatal conductance (g_s) [8], may be responsible to varying extents in determining how a genotype/rootstock combination responds to drought. A significant genotypic influence of apple rootstock tolerance to water stress has been reported [8–10]. Considerable variation among different apple genotypes has also been identified for a range of physiological and biochemical characteristics including responses to water stress [11,12]. Vegetative growth is drastically reduced as a result of water deficit stress. This growth reduction has been attributed to reduced Pn, g_s, and leaf Ψ_w [7]. In contrast, moderate water deficits have been shown to improve apple fruit quality with increased mean fruit weight, return bloom, flesh firmness, red skin color intensity [13], and total soluble solids content [11].

Deficit irrigation strategies can be useful for reducing water use and for managing fruit tree growth and yield [14]. Deficit irrigation has reduced apple tree growth without reducing fruit yield in some studies [15–17], but the effect may not be consistent year to year, as fruit yield has declined in other work [18,19]. As water deficit stress increases, negative impacts on yield are more likely, but finding the optimum deficit irrigation strategy may be challenging as environmental conditions (i.e., ambient temperature, relative humidity/vapor pressure deficit, sunny versus cloudy conditions) vary from season to season.

Numerous compounds have been studied as tools to alleviate water deficit stress by reducing the rate of leaf water loss. Foliar application of kaolin mineral particle film (PF) was reported to protect plants against drought stress [20–23]. Seasonal use in apple orchards increased Pn and T by reducing heat stress on leaves [20,21]. Abscisic acid (ABA) application reduces transpiration and stimulates osmotic adjustment in plants [24]. Exogenous application of ABA has reduced T and enhanced drought tolerance in several plant species [25–29]. Increasing the endogenous foliar ABA content in apple by inhibiting its hydroxylation improved physiological and biochemical aspects of tree drought tolerance [30].

Many studies of the effects of drought stress and antitranspirant application have been conducted for a relatively short duration, only a few weeks [12,31,32]. This approach can successfully expose incipient physiological and biochemical drought stress effects and drought tolerance responses. Fewer studies have looked at longer-term effects on physiological and biochemical responses. As trees develop drought avoidance modifications over longer periods, it is not clear if drought tolerance responses might also be adapting to the modified status. Whether deficit irrigation strategies are by choice to manage tree growth or are imposed by limited water availability, antitranspirants could be useful for alleviating some of the effects of drought stress. PF use in wine grapes has reduced the effects of moderate water stress during deficit irrigation [33,34], but there have been no reports examining this possibility with apple using either PF or ABA.

Results of container [25] and parallel field [11] experiments, in addition to those of other investigators [35], clearly demonstrated that water deficit stress caused plants to reduce Pn and, ultimately, growth. The response was strongly affected by the rapidity, severity and duration of the water shortage. In view of these results, and the need to minimize water stress effects during deficit irrigation of apple, the objective of this work was to study responses of three apple genotypes, "Granny Smith", "Royal Gala" and "Golden Delicious", grafted on MM106 rootstock, all common in Jordan, to scheduled use of PF and ABA during deficit irrigation, hypothesizing that the antitranspirants would moderate water deficit stress effects on leaf water relations, gas exchange, chlorophyll fluorescence properties, mineral composition, and leaf production.

2. Materials and Methods

This study was carried out in a greenhouse located in the Department of Horticulture, College of Agriculture, University of Kentucky, Lexington, KY, USA, from November 2010 to June 2011. The greenhouse was 22 $\pm$ 6 °C with ambient light levels during the investigation. Uniform, one year-old, dormant "Granny Smith", "Royal Gala" and "Golden Delicious" apple scions grafted on MM106 rootstock (Vaughn Nursery, McMinnville, TN, USA), were used. The leader shoots were pruned to a height of 90 cm, lateral branches were removed, the roots were pruned to a uniform length, and the trees were planted in 14.4 L black poly containers filled with Barky Beaver medium (composed of hardwood bark fines, pine park fines and peat moss, pH 5.5–6.5) (Barky Beaver Mulch & Soil Mix, Inc., Moss, TN, USA). The few floral buds present were removed when first evident during bud swell. The trees were irrigated uniformly, and were supplied with macro- and microelements (Peters Professional, Scotts-Sierra Horticultural Products Co., Marysville, OH, USA).

The trees were grown for two weeks prior to initiating the treatments. There were three replicate trees per irrigation level by antitranspirant treatment within each genotype, laid out in a randomized complete block design with three blocks (each on a separate greenhouse bench), each block with one replicate tree per treatment-genotype combination. Trees were subjected to one of the following irrigation treatments: Irrigation to runoff after 30%, 60%, or 80% depletion of available water (DAW). The irrigation treatments were derived after field capacity and wilting point were determined by generating a pressure-volume curve. Based on these two points, the total volume of available water (between field capacity and the wilting point) was calculated, and the weight derived. Each container was then weighed every two days, and when the depletion of available water was 30% $\pm$ 3%, 60% $\pm$ 3%, or 80% $\pm$ 3% of the total, each was irrigated to field capacity. Kaolin-based particle film (Surround WP Crop Protectant, Engelhard Corporation, Iselin, NJ, USA) and abscisic acid (ABA; ProTone SG, Valent BioSciences Corporation, Libertyville, IL, USA, 20% concentration, w/w), were used as antitranspirants. PF was applied at 6% w/v and ABA was applied at 5 mM as sprays to run off of half of the trees of each irrigation treatment on three dates: At 0, 30, and 60 days after the initiation of the study. Untreated trees at each % DAW constituted the controls. The trees were maintained under these conditions for 150 days.

Gas exchange measurements were taken when the assigned DAW was reached at or after 120 days after beginning the deficit irrigation treatments, between 9 a.m. and 1 p.m. The exact measurement date for each tree varied as the rate and severity of depletion differed. Measurements of net photosynthetic (CO_2 assimilation) rate (Pn), transpiration rate (T), and stomatal conductance (g_s) were conducted on both the second and third intact, fully-expanded leaf from the apex of the terminal shoot of each replicate tree using a portable steady state infrared gas analyzer (LI-6400, LI-COR, Lincoln, NE, USA), and the means per tree derived for data analysis. The leaf surface area within the measurement chamber of the gas analyzer was 6 cm^2, and the ambient CO_2 concentration was adjusted to 352 μmol·mol^{-1}. Water use efficiency (WUE) was calculated as the instantaneous ratio of Pn/T.

Leaf water potential (Ψ_w) of two recent fully expanded, mature leaves on each replicate tree was measured using a pressure bomb (Model 610, PMS Instruments, Corvallis, OR, USA) as described by Zhang and Archbold [36]. Measurements were taken between 9 a.m. and 2 p.m. when the assigned DAW was reached as described above, at or after 120 days after beginning the deficit irrigation treatments. Each leaf was cut at the petiole base and immediately placed in the pressure chamber with about 5 mm of the cut end of the petiole protruding through the compressible rubber gland used to seal the chamber. Compressed N_2 was released into the chamber until the sap returned to the cut end of the xylem, and the pressure required was recorded. For measuring leaf osmotic potential (Ψ_π), the leaves were then sealed between two layers of parafilm and frozen at −20 °C. Leaf discs with a diameter of the sample cups were taken with a cork borer, placed into a sample cup, and thawed for 1 h inside the sealed cup at room temperature. The cup was then placed in a chilled mirror dewpoint sensor (Model WP4-T, Decagon Devices, Pullman, WA, USA). Each sample was processed for ~20 min at room temperature. Leaf pressure potential (Ψ_p) was calculated as the difference between Ψ_w and Ψ_π. The means per tree of each component of water potential were derived for data analysis.

Chlorophyll fluorescence, as a measure of injury to the photosynthetic apparatus, was obtained when the assigned DAW was reached, as described above, using a Plant Efficiency Analyzer (PEA; Hansatech Instruments Ltd., Kings Lynn, UK). Chlorophyll fluorescence (as the Fv/Fm ratio) was measured on three dark-adapted leaves from each replicate tree. Dark-adaptation was achieved by covering the leaves for 30 min under plastic clips provided with the PEA. The mean per tree was derived for data analysis.

Two leaves were randomly collected from each replicate tree, when the assigned DAW was reached. The leaves were oven dried at 70 °C until a constant weight was obtained, combined, and ground to pass through a 20 mesh screen. Leaf mineral composition (N, P, K, Ca, and Mg) was determined as % dry weight on samples collected at 120 days as described by Tandon [37].

The number of new leaves produced on each replicate tree during the course of the study was measured 150 days after beginning the deficit irrigation treatments as a measure of the treatment effects on tree growth.

Using means per replicate tree, the data were subjected to analysis of variance using MSTATC software. Significant effects and interactions were determined at a $P \leq 0.05$, and means were compared using the Least Significant Difference (LSD) test.

3. Results

3.1. Main Effects

Genotype and deficit irrigation but not antitranspirant application affected nearly all leaf traits (Table 1). The most notable exceptions were the lack of genotype differences in Pn, and of any main effects or interactions on leaf Ca or Mg content.

Table 1. *P*-values from analysis of variance for the main effects of genotype (G), deficit irrigation (I), and antitranspirant (A) application and their interactions on leaf traits. Leaf traits are: Net photosynthetic rate (Pn), transpiration rate (T), stomatal conductance (g_s), water use efficiency (WUE), water, osmotic, and pressure potential (Ψ_w, Ψ_π and Ψ_p, respectively), ratio of variable-to-minimum chlorophyll fluorescence (Fv/Fm), N, P, K, Ca, and Mg content, and leaf number per tree. Measurements were taken at the completion of a final depletion cycle ending 120 or more days after starting the study but prior to re-watering. NS indicates the *P*-value was greater than 0.05.

Leaf Trait	Main Effects and Interactions						
	G	**I**	**A**	**G × I**	**G × A**	**I × A**	**G × I × A**
Pn	NS	$<10^{-4}$	NS	NS	NS	0.0082	NS
T	$<10^{-4}$	$<10^{-4}$	NS	0.0002	NS	NS	NS
g_s	0.0037	$<10^{-4}$	NS	NS	NS	NS	NS
WUE	0.0001	$<10^{-4}$	NS	0.0180	NS	NS	NS
Ψ_w	0.0091	$<10^{-4}$	NS	NS	NS	NS	NS
Ψ_π	0.0321	$<10^{-4}$	NS	0.0189	NS	NS	NS
Ψ_p	0.0346	$<10^{-4}$	NS	NS	NS	NS	0.0255
Fv/Fm	0.0091	$<10^{-4}$	NS	NS	NS	NS	NS
Leaf Number	$<10^{-4}$	$<10^{-4}$	NS	NS	NS	NS	NS
N	0.0002	$<10^{-4}$	NS	NS	NS	NS	NS
P	$<10^{-4}$	0.0001	NS	0.0025	NS	NS	NS
K	0.0051	$<10^{-4}$	NS	NS	NS	NS	NS
Ca	NS	NS	NS	NS	NS	NS	NS
Mg	NS	NS	NS	NS	NS	NS	NS

3.2. Leaf Gas Exchange

Differences among genotypes were evident in T, g_s, and WUE, but not Pn (Tables 1 and 2, Figure 1). The transpiration rate was the lowest and the WUE was the highest for "Granny Smith", and "Royal Gala" had the highest stomatal conductance. Deficit irrigation reduced Pn, T, and g_s as % DAW increased. Pn and T at 80% DAW were reduced 64.9% and 67.5%, respectively, compared to 30% DAW, and were also lower than those at 60% DAW. WUE was greater at 60% DAW than at 30% and 80% DAW.

Table 2. Main effect means for genotype and for deficit irrigation as dependent on reaching 30%, 60%, or 80% depletion of available water (DAW) prior to re-watering on leaf gas exchange. Measurements were taken at the completion of a final depletion cycle ending 120 or more days after starting the study but prior to watering.

Leaf Trait	Genotype			Deficit Irrigation (% DAW)		
	GS	RG	GD	30	60	80
Photosynthesis (Pn) ($\mu mol \cdot CO_2 \cdot m^{-2} \cdot s^{-1}$)	10.0 NS	9.7	9.4	13.6 a	11.2 b	4.3 c
Transpiration (T) ($mmol \cdot H_2O \cdot m^{-2} \cdot s^{-1}$)	2.50 c	3.06 a	2.92 b	4.14 a	2.67 b	1.67 c
Stomatal conductance (g_s) ($mmol \cdot H_2O \cdot m^{-2} \cdot s^{-1}$)	0.217 b	0.254 a	0.217 b	0.261 a	0.265 a	0.162 b
Water use efficiency (Pn/T)	4.04 a	3.20 b	3.29 b	3.34 b	4.37 a	2.81 b

Means followed by different letters for each trait within genotype or irrigation timing were significantly different by Fisher's Least Significant Difference (LSD) test at $P \leq 0.05$.

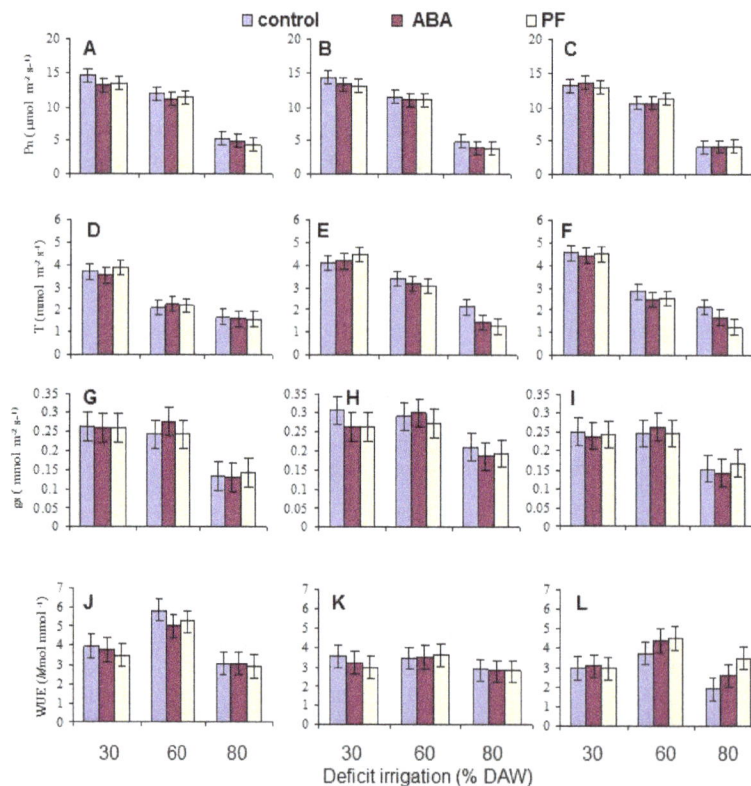

Figure 1. Effects of deficit irrigation as % depletion of available water (DAW) before re-watering and antitranspirant (ABA or PF) application on "Granny Smith" (**A,D,G,J**), "Royal Gala" (**B,E,H,K**), and "Golden Delicious" (**C,F,I,L**) apple net photosynthetic rate (Pn) (**A–C**), transpiration rate (T) (**D–F**), stomatal conductance (g_s) (**G–I**), and water use efficiency (WUE) (**J–L**). Values are the means ± standard error of the mean of three replicate measurements taken at the completion of a final depletion cycle ending 120 or more days after starting the study but prior to re-watering.

A significant genotype by deficit irrigation interaction was evident on T and WUE (Table 3). At 30% DAW, "Royal Gala" and "Golden Delicious" had higher T values than "Granny Smith", and the genotypes differed from one another at 60% DAW with "Royal Gala" the highest, followed by "Golden Delicious" and with "Granny Smith" the lowest, while they were all low at 80% DAW. Differences among genotypes were also evident in WUE, although in the opposite pattern. "Granny Smith" had the highest WUE at 30% and 60% DAW, and was equal to "Royal Gala" but higher than "Golden Delicious" at 80% DAW. WUE increased from 30% to 60% DAW for both "Granny Smith" and "Golden Delicious", but not for "Royal Gala", and it declined from 60% to 80% DAW for all genotypes. Reductions in Pn and/or T and an increase in WUE can be effective responses to avoid water deficit stress [11,12,31]. These reductions in Pn and T have been ascribed to increased mesophyll resistance and decreased leaf cell metabolic activity [38].

Averaged across genotypes, there was an interaction of the antitranspirant treatment with the irrigation level on Pn (Table 4). ABA and PF reduced Pn at 30% DAW, and PF but not ABA reduced it at 80% DAW, though all of the effects were small compared to the main effects of the irrigation level. Failure of ABA to affect T was observed in mulberry (*Morus alba* L.) [39], and PF-treated apple [40,41], and pecan (*Carya illinoinensis*) [42]. However, the reflective properties of the kaolin PF reduced leaf temperature by increasing leaf reflectance in field-grown trees, thus reducing leaf temperature, T, and the leaf-to-air vapor pressure gradient in other studies [20–22,43,44]. Gas exchange responses to PF were only evident at vapor pressure deficits less than 2 kPa [21], which could have contributed to the inconsistencies among studies.

Table 3. Interaction of genotype with deficit irrigation as dependent on reaching 30%, 60%, or 80% depletion of available water (DAW) prior to re-watering on transpiration rate, water use efficiency, and leaf osmotic potential (Ψ_π). Measurements were taken at the completion of a final depletion cycle ending 120 or more days after starting the study but prior to watering.

Genotype	Deficit Irrigation (% DAW)		
	30	60	80
Transpiration (mmol·H$_2$O·m^{-2}·s^{-1})			
Granny Smith	3.70 b	2.17 e	1.63 f
Royal Gala	4.23 a	3.23 c	1.70 f
Golden Delicious	4.50 a	2.60 d	1.67 f
Water use efficiency (net photosynthetic rate/transpiration rate)			
Granny Smith	3.74 c	5.37 a	3.01 ef
Royal Gala	3.26 de	3.54 cd	2.78 fg
Golden Delicious	3.00 ef	4.22 b	2.66 g
Leaf osmotic potential (Ψ_π, MPa)			
Granny Smith	−2.14 d	−2.97 c	−3.71 a
Royal Gala	−1.79 e	−3.23 b	−3.41 b
Golden Delicious	−1.89 e	−2.91 c	−3.38 b

Means followed by different letters among all values within an interaction set were significantly different by Fisher's Least Significant Difference (LSD) test at $P \leq 0.05$.

Table 4. Interaction of antitranspirant with deficit irrigation as dependent on reaching 30%, 60%, or 80% depletion of available water (DAW) prior to re-watering on net photosynthetic rate and water use efficiency. Measurements were taken at the completion of a final depletion cycle ending 120 or more days after starting the study but prior to watering.

Antitranspirant	Deficit Irrigation (% DAW)		
	30	60	80
Photosynthetic rate (μmol·CO$_2$·m^{-2}·s^{-1})			
None	14.1 a	11.4 c	4.6 d
Abscisic acid	13.4 b	11.0 c	4.3 d
Particle film	13.2 b	11.3 c	4.1 e

Means followed by different letters among all values were significantly different by Fisher's Least Significant Difference (LSD) test at $P \leq 0.05$.

3.3. Leaf Water Relations

A significant influence of genotype and deficit irrigation on leaf water relations was observed, but antitranspirants had no effect (Tables 1 and 5, Figure 2). While there were small differences among genotypes for leaf Ψ_w, Ψ_π and Ψ_p, the values changed considerably under water deficit conditions. Leaf Ψ_w and Ψ_π declined as the % DAW increased, but leaf Ψ_p declined at 80% DAW only. The only interaction of note on leaf water relations was a genotype by deficit irrigation effect on leaf Ψ_π (Table 3). "Granny Smith" had the lowest leaf Ψ_π at 30% and 80% DAW, while "Royal Gala" had the lowest at 60% DAW. Though not directly measured in the present work, osmotic adjustment under reduced irrigation has been observed in apple [12,32]. The low leaf Ψ_π for specific genotypes at each % DAW may have been evidence of genotypic differences in osmoregulation. The ability of apple trees to acclimate by active osmoregulation would be an advantage, providing increased flexibility in response to changing environmental conditions. The reduction in leaf Ψ_w as the moisture level decreased (i.e., from 30% to 80% DAW) in this investigation was correlated to lower T ($r = 0.87$, $P < 0.05$).

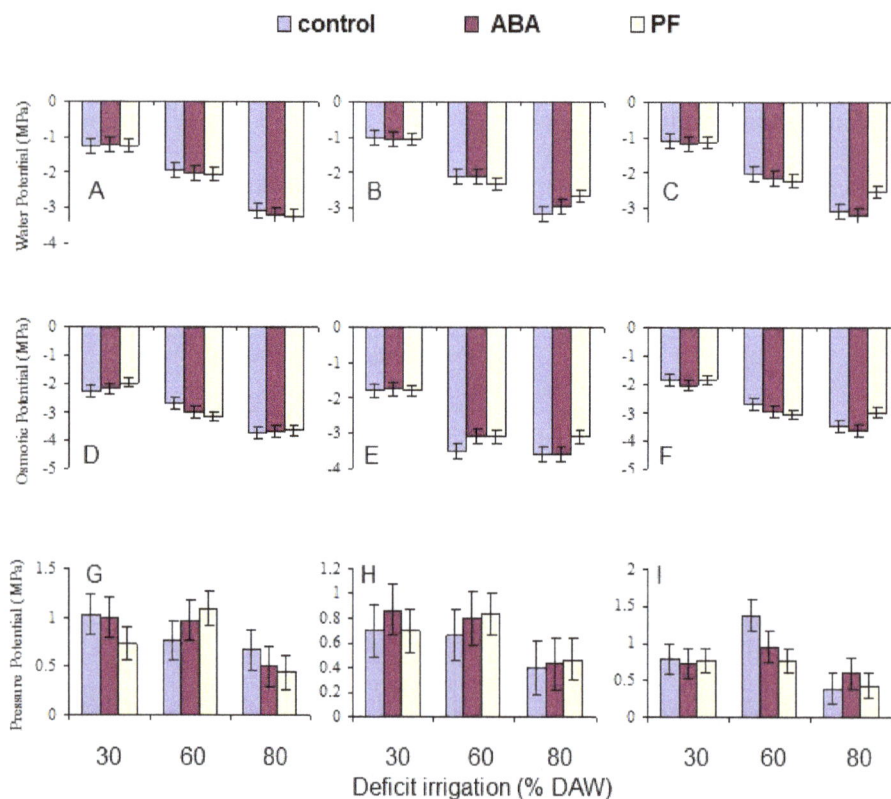

Figure 2. Effects of deficit irrigation as % depletion of available water (DAW) before re-watering and antitranspirant (ABA or PF) application on "Granny Smith" (**A,D,G**), "Royal Gala" (**B,E,H**), and "Golden Delicious" (**C,F,I**) apple on leaf water potential (Ψ_w) (**A–C**), osmotic potential (Ψ_π) (**D–F**), and pressure potential (Ψ_p) (**G–I**). Values are the means ± standard error of the mean of three replicate measurements taken at the completion of a final depletion cycle ending 120 or more days after starting the study but prior to re-watering.

Table 5. Main effect means for genotype and for deficit irrigation as dependent on reaching 30%, 60%, or 80% depletion of available water (DAW) prior to re-watering on leaf water relations components. Measurements were taken at the completion of a final depletion cycle ending 120 or more days after starting the study but prior to watering.

Leaf Water Relations	Genotype			Deficit Irrigation (% DAW)		
Component (MPa)	GS	RG	GD	30	60	80
Water potential (Ψ_w)	−2.14 a	−2.05 b	−2.07 b	−1.12 c	−2.12 b	−3.02 a
Osmotic potential (Ψ_π)	−2.94 a	−2.81 b	−2.72 c	−1.94 b	−3.04 a	−3.50 a
Pressure potential (Ψ_p)	0.80 a	0.76 a	0.65 b	0.81 a	0.92 a	0.48 b

Granny Smith (GS); Royal Gala (RG); Golden Delicious (GD); Means followed by different letters for each trait within genotype or irrigation timing were significantly different by Fisher's Least Significant Difference (LSD) test at $P \leq 0.05$.

3.4. Leaf Fv/Fm and Number

Chlorophyll fluorescence arises from overexcitation of the photochemical system, acting as a protective mechanism allowing plants under stressed conditions to dissipate light energy and survive, and the Fv/Fm ratio indicates photosystem II efficiency (PSII) [45]. The maximum quantum yield of PSII, as measured by the Fv/Fm ratio, differed among genotypes and declined with increasing DAW (Tables 1 and 6, Figure 3). As a result, there was a correlation ($r = 0.77$, $P < 0.05$) between leaf Ψ_w and Fv/Fm values. Fernandez et al. [9] observed drought stress effects on chlorophyll fluorescence in apple as well, though they occurred after effects on Pn were evident in contrast to our results where both declined as the % DAW increased. The influence of water deficit stress on chlorophyll fluorescence reflects effects on thylakoid membrane integrity and the relative efficiency of electron transport from PSII to photosystem I [46].

Although there were differences among genotypes for leaf number per tree (Tables 1 and 6, Figure 3), only 80% DAW reduced the number, indicating a significant impact of deficit irrigation on tree growth over the duration of the study. The reduction in leaf number at 80% DAW could be attributed to the reduction in the assimilate supply as Pn was lowest at 80% DAW, to a decline in cell division and enlargement and more leaf senescence [47], and/or to low leaf Ψ_p [48].

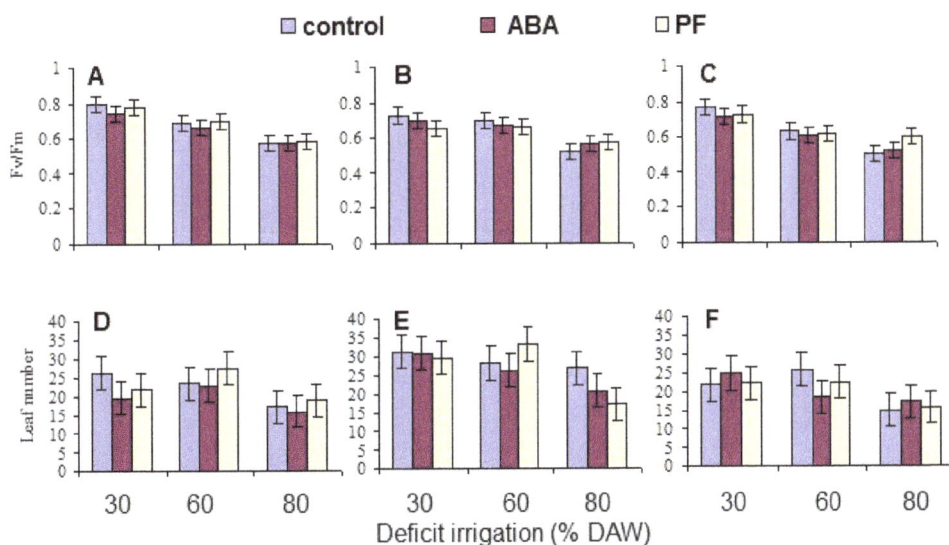

Figure 3. Effects of deficit irrigation as % depletion of available water (DAW) before re-watering and antitranspirant (ABA or PF) application on "Granny Smith" (**A,D**), "Royal Gala" (**B,E**), and "Golden Delicious" (**C,F**) apple on leaf chlorophyll fluorescence (**A–C**) and leaf number per tree (**D–F**). Values are the means ± standard error of the mean of three replicate measurements taken at the completion of a final depletion cycle ending 120 or more days after starting the study but prior to re-watering.

Table 6. Main effect means for genotype and for deficit irrigation as dependent on reaching 30%, 60%, or 80% depletion of available water (DAW) prior to re-watering on leaf chlorophyll fluorescence, leaf number, and leaf N, P, and K content (as % dry weight). Measurements were taken at the completion of a final depletion cycle ending 120 or more days after starting the study but prior to watering.

Leaf Trait	Genotype			Deficit Irrigation (% DAW)		
	GS	RG	GD	30	60	80
Fv/Fm	0.674 a	0.643 b	0.634 c	0.737 a	0.661 b	0.558 c
Leaf number	21.6 b	27.2 a	20.5 b	25.5 a	25.5 a	18.4 b
N content	2.24 bc	2.34 ab	2.18 c	2.37 a	2.28 a	2.12 b
P content	0.40 a	0.33 c	0.36 b	0.38 a	0.37 b	0.34 c
K content	2.09 a	1.91 b	2.10 a	2.13 a	2.11 a	1.86 b

Granny Smith (GS); Royal Gala (RG); Golden Delicious (GD); Means followed by different letters for each trait within genotype or irrigation timing were significantly different by Fisher's Least Significant Difference (LSD) test at $P \leq 0.05$.

3.5. Leaf N, P, and K Content

The contents of N, P, and K were the only nutrients that varied among genotypes and deficit irrigation levels (Tables 1 and 6, Figure 4). Increasing deficit irrigation reduced leaf N, P, and K contents, with the most notable effects at 80% DAW. Reductions in leaf N, P, and K contents in the stressed trees were likely affected by the decrease in T [31]. A low efficiency of nutrient absorption and transport within a plant has also been attributed to the slow movement of minerals in drying soil as well as to the decrease in root extension and root permeability [49].

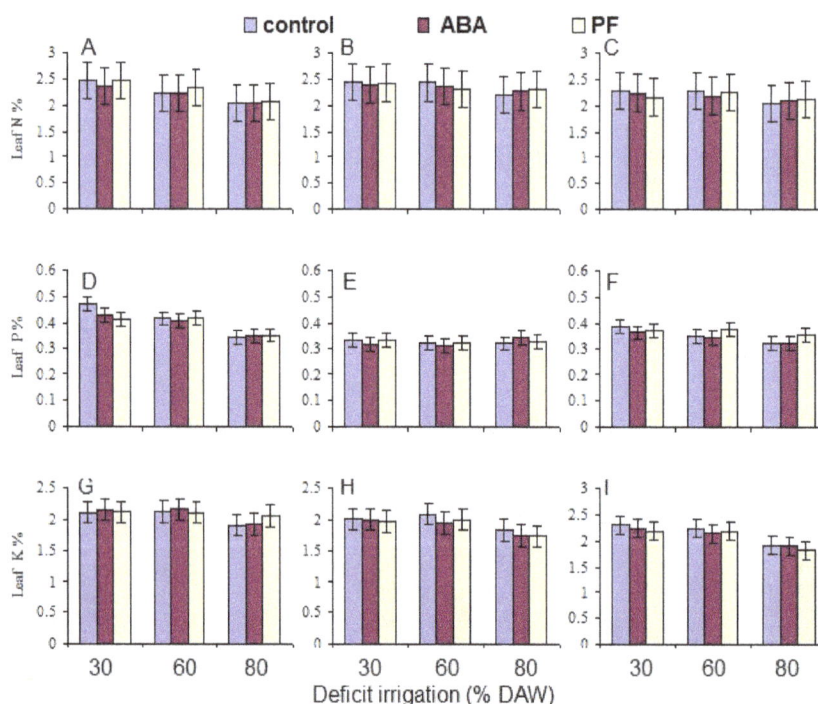

Figure 4. Effects of deficit irrigation as % depletion of available water (DAW) before re-watering and antitranspirant (ABA or PF) application on "Granny Smith" (**A,D,G**), "Royal Gala" (**B,E,H**), and "Golden Delicious" (**C,F,I**) apple on leaf N (**A–C**), P (**D–F**), and K (**G–I**) content. Values are the means ± standard error of the mean of three replicate measurements taken at the completion of a final depletion cycle ending 120 or more days after starting the study but prior to re-watering.

4. Discussion

Overall, the water deficit stress at increasing % DAW reduced Pn, T, leaf Ψ_w and Ψ_π, the Fv/Fm ratio, and leaf N, P and K contents. WUE increased for two of the genotypes from 30% to 60% DAW because Pn did not change, but g_s declined, and it decreased in all genotypes from 60% to 80% DAW. Apart from genotypic variation, deficit irrigation had the most important effect. PF affected leaf Pn but not the other gas exchange traits, suggesting it did not alter leaf physiology or development. The lack of response to ABA was likely due to the fact that only three applications were made to the trees. ABA metabolism would occur in leaves shortly after each application [50], and measurements reported here were taken 60 or more days after the last application, so ABA effects on stomatal closure or other physiological parameters would likely have subsided much earlier. More frequent ABA application might produce beneficial effects on gas exchange responses and WUE under stress conditions [27]. Nonetheless, there were no long-term effects from the ABA applications in the present work.

Under conditions of water scarcity, irrigation should be scheduled at a level of no less than 60% DAW, as this optimized WUE and maintained growth, as indicated by the leaf number, while detrimental effects on leaf gas exchange and growth occurred at 80% DAW. Even though there was a residue of kaolin PF on the leaves throughout the period of the study because the trees were only watered at their base, and the residue was never washed off as may occur in environments with periodic rainfall, there were no sustained effects on any measured trait other than leaf Pn mitigating the impact of reduced water availability. This was also evident in tree growth (as indicated by leaf production) which did not respond to PF. The effects of PF film on apple have been most notable in environments with high vapor pressure deficits and temperature, and with irrigated trees [20–23], quite different than the conditions of the present work. Although the trees in the present work were young and had limited root development and container volume which may have impacted the results, the application of PF to orchard-grown walnut (*Juglans regia*) also failed to mitigate the effects of water stress [43], so it is likely to have a similar impact on mature apple trees in the field.

5. Conclusions

Periodic use of PF or ABA during deficit irrigation did not alleviate water deficit stress effects from deficit irrigation strategies that impact apple tree growth under the conditions of this study.

Acknowledgments: This study was performed as a Fulbright Scholar during a sabbatical leave from University of Mu'tah, Jordan, and spent at the Department of Horticulture, University of Kentucky, KY, USA. The information reported in this paper (No. 16-11-093) is part of a project of the Kentucky Agricultural Experiment Station and is published with the approval of the Director.

Author Contributions: K.M.A. and D.D.A. conceived and designed the experiment; K.M.A. performed the experiment and analyzed the data; K.M.A. and D.D.A. wrote the paper.

Conflicts of Interest: The authors declare no conflict of interest.

References

1. Lobell, D.B.; Burke, M.B.; Tebaldi, C.; Mastrandrea, M.D.; Falcon, W.P.; Naylor, R.L. Prioritizing climate change adaptation needs for food security in 2030. *Science* **2008**, *319*, 607–610. [CrossRef] [PubMed]
2. Arar, A. Wastewater reuse for irrigation in the near east region. *Water Sci. Technol.* **1991**, *23*, 2127–2134.
3. Geerts, S.; Raes, D. Deficit irrigation as an on-farm strategy to maximize crop water productivity in dry areas. *Agric. Water Manag.* **2009**, *96*, 1275–1284. [CrossRef]
4. Aroca, R. *Plant Responses to Drought Stress: From Morphological to Molecular Features*; Springer: Berlin, Germany, 2012.
5. Ludlow, M.M. Strategies of response to water stress. In *Structural and Functional Responses to Environmental Stress*; Kree, K.H., Richter, H., Minckley, T.M., Eds.; SPB Academic: The Hague, The Netherlands, 1989; pp. 269–281.
6. Hanson, A.D.; Hitz, W.D. Metabolic responses of mesophytes to plant water deficits. *Ann. Rev. Plant Physiol.* **1982**, *33*, 163–203. [CrossRef]

7. Li, F.; Cohen, S.; Naor, A.; Shaozong, K.; Erez, A. Studies of canopy structure and water use of apple trees on three rootstocks. *Agric. Water Manag.* **2002**, *55*, 1–14. [CrossRef]

8. Atkinson, C.J.; Policarpo, M.; Webster, A.D.; Kingswell, G. Drought tolerance of clonal *Malus* determined from measurements of stomatal conductance and leaf water potential. *Tree Physiol.* **2000**, *20*, 557–563. [CrossRef] [PubMed]

9. Fernandez, R.T.; Perry, R.L.; Flore, J.A. Drought response of young apple trees on three rootstocks. II. Gas exchange, chlorophyll fluorescence, water relations, and leaf abscisic acid. *J. Am. Soc. Hortic. Sci.* **1997**, *122*, 841–848.

10. Lombardini, L.; Toselli, M.; Flore, J.A. Carbon translocation and root respiration in potted apple trees during conditions of moderate drought. *Acta Hortic.* **2001**, *557*, 413–420. [CrossRef]

11. Mohawesh, O.E.; Al-Absi, K.M. Physiological response of two apple genotypes to different water regimes under semiarid conditions. *Adv. Hortic. Sci.* **2009**, *23*, 158–165.

12. Sircelj, H.; Tausz, M.; Grill, D.; Batic, F. Detecting different levels of drought stress in apple trees (*Malus domestica* Borkh.) with selected biochemical and physiological parameters. *Sci. Hortic.* **2007**, *113*, 362–369. [CrossRef]

13. Kilili, A.W.; Behboudian, M.H.; Mills, T.M. Composition and quality of "Braeburn" apples under reduced irrigation. *Sci. Hortic.* **1996**, *67*, 1–11. [CrossRef]

14. Behboudian, M.H.; Mills, T.M. Deficit irrigation in deciduous orchards. *Hortic. Rev.* **1997**, *21*, 105–130.

15. Ebel, R.C.; Proebsting, E.L.; Evans, R.G. Deficit irrigation to control vegetative growth in apple and monitoring fruit growth to schedule irrigation. *HortScience* **1995**, *30*, 1229–1232.

16. Irving, D.E.; Drost, J.H. Effects of water deficit on vegetative growth, fruit growth and fruit quality in Cox's Orange Pippin apple. *J. Hortic. Sci.* **1987**, *62*, 427–432. [CrossRef]

17. Mpelasoka, B.; Behboudian, M.; Green, S. Water use, yield and fruit quality of lysimeter-grown apple trees: Responses to deficit irrigation and to crop load. *Irrig. Sci.* **2001**, *20*, 107–113.

18. Leib, B.G.; Caspari, H.W.; Redulla, C.A.; Andrews, P.K.; Jabro, J.J. Partial rootzone drying and deficit irrigation of "Fuji" apples in a semi-arid climate. *Irrig. Sci.* **2006**, *24*, 85–99. [CrossRef]

19. O'Connell, M.G.; Goodwin, I. Responses of "Pink Lady" apple to deficit irrigation and partial rootzone drying: Physiology, growth, yield, and fruit quality. *Aust. J. Agric. Res.* **2005**, *58*, 1068–1076. [CrossRef]

20. Glenn, D.M. Particle film mechanisms of action that reduce the effect of environmental stress in "Empire" apple. *J. Am. Soc. Hortic. Sci.* **2009**, *134*, 314–321.

21. Glenn, D.M. Canopy gas exchange and water use efficiency of "Empire" apple in response to particle film, irrigation, and microclimatic factors. *J. Am. Soc. Hortic. Sci.* **2010**, *135*, 25–32.

22. Glenn, D.M.; Erez, A.; Puterka, G.J.; Gundrum, P. Particle films affect carbon assimilation and yield in "Empire" apple. *J. Am. Soc. Hortic. Sci.* **2003**, *128*, 356–362.

23. Glenn, D.M.; Puterka, G.J.; Drake, S.R.; Unruh, T.R.; Knight, A.L.; Baherle, P.; Prado, E.; Baugher, T.A. Particle film application influences apple leaf physiology, fruit yield, and fruit quality. *J. Am. Soc. Hortic. Sci.* **2001**, *126*, 175–181.

24. Davies, W.J.; Zhang, J. Root signals and the regulation of growth and development of plants in drying soil. *Ann. Rev. Plant Physiol. Plant Mol. Biol.* **1991**, *42*, 55–76. [CrossRef]

25. Al-Absi, K.M. Gas exchange, chlorophyll and growth response of three orange genotypes (*Citrus sinensis* [L.] Osbeck) to abscisic acid under progressive water deficit. *Jordan J. Agric. Sci.* **2009**, *5*, 421–433.

26. Du, Y.-L.; Wang, Z.-Y.; Fan, J.-W.; Turner, N.C.; He, J.; Wang, T.; Li, F.-M. Exogenous abscisic acid reduces water loss and improves antioxidant defence, desiccation tolerance and transpiration efficiency in two spring wheat genotypes subjected to a soil water deficit. *Func. Plant Biol.* **2013**, *40*, 494–506. [CrossRef]

27. Ma, X.; Ma, F.; Mi, Y.; Ma, Y.; Shu, H. Morphological and physiological responses of two contrasting *Malus* species to exogenous abscisic acid application. *Plant Growth Regul.* **2008**, *56*, 77–87. [CrossRef]

28. Wang, Z.; Huang, B.; Xu, Q. Effects of abscisic acid on drought responses of Kentucky bluegrass. *J. Am. Soc. Hortic. Sci.* **2003**, *128*, 36–41.

29. Waterland, N.L.; Campbell, C.A.; Finer, J.J.; Jones, M.L. Abscisic acid application enhances drought stress tolerance in bedding plants. *HortScience* **2010**, *45*, 409–413.

30. Kondo, S.; Sugaya, S.; Sugawa, M.; Ninomiya, S.; Kittikorn, M.; Okawa, K.; Ohara, H.; Ueno, K.; Todoroki, Y.; Mizutani, M.; et al. Dehydration tolerance in apple seedlings is affected by an inhibitor of ABA 8-hydroxylase CYP707A. *J. Plant Physiol.* **2012**, *169*, 234–241. [CrossRef] [PubMed]

31. Jie, Z.; Yuncong, Y.; Streeter, J.G.; Ferree, D.C. Influence of soil drought stress on photosynthesis, carbohydrates and the nitrogen and phosphorus absorb in different section of leaves and stem of Fugi/M.9EML, a young apple seedling. *Afr. J. Biotechnol.* **2010**, *9*, 5320–5325.

32. Wang, Z.; Quebedeaux, B.; Stutte, G.W. Osmotic adjustment: Effect of water stress on carbohydrates in leaves, stems and roots of apple. *Aust. J. Plant Physiol.* **1995**, *22*, 747–754. [CrossRef]

33. Shellie, K.; Glenn, D.M. Wine grape response to foliar particle film under differing levels of preveraison water stress. *HortScience* **2008**, *43*, 1392–1397.

34. Shellie, K.C.; King, B.A. Kaolin-based foliar reflectant and water deficit influence Malbec leaf and berry temperature, pigments, and photosynthesis. *Am. J. Enol. Vit.* **2013**, *64*, 223–230. [CrossRef]

35. Lawlor, D.W.; Cornic, G. Photosynthetic carbon assimilation and associated metabolism in relation to water deficits in higher plants. *Plant Cell Environ.* **2002**, *25*, 275–294. [CrossRef] [PubMed]

36. Zhang, B.; Archbold, D.D. Water relations of a *Fragaria chiloensis* and a *F. virginiana* selection respond to water deficit stress. *J. Am. Soc. Hortic. Sci.* **1993**, *118*, 280–285.

37. Tandon, H.L.S. *Methods of Analysis of Soils, Plants, Waters and Fertilizers*; Fertilizer Development and Consultation Organization: New Delhi, India, 1995; pp. 49–82.

38. Reddy, A.R.; Chaitanya, K.V.; Vivekanandan, M. Drought-induced responses of photosynthesis and antioxidant metabolism in higher plants. *J. Plant Physiol.* **2004**, *161*, 1189–1202. [CrossRef]

39. Misra, A.K.; Das, B.K.; Datta, J.K.; De, G.C. Influence of antitranspirants on photosynthesis, leaf dry wt., nitrate reductase activity and leaf yield of mulberry (*Morus alba* L.) under water stress condition. *Indian J. Agric. Res.* **2009**, *43*, 144–147.

40. Grange, M.; Wand, S.J.; Theron, K.L. Effect of kaolin applications on apple fruit quality and gas exchange of apple leaves. *Acta Hortic.* **2004**, *636*, 545–550. [CrossRef]

41. Prive, J.P.; Russell, L.; LeBlanc, A. The impact of kaolin clay sprays on leaf gas exchange of Ginger Gold apple trees in New Brunswick. *Can. J. Plant Sci.* **2006**, *86*, 1377–1381. [CrossRef]

42. Lombardini, L.; Glenn, D.M.; Harris, M.K. Application of kaolin-based particle film on pecan trees: Consequences on leaf gas exchange, stem water potential, nut quality, and insect populations. *HortScience* **2004**, *39*, 857–858.

43. Rosati, A.; Metcalf, S.G.; Buchner, R.P.; Fulton, A.E.; Lampinen, B.D. Physiological effects of kaolin applications in well-irrigated and water-stressed walnut and almond trees. *Ann. Bot.* **2006**, *98*, 267–275. [CrossRef] [PubMed]

44. Wunsche, J.N.; Lombardini, L.; Greer, D.H. Surround particle film applications—Effects on whole canopy physiology of apple. *Acta Hortic.* **2004**, *636*, 565–571. [CrossRef]

45. Demmig-Adams, B.; Adams, W.W.; Grace, S.C. Physiology of light tolerance in plants. *Hortic. Rev.* **1997**, *18*, 215–246.

46. Johnson, J.D.; Tognetti, R.; Paris, P. Water relations and gas exchange in poplar and willow under water stress and elevated atmospheric CO_2. *Physiol. Plant.* **2002**, *115*, 93–100. [CrossRef] [PubMed]

47. Shao, H.; Chu, L.; Jaleel, C.A.; Zhao, C. Water-deficit stress-induced anatomical changes in higher plants. *Plant Biol. Path.* **2008**, *331*, 215–225. [CrossRef] [PubMed]

48. Cabuslay, G.S.; Ito, O.; Alejar, A.A. Physiological evaluation of responses of rice (*Oryza sativa* L.) to water deficit. *Plant Sci.* **2002**, *63*, 815–827. [CrossRef]

49. Nilsen, E.T.; Orcutt, D.M. *The Physiology of Plant under Stress*; John Wiley and Sons: New York, NY, USA, 1996.

50. Zeevaart, J.A.D.; Creelman, R.A. Metabolism and physiology of abscisic acid. *Ann. Rev. Plant Physiol. Plant Mol. Biol.* **1998**, *39*, 439–473. [CrossRef]

14

Estimating the Leaf Area of Cut Roses in Different Growth Stages Using Image Processing and Allometrics

Ana Patrícia Costa [1], Isabel Pôças [2,3] and Mário Cunha [1,3,4,*]

[1]   Faculty of Sciences, Universidade do Porto, Porto 4169-007, Portugal; patricia.malva.costa@gmail.com
[2]   Linking Landscape, Environment, Agriculture and Food, Instituto Superior de Agronomia, Universidade de Lisboa, Lisboa 1349-017, Portugal; ipocas@mail.icav.up.pt
[3]   Geo-Space Sciences Research Centre, Universidade do Porto, Porto 4169-007, Portugal
[4]   CITAB—Center for the Research and Technology of Agro-Environmental and Biological Sciences, Universidade de Trás-os-Montes e Alto Douro, Quinta de Prados, Ap. 1013, Vila Real 5001-801, Portugal
*   Correspondence: mcunha@mail.icav.up.pt

Abstract: Non-destructive, accurate, user-friendly and low-cost approaches to determining crop leaf area (LA) are a key tool in many agronomic and physiological studies, as well as in current agricultural management. Although there are models that estimate cut rose LA in the literature, they are generally designed for a specific stage of the crop cycle, usually harvest. This study aimed to estimate the LA of cut "Red Naomi" rose stems in several phenological phases using morphological descriptors and allometric measurements derived from image processing. A statistical model was developed based on the "multiple stepwise regression" technique and considered the stem height, the number of stem leaves, and the stage of the flower bud. The model, based on 26 stems (232 leaves) collected at different developmental stages, explained 95% of the LA variance ($R^2 = 0.95$, $n = 26$, $p < 0.0001$). The mean relative difference between the observed and the estimated LA was 8.2%. The methodology had a high accuracy and precision in the estimation of LA during crop development. It can save time, effort, and resources in determining cut rose stem LA, enhancing its application in research and production contexts.

Keywords: allometric descriptors; leaf area modeling; non-destructive measurements; stem morphology; *Rosa hybrida* L.

1. Introduction

Productivity and crop growth are directly related to leaf area (LA). Therefore, accurate and prompt determination of this parameter is of great importance. Radiation and photosynthetic efficiency, crop transpiration, crop water use, and nutrient use are some of the processes impacted by the LA of a crop [1,2].

The implementation of tools for measuring and estimating crop LA has long been a concern for researchers. There are currently several approaches for LA determination, which include direct and indirect methods. Direct methods include planimetric or gravimetric analyses of leaves, harvested directly or indirectly [3,4]. Portable scanning planimeters (e.g., LI-3000, Licor, NE, USA) are often used as a reference method for obtaining the LA. Direct methods are more accurate but have the disadvantages of being very time-consuming, not user-friendly, and having constraints regarding equipment acquisition, price, and operation [4]. Moreover, direct methods can be destructive, not allowing successive measurements of LA [5].

One of the most frequently used indirect methods for LA estimation is based on observations and measurements of allometric parameters of the plants, which are used as inputs in mathematical models [6]. Such mathematical models are based on the correlation between the allometric measures of plants and the area of the leaves. These methods are non-destructive and allow for faster LA determination, eventually being suitable for automation. Nevertheless, an adequate parameterization and calibration of such methods is necessary. Stereo and time-of-flight image combination [7] and light curtain arrays [8] have been used for non-destructive automatic leaf area measurements. Despite promising results in experimental conditions, the costs and operability of these approaches are, for now, still unknown.

Considering the advantages of indirect methods, they are now assuming a particular relevance when compared with direct methods [4]. In 1911, scientists suggested for the first time estimating plant LA through its relation with allometric leaf measurements, using the width and length of the leaf [9]. Since then, this method has been applied to different crops such as tomatoes and cucumbers [10], grapevines [11], peppers [12], and soybeans [13], as well as to fruit trees, such as hazelnut [14], and to ornamental crops such as anthurium [15], begonia [16], and cut roses [1,5]. According to Zhang and Liu [16], this indirect and non-destructive approach can provide precise and in situ LA estimations.

Several models are currently available for rose crop to determine the area of individual leaves [1,17,18], leaflets [5], and stems [19] using allometric measures such as the length and width of leaves or leaflets, the number of leaflets, or the height of the stem. However, despite the economic importance of the rose as a major ornamental crop [1,5], these methods for estimating LA still have several limitations. For example, not all models that have been developed take into account the diversity of existing genotypes [17,19]. The limitations also extend to the architecture of the plant, with some models being developed using material exclusively from erect stems, disregarding samples resulting from the bending of weak stems of the plant [1,19]. Additionally, most models only consider a phenological stage, usually harvest, which limits its use along the cultural cycle of the crop and consequently the continuous monitoring of crop growth and leaf expansion.

Consequently, LA measurements along the crop cultural cycle are often not possible, although they have been empirically simulated in complex crop-specific growth models when available [20]. Most of the models described in the literature are based on allometric measurements of compound leaves. However, this methodology, while allowing non-destructive LA estimation, can become time-consuming depending on the number of leaves whose area has to be determined.

In this context, the goal of this study was to develop a non-destructive and expeditious method and mathematical model for estimating LA in the cut rose cv. "Red Naomi" at different phenological stages. The specific goals included (i) the development of an estimation methodology based on morphological descriptors of the stems and biometric measurements of the leaves using image processing; and (ii) the development of a dynamic mathematical model that estimates the LA of the crop stems throughout the cultural cycle of the rose.

2. Materials and Methods

2.1. Study Area

This case study was carried out in a commercial glass "Venlo" greenhouse of the company Floralves, located in the Vila do Conde (41°19'40.8"N 8°42'17.4"W), in the North of Portugal. The greenhouse has a North–South orientation and an area of approximately 1 ha.

The greenhouse is exclusively occupied by the production of cut roses (*Rosa hybrida* L.) grown in a substrate cultivation system (coir). The cultivar "Red Naomi," which occupies about 30% of the greenhouse area, was used for the case study. The plants used in this study were transplanted in 2011 with a density of eight plants per m^2.

The culture was irrigated by a closed drip system with a standard nutrient solution for cut roses.

The irrigation water was reused after disinfection with ultraviolet light. The limit values for electrical conductivity and pH were 1.5 dS· m^{-1} and 5.4, respectively.

The crop was managed by the producer following standard cultural practices, using the stem-bending system, where nonproductive and lesser quality stems are bent into the canopy or aisle.

2.2. Field Data

The data used for the development of the LA estimation model were collected in an 11-day period between mid-March and early May of 2014.

A total of 26 erect stems (accounting for 232 leaves) in different stages of development, from randomly-selected plants located in the same sector of the greenhouse, were sampled. The stems were collected randomly among healthy plants and within the plant. Samples from bended stems and from stems with no floral bud formation ("blind stems") were also collected. Data for several morphological/allometric descriptors were recorded on each date.

The effect of the phenological stage on LA estimation was evaluated through the qualitatively variable stages of flower buds (SB). This variable could assume three categories defined according to the level of aggregation of the floral bud stages—SB_A1, SB_A2, and SB_A3—which include 2, 4, and 3 classes, respectively, as shown in Figure 1 and defined in Table 1. These three flower bud categories were tested as potential predictors in the model for LA estimation. The classes within the SB_A1, SB_A2, and SB_A3 categories were based on the phenological stages of the rose reproductive period described by Flórez and Rodríguez [21].

(a)   (b)   (c)   (d)

Figure 1. Classification of flower bud development stages: (a) stem without flower bud; (b) stem whose flower bud is closed with petal color not visible; (c) stem with flower bud partially closed but petal color visible; (d) stem with petal color visible and ready for harvest.

Table 1. Model input data for each aggregation level of flower bud stage tested as potential descriptors in the model for leaf area estimation.

Stages of Bud Development	Model Input Data for Each Aggregation Level		
	SB_A1	SB_A2	SB_A3
No flower bud (flower bud nonvisible or "blind stems") (Figure 1a)	0	0	0
Flower bud is closed and color of the petals is not visible (Figure 1b)	1	1	1
Flower bud partially closed but color of the petals is visible (Figure 1c)	1	2	2
Flower bud ready to be harvest (Figure 1d)	1	3	2

Additionally, images of the leaves from the selected stems were collected in the greenhouse using a digital camera (Sony Alpha Nex F3, New York, NY, USA) with a maximum resolution of 4912 × 3264 pixels in order to calibrate and further validate the model for the estimation of LA based on the morphological/allometric descriptors. The leaves were placed on a white color support plate suitable for the leaf size. The plates had a transparent acrylic cover in order to flatten the leaves, avoiding perspective distortion and allowing greater accuracy in the attainment of the total leaf area (Figure 2). The plate also had a graduated scale and a mark at each corner forming a parallelogram with right angles. By combining the corner marks on the plate with the digital camera grid, it was possible to avoid perspective distortion and enhance the accuracy of the measurements.

Figure 2. Nondestructive image acquisition for LA measurements for image processing using a plate with a graduated scale and mark in each corner.

The digital images were processed using Photoshop CS6 (San Jose, CA, USA) and included three main steps: (i) the assignment of the number of pixels that corresponded to one cm of the graduated scale in the image by using the Measurement Scale option in the Analysis feature; (ii) the selection of the area of interest within the image, i.e., the area to measure by using the Quick Selection Tool; and (iii) the acquisition of the area previously selected by using the Record Measurements option in the Analysis feature.

The accuracy of LA measurements based on image processing was compared with the LA obtained with a portable LA meter (AM350, ADC BioScientific Ltd., Hoddesdon, UK). This comparison was assessed by the measurement of four stems for a total of 36 leaves, from plants cultivated under the same conditions and from the same cultivar. Only open leaves with a minimum of three leaflets were considered. After the leaves were photographed in the greenhouse, the stems were cut and immediately placed in a container with water. After the stem harvest, they were rapidly transported to the laboratory for LA determination with the portable LA meter [17].

2.3. Development of the Model for Stem LA Estimation

The model for stem LA estimation based on morphological and allometric descriptors was developed through a multiple stepwise regression. The descriptors considered were stem height (SH), leaves per stem (SLN), flower bud stage (SB), leaf length (LL), leaf width (LW), and leaflet number per leaf (LNL). The model development took into consideration potential allometric descriptor combinations such as the square of the SH. For each stem, the length and width of the leaf with the maximum LA, the length and width of the first and last leaf, the number of leaflets on the first and last leaf, the average number of leaflets, and the sum of the leaflets were also considered.

Multiple stepwise regression allows, at each step in calculation, the evaluation of the significance level ($p < 0.05$) of each independent variable to the dependent variable. Assumptions of normality, homocedasticity, and the existence of multicollinearity among the independent variables were previously verified. In order to detect the existence of collinearity between the model variables, the value of the variance inflation factor (VIF) and tolerance (T) were calculated. The variables, selected by multiple stepwise regression, with VIF > 10 and T < 0.1, were excluded from the model [22]. The statistical significance of the parameters of the model was evaluated by the Student's t-test.

The model adequacy was assessed by the percentage of variance explained by the model, expressed by the R-square. A set of indicators of residual estimation errors was also computed for model fitting analysis: the root mean square error (RMSE), the mean relative error (MRE) and the mean absolute error (MAE). Additionally, the analysis of uncertainty based on the model reliability was considered, which encompasses both accuracy and precision. The regression mean prediction interval for the 95% probability level was illustrated graphically [22].

We performed all analyses using IBM SPSS statistical software (Version 23.0., IBM Corp., Armonk, NY, USA).

2.4. Model Validation and Prediction Accuracy

An external validation of the model for estimating LA based on allometric measurements was performed to evaluate the prediction reliability. The external validation was performed by analyzing the model performance in predicting the leaf area of three stems (total of 24 leafs) not used in the model parameter estimation.

An additional validation was applied over the full set of data ($n = 26$) using the "leave-one-out" (LOO) cross-validation method [23]. The LOO cross-validation evaluates the model performance for observations not considered in the estimation step, thus providing independent estimates of the predictive capability of the selected models. This technique consists in the removal of one observation from the dataset used, and the estimation of a new regression model with the remaining observations. This new regression model is then used to estimate the stem LA.

Residual errors (RMSE, MAE, and MRE) were also used to estimate the goodness of fit of both external and cross-validation results.

3. Results and Discussion

The comparison of the LA results obtained by the destructive method (portable leaf area meter) and by the newly-developed non-destructive image processing method (Figure 3) showed a strong correlation ($R^2 = 0.95$; $n = 36$) with the slope very close to one (0.995), which was strong evidence for the reliability of the image processing method. For the data set of 36 leaves, the descriptive statistics showed that 53% of cases of LA estimates had between-methods differences below 5%, with an average deviation of 5.7%.

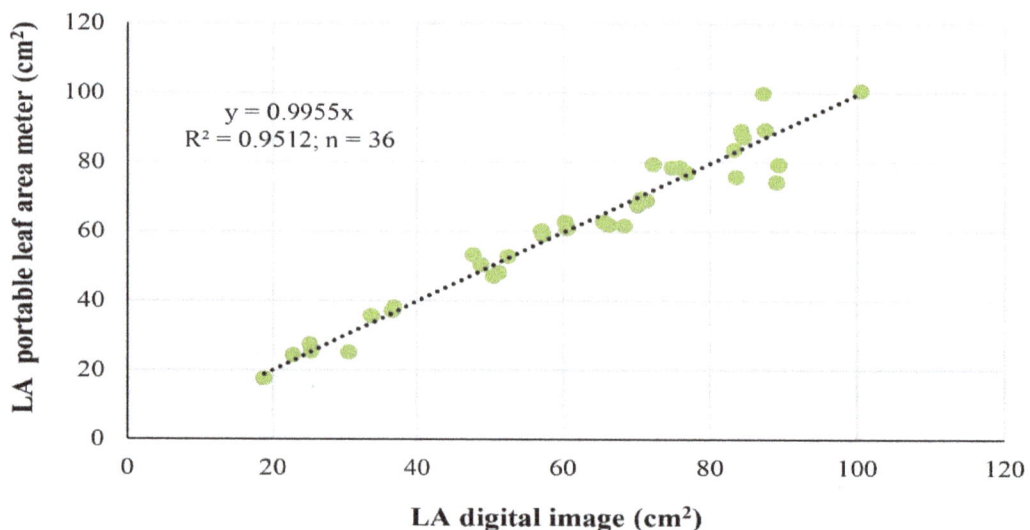

$$y = 0.9955x$$
$$R^2 = 0.9512; n = 36$$

Figure 3. Regression through the origin relating leaf area (LA) obtained by the destructive method (portable leaf area meter, *AM350; BioScientific ADC*) and the non-destructive method (processed digital images).

The results were consistent with those for tomato and corn crops [24], where the accuracy of LA determination methods using digital photographs processed by two different software programs against a standard method were compared. Results using 20 leaves showed no significant differences between methods, with a coefficient of correlation above 99%.

The image processing method used in the present study revealed an accuracy and precision that permits estimating a cut rose crop LA without the need to damage or replace the plants, although when compared to the reference method using a portable area meter, it was more time-consuming. The high correlation achieved supports the use of LA values estimated by the image processing method as a benchmark for the comparison with values estimated by models developed from morphologic and allometric descriptors.

The analysis of the range of values of the variables considered for the LA estimation model showed that a large set of morphological conditions were sampled (Table 2). The values of SH ranged from 11.4 to 68.5 cm, and the values of SLN ranged from 4 to 15 (Table 2). The SH and SLN coefficients of variation were 52% and 28%, respectively (Table 2). There was a marked variability in the selected descriptors, which allowed for the formulation of LA estimates with a wide validation interval.

Table 2. Morphological and allometric descriptors used as a predictor candidate in the leaf area model.

Descriptors	Units	Minimum	Maximum	Average	Coefficient of Variation
Stem height [1] (SH)	cm	11.4	68.5	38.9	52.2
Leaves per stem [2] (SLN)	-	4	15	9	28.2
Flower bud stage (SB)	-	–	–	–	–
Leaf length [3] (LL)	cm	4.7	20.3	13.3	21.3
Leaf width [4] (LW)	cm	3.8	13.1	8.7	22.6
Leaflet number per leaf (LNL)	-	3	7	5.3	24.6

[1] Distance from the base of the stem to the base of the flower bud. In stems without a visible flower bud, stem height was obtained by measuring the distance from the base of the stem to the last open leaf on the stem. [2] Only open leaves with a minimum of three leaflets were considered. [3] Distance between the terminal leaflet and the intersection point of the leaf with the stem. [4] Distance between the extremities of the first pair of leaflets below the terminal leaflet.

In the application of the stepwise regression technique, the variables selected for the LA estimation model were aggregation level SB_A1, SH, and SLN. Aggregation levels SB_A2 and SB_A3, LL, LW, and LNL were not statistically significant by the stepwise regression method. Table 3 shows the estimated model parameters for the selected variables, adjustment and diagnostic tests, and the validation model results.

Table 3. Estimation of model parameter coefficients and measures of model adequacy and validation.

Selected Variables *	Parameters		Adjustment and Diagnostic Tests				Residual Analysis			
	$\beta \pm SE$	Student T **	R^2	SE	T	VIF	Test	Model	Cross Validation	External Validation
Constant	30.16 ± 34.50						n	26	26	3
SH	10.01 ± 0.99	<0.001	0.80	72.6	0.19	5.26	MAE	31.71	38.0	41.33
SB_A1	158.89 ± 32.22	<0.001	0.93	47.6	0.32	3.12	MRE	8.22	9.8	16.12
SLN	11.59 ± 5.39	0.0426	0.95	44.25	0.43	2.34	RMSE	40.20	49.0	50.72

* Selected variables using the stepwise regression method ($p < 0.05$) and the value of tolerance value (T) and variance inflation factor (VIF) were stem height (SH), flower bud stage with aggregation 1 (SB_A1), and leaves per stem (SLN). The variables representing aggregation levels SB_A2 and SB_A3, leaf length, leaf width, and leaflets per leaf were not statistically significant by the stepwise regression method. ** Probability associated with the Student's t-test.

The selection by the model of aggregation level SB_A1 indicated that, in addition to the variables directly related to the stem (height and number of leaves), the phenological stage also influenced stem LA. This could be due to the distribution of assimilates between the vegetative and reproductive parts of the plant. On stems without a terminal flower bud, there will be a distribution of assimilates to other parts of the plant (leaves and roots). This influences the development of the stems and promotes leaf development [25]. According to Matloobi et al. [25], the practice of debudding favors leaf development and subsequently supplies assimilates to other organs, because this cultural practice suppresses terminal flower bud formation, a strong sink for assimilates. Yassin et al. [26] also found that flower bud removal treatments had highly significant ($p < 0.01$) effects on LA of anchote (*Coccinia abyssinica*), a tuberous root crop, increasing its LA by 6.69%.

The final model, based on the variables SH, SB_A1, and SLN, explained 95% ($R^2 = 0.95$, $n = 26$, $p < 0.0001$) of the LA variability for different stages of leaves development (Table 3). VIF values were lower than 10 and T values were higher than 0.1, indicating that there was no collinearity between the selected variables.

According to Suay et al. [19], the estimation of stem LA in cut roses of the cv. "Dallas" can be done using a linear function based only on stem height. The model developed in our study presented a higher R^2 ($R^2 = 0.95$, $n = 26$) than the value obtained by Suay et al. [19], although a lower number of observations were used. These authors also concluded that only 11 stems were needed to estimate the LA of the flowering shoots with an error of less than 5%. Contrary to the Suay et al. [19] study, the model developed in our study considered successive measurements of the LA of all erect stems of one plant through its crop cycle, enabling accommodation of a range of different development stages, which increases its potential applicability.

When the observed stem LA values obtained by the non-destructive image processing method were plotted against the LA values estimated by the model (Figure 4), the slope was very close to one (0.99) and the coefficient of determination was 94%, showing that the model produced stem LA values with high accuracy and precision at different plant development stages. Moreover, the MRE between the observed and estimated LA was 8.2%, corresponding to a MAE of 31.71 cm^2 and RMSE of 40.20 cm^2, thus supporting the good performance of the model.

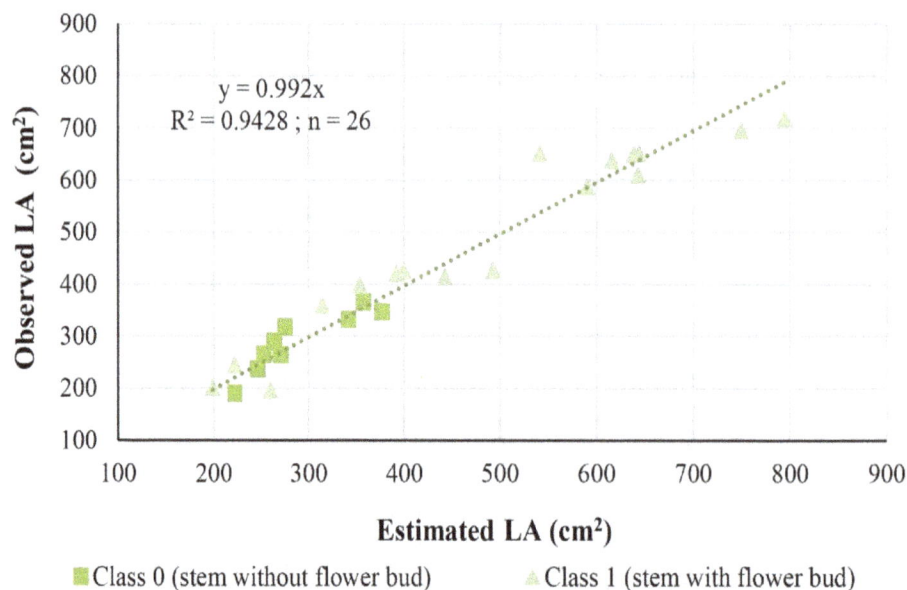

Figure 4. Regression through the origin between observed and modeled (estimated or predicted) stem leaf area and its distribution according to the classes (0—stem without flower bud; 1—stem with flower bud) of the category flower bud stage at aggregation level 1 (SB_A1).

When applying the model to a dataset of three independent observations (external validation), a value of 50.7 cm^2 and 41.3 cm^2 were obtained for RSME and MAE, respectively. Similar values were obtained in the cross-validation with an RSME of 49.0 cm^2 and MAE of 38.0 cm^2. The MRE value of the external validation was 9.8%, lower than the value obtained by the cross-validation (MRE = 16.1%). The MRE values of the calibration process indicated model robustness.

Observed and modeled LA values were compared using the leave-one-out cross validation procedure for 26 plant stems (Figure 5). The results showed a good similarity between the sets of data, which indicate good performance of the predictive model.

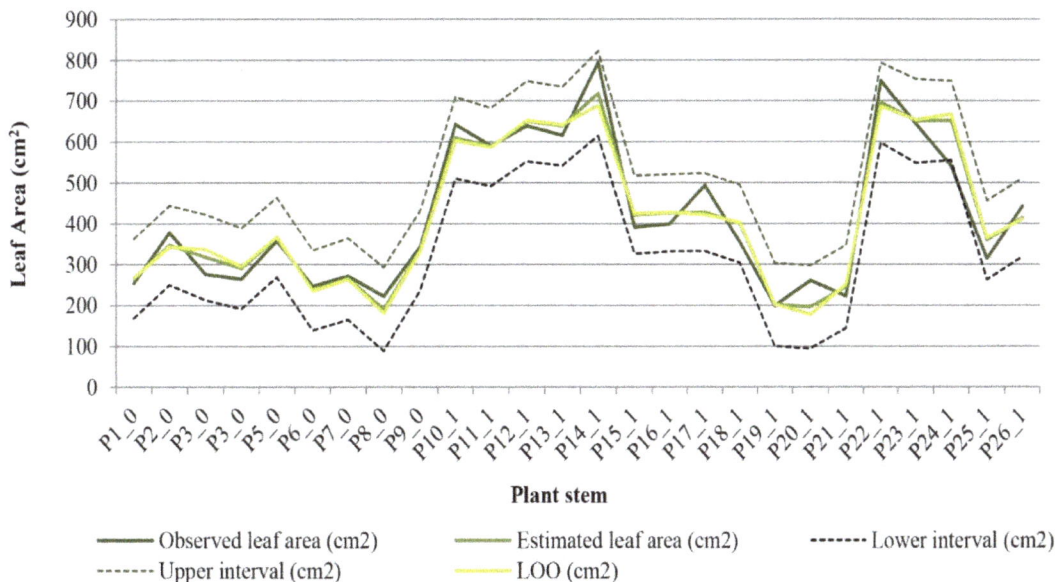

Figure 5. Overall comparison between observed and modeled (estimated and predicted) leaf area for 26 plant stems (Pn_0–Class 0: stem without flower bud; Pn_1-Class 1: stem with flower bud). The predicted was obtained from the leave-one-out cross validation procedure (LOO). Dashed lines are the upper and lower prediction intervals ($\alpha = 95\%$).

Additionally, 80% (cross validation) to 90% (estimation) of the errors between observed and modeled values were less than 15% (Figure 6), all consistently inside the prediction interval ($\alpha = 5\%$).

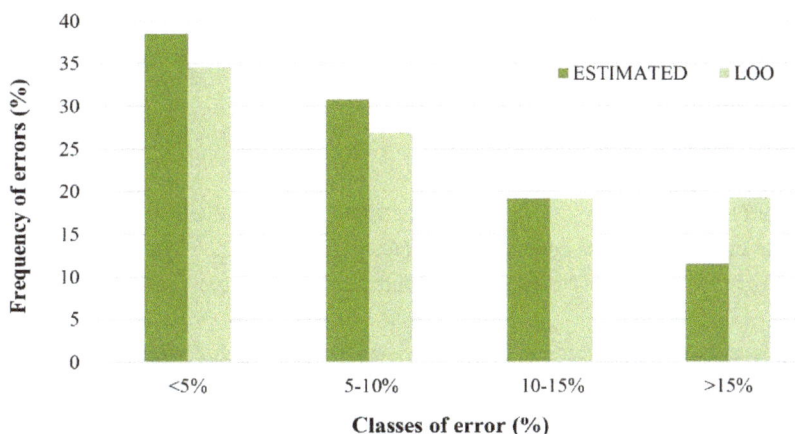

Figure 6. Frequency levels of errors between each pair of observed and modeled (estimated or predicted) leaf areas.

4. Conclusions

In this work, the leaf area of cut "Red Naomi" roses were modeled from simple non-destructive allometric measurements. The LA estimation model based on in situ descriptors such as stem height, number of leaves, and flower bud stage provided accurate and rigorous estimates for this cut rose cultivar. The model presented in this study was unbiased and robust, allowing its use throughout the entire crop cycle instead of being limited to a specific stage of crop development, as has limited most models developed to date. Moreover, it allowed more rapid estimates of cut rose stem LA, potentiating its application in support of a decision-making process. The fact that this method is non-destructive, is user-friendly, and avoids the acquisition of expensive equipment also contributes important advantages when compared with other indirect methods.

A desirable model of non-destructive LA estimation must be able to predict LA independent of genotype and growing conditions. Future studies must be undertaken to test the robustness of the model for other cut rose cultivars and growing conditions.

Acknowledgments: This research was financially supported by the European Regional Development Funds (ERDF), the program COMPETE, national funds from FCT-Foundation for Science and Technology, the Sabbatical Leave Grants provided to Mario Cunha (SFRH/BSAB/113908/2015), and the FCT EXPL/AGR-PRO/1559/2012 project, in particular the fellowship attributed to the first author. The authors thank Susana Carvalho, Sara Gonçalves, and Manuela Pereira for their help. We also thank Floralves Company and Sérgio Alves.

Author Contributions: This study was conceptualized and designed by Mário Cunha, Isabel Pôças, and Ana Patrícia Costa performed the experiment, collected the data, and wrote the manuscript with major assistance and revision from the other authors.

Conflicts of Interest: The authors declare no conflict of interest.

Abbreviations

The following abbreviations are used in this manuscript:

LA	leaf area
LL	leaf length
LNL	leaf number of leaflets
LOO	cross-validation method "leave-one-out"
LW	leaf width
MAE	mean absolute error
MRE	mean relative error
N	number of observations
RMSE	root mean square error
SB	stage of the flower bud
SB_A1	stage flower bud 1
SB_A2	stage flower bud 2
SB_A3	stage flower bud 3
SH	stem height
SLN	stem leaves number
VIF	variance inflation factor
T	tolerance value

References

1. Gao, M.; van Der Heijden, G.; Vos, J.; Eveleens, B.A.; Marcelis, L.F.M. Estimation of leaf area for large scale phenotyping and modeling of rose genotypes. *Sci. Hortic.* **2012**, *138*, 227–234. [CrossRef]

2. Pandey, S.K.; Singh, H. A Simple, Cost-Effective Method for Leaf Area Estimation. *J. Bot.* **2011**, *2011*, 1–6. [CrossRef]

3. Breda, N.J.J. Ground-based measurements of leaf area index: A review of methods, instruments and current controversies. *J. Exp. Bot.* **2003**, *54*, 2403–2417. [CrossRef] [PubMed]

4. Jonckheere, I.; Fleck, S.; Nackaerts, K.; Muys, B.; Coppin, P.; Weiss, M.; Baret, F. Review of methods for in situ leaf area index determination—Part I. Theories, sensors and hemispherical photography. *Agric. For. Meteorol.* **2004**, *121*, 19–35. [CrossRef]

5. Rouphael, Y.; Mouneimne, A.H.; Ismail, A.; Mendoza-De Gyves, E.; Rivera, C.M.; Colla, G. Modeling individual leaf area of rose (*Rosa hybrida* L.) based on leaf length and width measurement. *Photosynthetica* **2010**, *48*, 9–15. [CrossRef]

6. Peksen, E. Non-destructive leaf area estimation model for faba bean (Vicia faba L.). *Sci. Hortic.* **2007**, *113*, 322–328. [CrossRef]

7. Song, Y.; Glasbey, C.A.; Polder, G.; van der Heijden, G.W.A.M. Non-destructive automatic leaf area measurements by combining stereo and time-of-flight images. *IET Comput. Vis.* **2014**, *8*, 391–403. [CrossRef]

8. Fanourakis, D.; Briese, C.; Max, J.F.J.; Kleinen, S.; Putz, A.; Fiorani, F.; Ulbrich, A.; Schurr, U. Rapid determination of leaf area and plant height by using light curtain arrays in four species with contrasting shoot architecture. *Plant Meth.* **2014**, *10*, 1–11. [CrossRef] [PubMed]

9. Leroy, C.; Saint-Andre, L.; Auclair, D. Practical methods for non-destructive measurement of tree leaf area. *Agrofor. Syst.* **2007**, *71*, 99–108. [CrossRef]

10. Blanco, F.F.; Folegatti, M.V. A new method for estimating the leaf area index of cucumber and tomato plants. *Hortic. Bras.* **2003**, *21*, 666–669. [CrossRef]

11. Buttaro, D.; Rouphael, Y.; Rivera, C.M.; Colla, G.; Gonnella, M. Simple and accurate allometric model for leaf area estimation in *Vitis vinifera* L. genotypes. *Photosynthetica* **2015**, *53*, 1–8. [CrossRef]

12. De Swart, E.A.M.; Groenwold, R.; Kanne, H.J.; Stam, P.; Marcelis, L.F.M.; Voorrips, R.E. Non-destructive estimation of leaf area for different plant ages and accessions of *Capsicum annuum* L. *J. Hortic. Sci. Biotechnol.* **2004**, *79*, 764–770. [CrossRef]

13. Bakhshandeh, E.; Kamkar, B.; Tsialtas, J.T. Application of linear models for estimation of leaf area in soybean *Glycine max* (L.) Merr. *Photosynthetica* **2011**, *49*, 405–416. [CrossRef]

14. Cristofori, V.; Rouphael, Y.; Gyves, E.; Bignami, C. A simple model for estimating leaf area of hazelnut from linear measurements. *Sci. Hortic.* **2007**, *113*, 221–225. [CrossRef]

15. Silva, S.H.; Lima, J.D.; Bendini, H.; Nomura, E.S.; Moraes, W. Estimating leaf area in anthurium with regression functions. *Cienc. Rural* **2008**, *38*, 243–246. [CrossRef]

16. Zhang, L.; Liu, X.S. Non-destructive leaf-area estimation for *Bergenia purpurascens* across timberline ecotone, southeast Tibet. *Ann. Bot. Fenn.* **2010**, *47*, 346–352. [CrossRef]

17. Costa, P.M.; Poças, I.; Alves, S.; Pereira, M.; Carvalho, S.M.P.; Cunha, M. Leaf area estimation in different growth stages based on allometric descriptors of cut roses "Red Naomi". In *Actas de Horticultura No. 68*; SECH: Valencia, Spain, 2014; pp. 150–156.

18. Katsoulas, N.; Baille, A.; Kittas, C. Influence of leaf area index on canopy energy partitioning and greenhouse cooling requirements. *Biosyst. Eng.* **2002**, *83*, 349–359. [CrossRef]

19. Suay, R.; Martinez, P.F.; Roca, V.; Martinez, M.; Herrero, J.M.; Ramos, C. Measurement and estimation of transpiration of a soilless rose crop and application to irrigation management. *Acta Hortic.* **2003**, *614*, 625–630. [CrossRef]

20. Heuvelink, E. Evaluation of a dynamic simulation model for tomato crop growth and development. *Ann. Bot.* **1999**, *83*, 413–422. [CrossRef]

21. Rodríguez, E.W.; Flórez, V.L. Phenological behavior of three red rose varieties according to temperature accumulation. *Agron. Colomb.* **2006**, *24*, 247–257.

22. Montgomery, D.; Peck, E.; Geoffrey, V. *Introduction to Linear Regression Analysis*, 5th ed.; Wiley: Adelaid, Australia, 2012.

23. Cunha, M.; Marcal, A.; Silva, L. Very early prediction of wine yield based on satellite data from vegetation. *Int. J. Remote Sens.* **2010**, *31*, 3125–3142. [CrossRef]

24. Rico-Garcia, E.; Hernandez-Hernandez, F.; Soto-Zarazua, G.M.; Herrera-Ruiz, G. Two new methods for the estimation of leaf area using digital photography. *Int. J. Agric. Biol.* **2009**, *11*, 397–400.

25. Matloobi, M.; Baille, A.; Gonzalez-Real, M.M.; Colomer, R.P.G. Effects of sink removal on leaf photosynthetic attributes of rose flower shoots (*Rosa hybrida* L., cv. Dallas). *Sci. Hortic.* **2008**, *118*, 321–327. [CrossRef]

26. Yassin, H.; Mohammed, A.; Fekadu, D.; Hussen, S. Effect of flower bud removal on growth and yield of anchote root (*Coccinia abyssinica* (Lam.) Cogn.) accessions at bishoftu. *Adv. Res. J. Plant Anim. Sci.* **2013**, *1*, 7–13.

15

Analysis of Barriers to Development of Malagasy Horticultural Microenterprises in Madagascar

Tamby Ramanankonenana *, Jules Razafiarijaona, Sylvain Ramananarivo and Romaine Ramananarivo

Agro-Management, Développement Durable et Territoires, Ecole Doctorale Gestion des Ressources Naturelles et Développement, Ecole Supérieure des Sciences Agronomiques, University of Antananarivo, Antananarivo 101, Madagascar; razafiarijaonaj@yahoo.fr (J.R.); sylaramananarivo@yahoo.fr (S.R.); s.ramananarivo@esmia-mada.com (R.R.)
* Correspondence: rtambymsn@gmail.com

Abstract: The malagasy rural environment and the development of agricultural microenterprises are closely linked. For Ambalavao Atsimondrano located in the suburban area of Antananarivo, Madagascar, the horticultural chain appears as a buoyant sector; many rural or urban households are dependent upon it. Despite the reputation of the region in this field and support from different organizations, Rural Microenterprises (RMEs) encounter problems in their development. This study highlights the factors blocking entrepreneurial development including education of horticultural entrepreneurs and their ability to deal with complex situations. The aim of the study was to identify the factors affecting the growth of RMEs. A focus group on 33 small farmers considered as RMEs was conducted taking into account their individual characteristics. A typology and value chain analysis resulted in their classification and in comprehension of their empowerment in management. The results indicated that there were 3 types of entrepreneurs or promoters: the experienced traditional (36%), the educated young (33%), and the professional young (31%). Ishikawa diagrams highlight the problems related to entrepreneurial development in funding and information systems. Our conclusions insist on the necessity of improving communications strategies among microentrepreneurs, guidance for entering the market, and professionalizing the horticultural trade, while emphasizing the importance of cooperation between producers.

Keywords: horticulture; entrepreneurship initiative; entrepreneur's education; cluster analysis

1. Introduction

In Madacasgar, the primary sector of 2.4 million agricultural households indicates 300,000 rural young farmers joining the labour market each year [1]. While the illiteracy rate has been less than 20% in urban areas in the last 15 years, it is 36% in rural populations and only 10% of them continue beyond elementary school [2]. In the study area located in the rural municipality of Ambalavao, District of Atsimondrano, Region Analamanga, Madacasgar, a multitude of Rural Microenterprises (RMEs) created by smallholder farmers operate in all rural activities. Youths and adults are looking for livelihoods. Thus, they are forced to be entrepreneurs, but the majority have not been educated to be successful.

Despite the large workforce of the district, 372,077 inhabitants in 2012, it has failed to increase its agricultural productivity. Although farmers have a huge agricultural and craftsmanship potential, they are slow to make progress [3]. Facing this difficulty, the aim of this research was to identify and explain the factors influencing performance, and the underperformance, of RMEs. The specific goals were: (i) to classify the different characteristics of horticultural microentrepreneurs; (ii) to evaluate RMEs' management levels; and (iii) to define influences of organizations and institutions in the horticultural fields.

2. Experimental Section

2.1. Materials

2.1.1. Site

The municipalities of Ambalavao, Ankadinandriana, and Anjeva are located in the southern and in the eastern parts, respectively, of the suburban area of Antananarivo, Madacasgar. They are among Antananarivo's suppliers (Figure 1), but the survey was focused on Ambalavao.

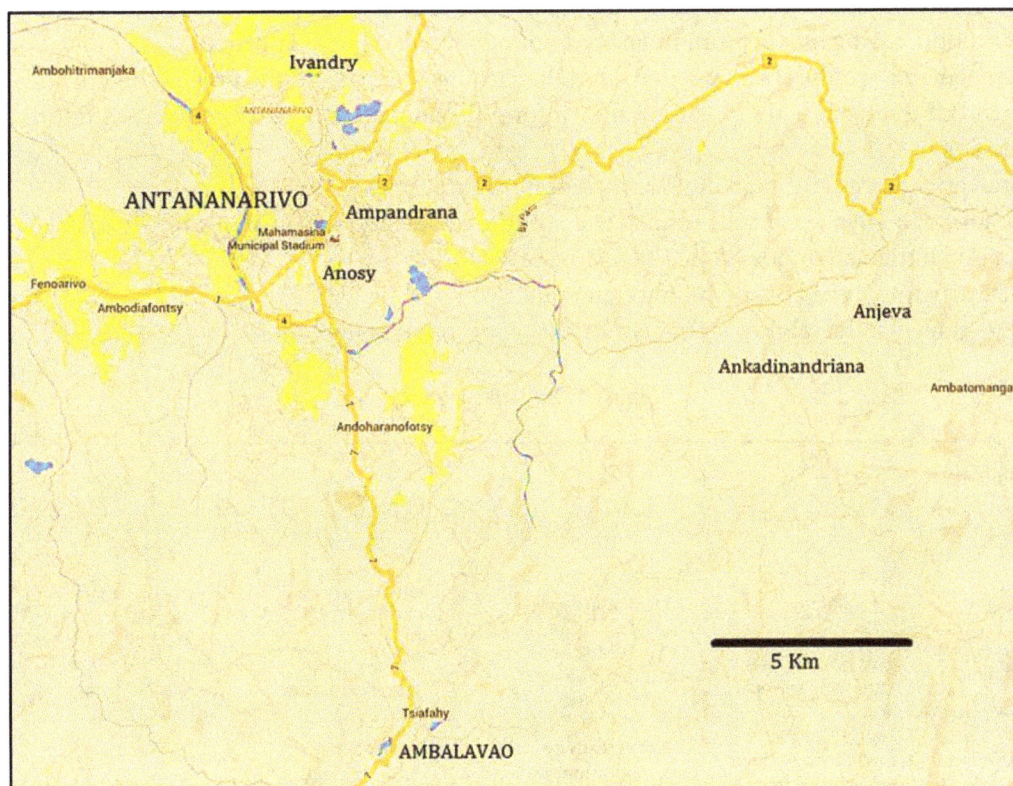

Figure 1. Map of Antananarivo suburban area.

Ambalavao has 2540 ha of land suitable for cultivation, but only 57% is exploited [4]. The RMEs in this area are specialized in raphia weaving, and among horticultural crops are mainly involved in the production of pineapple, strawberry, and ornamental plants and nurseries [3].

2.1.2. Data Collection

Data were collected in two steps: (i) in order to get an overview of RMEs, some discussions were held with PROSPERER s (PROSPERER (Programme de Soutien aux Pôles des Microentreprises Rurales et aux économies Régionales) is a program funded by IFAD in 2008 for a 7 years duration) agents and senior managers, and with the municipality's mayor; and then, (ii) data from a focus group of 33 horticultural farmers spread across 3 points of sale alongside the 7th National Road were collected. Most of analyses were based on qualitative data.

2.2. Methods

2.2.1. Microentrepreneur Characterization

The collected data focused on education, socioeconomic surroundings, experience, innovation, creativity, and entrepreneurial culture of the surveyed. Data analysis and their processing occurred in 3 steps: (i) a classification for identifying types of farmers by Agglomerative Hierarchical Clustering (AHC); (ii) reduction of the variables by Discriminant Analysis (DA) to confirm the results of Agglomerative Hierarchical Clustering (AHC); and (iii) identification of the characteristics of each class by Multiple Correspondence Analysis (MCA).

For step 1, a scoring model was necessary in order to codify the results. Selected variables and their importance on a 1 to 5 scale were used (Table 1), and the variables were codified (Table 2). The higher the score of a variable, the more significant was the variable. The variables studied were:

Duration of practice: the years increase in intervals of five.

Education: taking into account illiteracy until or beyond secondary school.

Management: considers on the one hand the farming system and the time involved, and on the other hand, if the main activity is formal or informal; for example, part-time versus informal full time farming.

Innovation: creativity, marketing and communication strategies.

Finance: if a member of a microfinance institution (MFI) and the level of capitalization.

Supply: if difficult, moderately difficult, or easy.

Infrastructure: based on the selling point quality.

Market level vulnerability to economic crisis; may be high, medium or low.

Table 1. Variable scoring.

Variables	Score				
	1	2	3	4	5
Duration of practice (years)	5 to 9	10 to 14	15 to 19	20 to 24	>25
Education	illiterate	Elementary school	Elementary school level 2	Elementary school level 3	more
Management	part-time farming/ informal/seasonal	part-time farming/informal/ full year	Main activity Informal/full year	Main activity/ formal/full year	
Innovation	Tough competition/ traditional/without market strategy	Average creativity skills/ communication strategy	Good packaging/ niche strategy		
Finance	MFI aversion/ Low capitalisation	MFI aversion/ medium capitalization	Possible member/ medium capitalization	MFI member/ large capitalization	
Supply	Hard	Medium	Easy		
Infrastructure	Market stall	Wooden house	Market pavilion		
Market level vulnerability to economic crisis	High vulnerability	Medium vulnerability	Low vulnerability		

Table 2. Variable codification.

Variables	Score				
	1	2	3	4	5
Duration of practice (years)	year-1	year-2	year-3	year-4	year-5
Education	education-1	education-2	education-3	education-4	education-5
Management	management-1	management-2	management-3	management-4	
Innovation	innovation-1	innovation-2	innovation-3		
Finance	finance-1	finance-2	finance-3	finance-4	
Supply	supply-1	supply-2	supply-3		
Infrastructure	infrastructure-1	infrastructure-2	infrastructure-3		
Market level vulnerability to economic crisis	market-1	market-2	market-3		

2.2.2. Assessment of the Importance of Management of RMEs

The goal was to prioritize the outlines of the blocking system. Then, an Ishikawa diagram was created by brainstorming with PROSPERER's agents in the *Atsimondrano* and *Avaradrano* districts. Generally, in agricultural fields, causes are categorized into 5 families (5M Methods), but the fish bone diagram adopted considers 6M Methods (Material, Machine, Method, Measurement, Man Power, and Environment for *Milieu*). Respectively, they are presented as follow:

1. Material: consumables, raw materials, documents and information, etc.
2. Machine: requires investment such as equipment, telephones, etc.
3. Method: organization, working methods, procedures, system and operating methods, law, regulations, specifications, etc.
4. Milieu (environment): work environment, working conditions, light, noise, temperature, pathogens, etc.
5. Man power: human activity, behavior, skills, work habits, management, staff, experience, etc.
6. Measurement (financial means): budget and price.

The resulting diagram shows a problem tree of the development of microenterprises.

2.2.3. The Influence of Organizations and A Horticultural RMEs' Level

According to the International Labor Organization's approach [5], the research focused on the different institutions belonging to the production chain. In addition, the analysis of the value chain allowed an understanding of the connection between horticultural producers as well as the opportunities and constraints that result from such connection. A map of the value chain was built. In each step of the chain, the support organizations were cited.

2.3. Data Analysis

Data were analyzed using XLSTATTM 2008 software manufactured by Addinsoft.

3. Results and Discussion

3.1. Individual Characteristics of Horticultural Entrepreneurs

3.1.1. Characteristics of Each Class

AHC led to three types of entrepreneurs, reclassified with discriminant analysis (Figure 2).

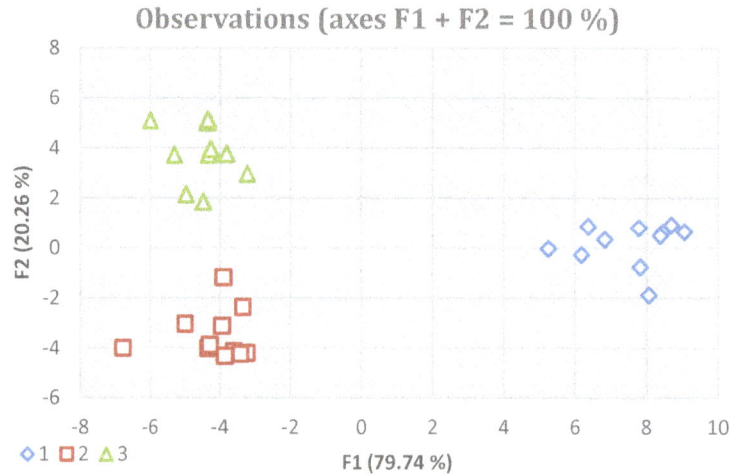

Figure 2. Entrepreneur/promoter classification based on discriminant analysis: Types 1, 2, and 3.

About 79.74% of the variability was explained by axis F1, while 20% of the variability was explained by axis F2. The promoters characteristics included the experienced traditional (Type 1; 36%), the educated young (Type 2; 37%), and the young professional promoters (Type 3; 31%) (Table 3).

Table 3. Entrepeneurs/promoters characteristics.

Type	1	2	3
Name	Experienced Traditional Promoter	Educated young Promoter	Professional young Promoter
Workforce (effectif)	12	11	10
Proportion (%)	36	33	31
Experience	High	Medium	High
Education	Elementary school	Elementary school level 2	Elementary school
Management	Informal and main activity	Member of cooperatives and main activity	Informal and part-time farming
Innovation	Medium	Medium	Low
Finance	Not member of microfinance with low capitalization	Member of microfinance and medium capitalization	Not member of microfinance with low capitalization
Supply	Very easy	Medium	Easy
Infrastructure	Depends on program supports	Depends on program supports	Depends on program supports
Market	Medium vulnerability to socio-economic crisis		Low vulnerability to socio-economic crisis

Type 1: Experienced traditional promoters were characterized by a long period of practice (>20 years); they were represented by year-4 and year-5. The individuals of this type had a good strategy (supply-3) and a solid resilience facing possible crises (market-3). However, these individuals were poorly educated (education-1) with low innovation skills (Innovation-1). They had a microfinance aversion (finance-1), and they operated in informal situations (management-1).

Type 2: Educated young promoters were the opposite of the Type 1. They were characterized by a relatively high level of education (education-4 and education-5). They had high innovation and management ability, represented, respectively, by (innovation-3) and (management-3). They were members of microfinance enterprises (finance-3) and were quite sensitive to market fluctuations (market-1) (Figure 3). Their lack of experience (<10 years) (year-1) was noted.

Type 3: Professional young entrepreneurs were an intermediate class between the two previous types (Figure 3). Nevertheless, they were much closer to Type 1 by their low innovation capacity (innovation-2) linked with their low education (education-2). They had strong reinvestment ability. They were not members of rural organizations (management-2). However, they were members of microfinance enterprises (finance-4). Their feature in common with Type-2 was the short duration of activities.

The 3 types of promoters obtained from the Multiple Correspondence Analysis graphic were considered as supplementary variables and were framed. The remaining variables were their characteristics. If they were closed, the promoters' features were identified by associating them with these remaining variables (Figure 3).

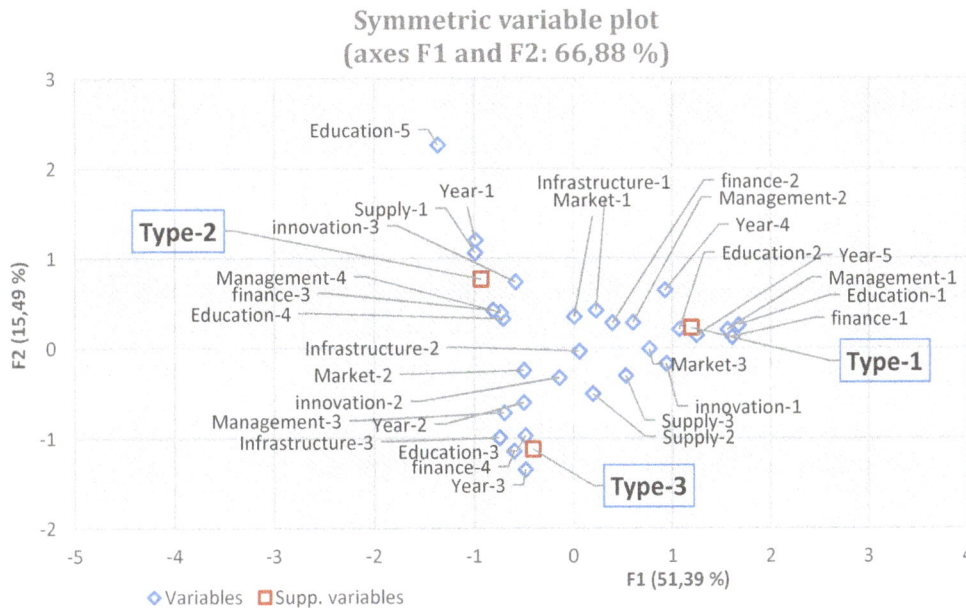

Figure 3. RMEs' features by multiple correspondence analysis.

3.1.2. RMEs' Analysis Axes

The analysis of each type of promoter was performed on the basis of education, experience, innovation and creativity (Table 2).

Education: Education is the basis of sustainable development [6]. It is an essential tool for evolving values and attitudes, abilities, behaviors and life styles. Type 2 was the most receptive; they can easily get a strong entrepreneurial culture.

Experience: Work experiences can be earned as a salaried employee or as a manager. Moreover, theoretical conceptions distinguish vocational training from experience which is more important; it makes managers at ease in task accomplishment [7]. Type 1 is the most experienced promoter; their experience helps them identify and assess the importance of their decision criteria [8]. So, their experience explains their solid resilience facing market crises.

Innovation and creativity: Creativity results from the application of new ideas linked to innovation [9]. The young educated (type 2) and the young vocational are characterized by higher innovation and creative abilities.

Entrepreneurial culture: FORTIN [10] asserts that entrepreneurial culture is the antidote against poverty. Here, all promoters have a desire to start a business. Though the Type 2 and the Type 3 as members in Microfinance Institutions (MFIs) have more opportunities, the Type 2 is the most able to expand their activities by exploiting the networks of farmers' organizations where they work.

3.2. The Horticultural Microenterprise Management

The aforementioned results hinder the development of RMEs. The Ishikawa diagram outlines the factors blocking RMEs (Figure 4), and Figure 5 shows the assessed elements for each variable.

1. Material. Among the identified materials, packaging remains the main issue of RMEs (51%).
2. Machine. In order to add value to their product, microenterprises face constraints concerning the organization of locales (46%) and the creation of infrastructures (39%). During the brainstorming, the access to water was also raised as a major issue (15%).
3. Methods. Microenterprises fail to assure their visibility. Moreover, the Government has yet to establish a RME policy.
4. Milieu (environment): Seasonality may have an impact on supply. Farmer organizations and market operators are competing.
5. Manpower. The low education level of the entrepreneurs is negatively impacting their entrepreneurial culture and their management skills. In addition, some weaknesses include the supervision, innovation, and creativity.
6. Measurement (financial means). They include external and internal financial difficulties. For external difficulties, the majority of microenterprises cannot support the high interest rates of MFI, and there are delays in the release of funds. Internal difficulties are the culture of savings leading to the lack of reinvestment.

Considering these variables, the problems cited by the RMEs were mainly financial means (82%), materials (76%), Method (18%), and Environment (24%) (Figure 6).

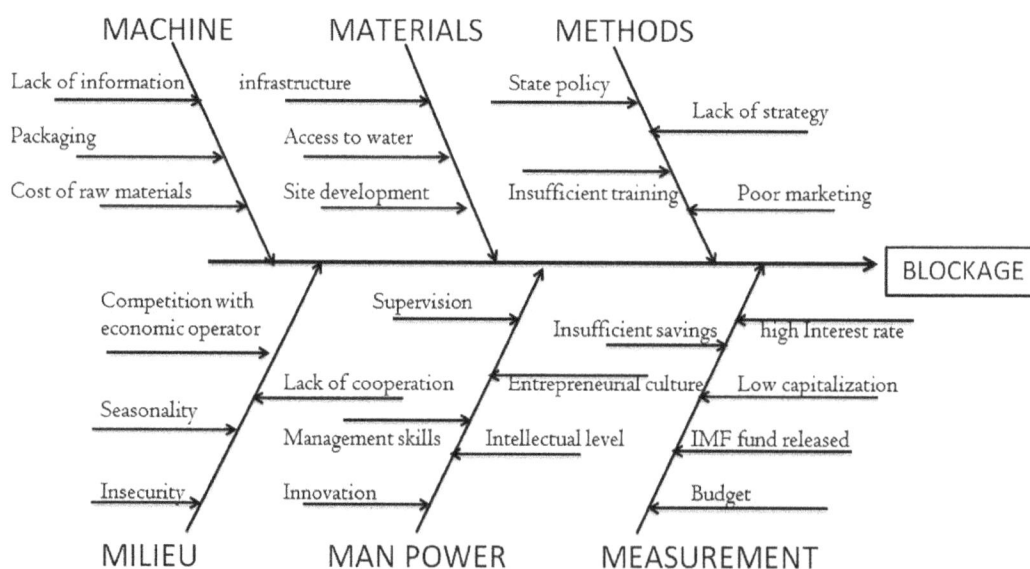

Figure 4. Ishikawa diagram.

Each main problem was comprised of associated problems. The whole constituted the blocking factor in the development of RMEs (Figure 5). As a synthesis, the RMEs' main problems were the Machine and the Measurement. So, they are common problems of the horticulturalists. Milieu and Methods seem to be tolerable in comparison to the others (Figure 6).

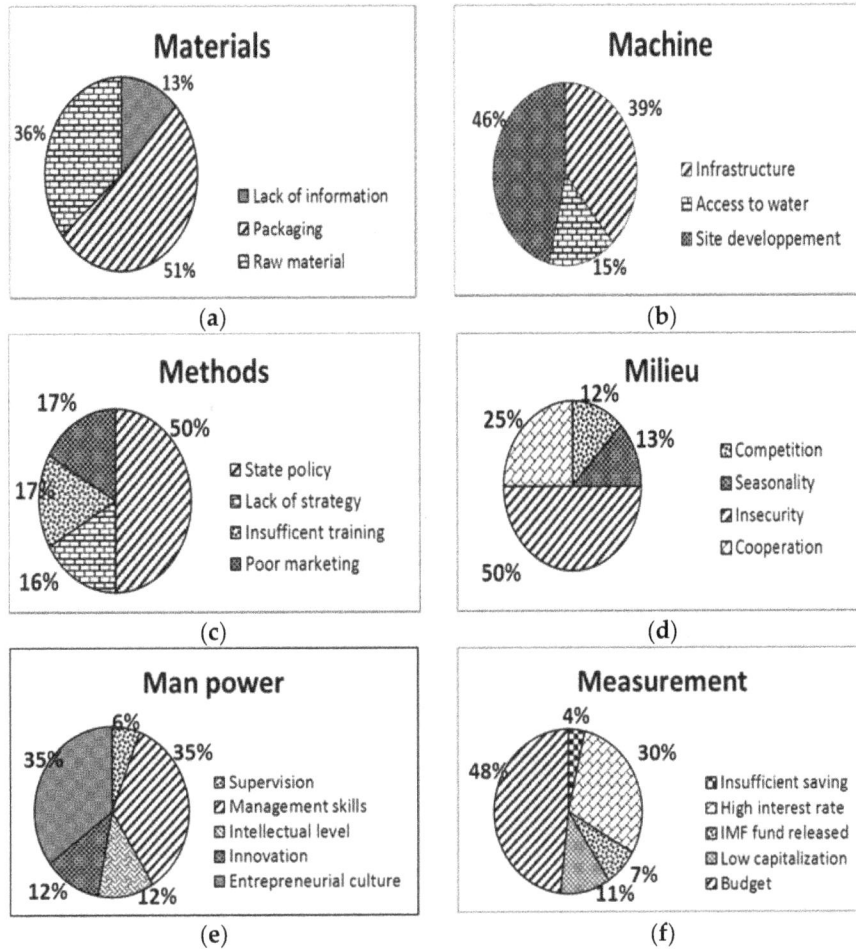

Figure 5. Detailed problems of RMEs.

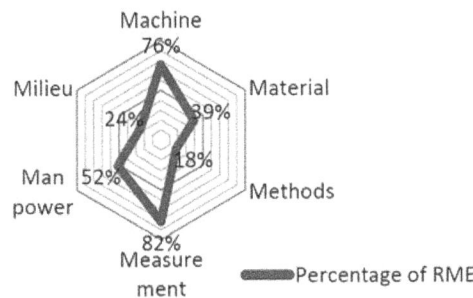

Figure 6. The primary problems for RMEs.

3.3. Institutions and Organism for Horticultural Sector Development

3.3.1. Value Chain

The public and private institutions considered as facilitators shared the same goal: develop the horticultural sector through support services (Figure 7). The facilitators were PROSPERER, MFI, Enterprise Advisor (CE), Municipality, Chamber of Commerce, Industry of Antananarivo (CCIA), and Telecom Malagasy (TELMA). PROSPERER supports microenterprises by providing infrastucture and facilitates the creation of agricultural cooperatives, by access to MFIs for RMEs to resolve gaps in cash flow and investment problems. It closely works with CCIA and CE as their interface for their know-how. CCIA is hosting PROSPERER, and CEs are service providers.

1. Supply. The seed suppliers sell inputs to the farmers. In parallel, horticulturalists may be self-sufficient. The facilitators are the MFI and CE which are PROSPERER's partners. MFIs contribute to the financial support of RMEs including the purchase of inputs, tools and related materials. The CE supports the choice of strategies and methods for approaching suppliers.
2. Production. Production mainly concerns RMEs, cooperatives and individual farmers. There are horticultural fields and seedling nurseries.
3. Processing. One part of the raw plant material may be sold directly by retailers or dealers. The remainder is processed for packaging.
4. Trading. This is the final step of the value chain. Market operators and retailers are those who work the most in the marketing of flowers and seedlings. They are closely linked to RMEs which are their suppliers. The products are intended for the urban market.

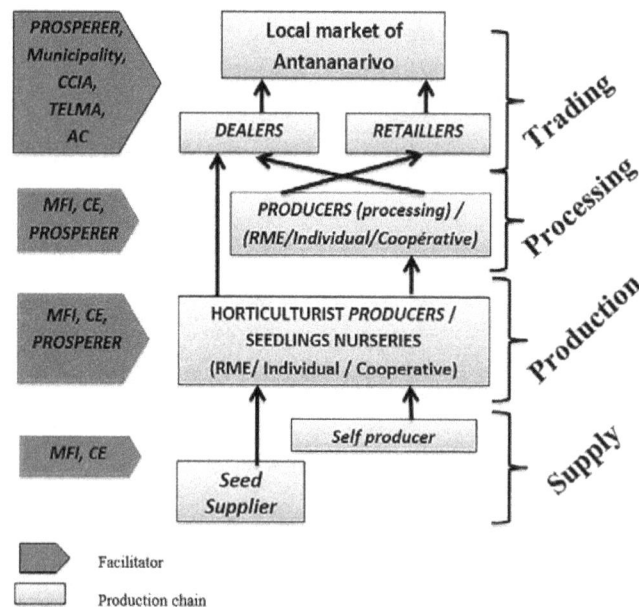

Figure 7. Value chain process.

3.3.2. Flow Map of Major Floral Crops

Figure 8 illustrates the flows of 3 types of flowers on the market in Antananarivo center and in the surroundings. Most flowers are intended for decoration rather than for crops or gardening. Farmers sell a variety of horticultural products, but *Pinus pinaster*, *Clarkia amoena*, and *Dianthus caryophyllus* are the flagship products. They are usually sold in bunches, reflecting consumer choice. The most productive period is between June and September. Daily average sales reach 4 bouquets per operator.

For *Pinus pinaster*, the majority of production comes from the rural municipality of Ambalavao and Tsiafahy. Indeed, its sales period peaked during reforestation campaigns between January and March. At a unit price of 400 ariary, a single point of sale in these 2 municipalities can provide 10,000 plants ordered a day in advance. Consumers make their purchases directly from producers in the municipality or from dealers.

For *Clarkia amoena*, production comes from the Anjeva municipality. This kind of flower were sold by bunch and one of them costs 1000 ariary. Producers make a wholesale daily delivery in the capital around 3 a.m. in the neighborhood of Anosibe, where traders from Anosy, Ivandry and Ampandrana, the three major selling points in the city of Antananarivo, buy their supply. They account for 50, 20, and 10 operators, respectively.

For *Dianthus caryophyllus*, the priamry production comes from Ankadinandriana municipality. A bouquet of *Dianthus caryophyllus* costs 2000 ariary.

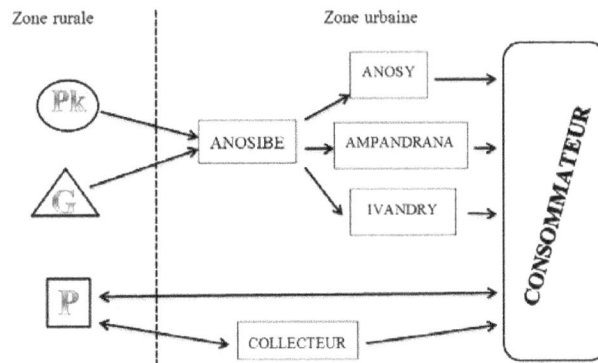

Figure 8. Flow map of floral crops in Antananarivo center and in the surroundings.

4. Conclusions

Rural microentreprises should put the focus on the restructuring of the trade system and should improve cooperation between producers. Future prospects must strengthen the profession. Pouillaude's conclusion [11], that an interactive policy of the government for the promotion of RMEs, is mandatory, the economic and social environment having a significant impact throughout the life of the RME.

Author Contributions: This work was a product of the combined effort of all authors. Tamby Misaina Ramanankonenana performed the experiments, gathered and analyzed the data, and wrote the manuscript. Romaine Ramananarivo, Sylvain Ramananarivo and Jules Razafiarijaona revised and improved the manuscript.

Conflicts of Interest: The authors declare no conflict of interest.

References

1. Plan d'Action pour le Développement Rural (PADR). *Cadre Référentiel et Processus pour le Développement Rural Durable à Madagascar*; NEPAD Planning and Coordinating Agency: Cotonou, Benin, 2013.

2. UNICEF. *Exclusion Scolaire et Moyen D'inclusion au Cycle Primaire à Madagascar*; Rohen D'aiglepierre en Collaboration Avec L'équipe de Focus Development Association: Antananarivo, Madagascar, 2012; p. 156.

3. Programme de Soutien aux Poles des Microentreprises Rurales et aux Economies Régionales (PROSPERER). *Rapport de Préévaluation*; Fonds International pour le Développement Agricole: Antananarivo, Madagascar, 2008.

4. Plan Communal de Développement (PCD). *Commune Rurale d'Ambalavao Atsimondrano*; Ministère de l'Intérieur: Antananarivo, Madagascar, 2010; p. 45.

5. ILO. Avaliable online: http://www.ilo.org/valuechains (accessed on 16 May 2014).

6. UNESCO. Available online: www.unesco.org. (accessed on 15 June 2014).

7. Duchéneault, B. *Les Dirigeants de PME: Enquête, Chiffres, Analyses pour Mieux les Connaître*; Maxima: Paris, France, 1996; p. 345.

8. Rabenilaina, H. Le Contrôle de Gestion, Système d'aide à la Décision Dans Une Organisation: Cas des Petites et Moyennes Entreprises Malgaches des Deux Districts d'Antsirabe. Ph.D. Thesis, en Science de Gestion, Faculté de Droit Economie Gestion et Sociologie, University of Antananarivo, Antananarivo, Madagascar, 2012.

9. Meier, O. *Le Dico du Manager*; Dunod: Paris, France, 2008; p. 228.

10. Fortin. *La Culture Entrepreneuriale un Antidote à la Pauvreté*; Collection Entreprendre: Paris, France, 2002.

11. Pouillaude, A. *Gouvernance et Développement des Micro-Entreprises: Approche Conceptuelle et Méthodologique*; Groupe d'Economie du Développement, Université Montesquieu Bordeaux IV: Bordeaux, France, 1998; p. 11.

16

An Overview of Fertilization and Irrigation Management in the Conventional and Certified Organic Production of Vegetable Crops in Florida

Eric Simonne [1,*] and Robert Hochmuth [2]

[1] Horticultural Sciences Department, Institute of Food and Agricultural Sciences, University of Florida, Gainesville, FL 32611, USA

[2] Suwannee Valley Agricultural Extension Center, Florida Cooperative Extension Service, Live Oak, FL 32060, USA; bobhoch@ufl.edu

[*] Correspondence: esimonne@ufl.edu

Abstract: The postharvest quality of vegetable crops from conventional and organic production systems depends on pre-harvest factors such as variety genetic potential, fertilization, and irrigation. The five principles of plant nutrition (plants absorb ions, not fertilizers; Leibeig's law of the minimum; nutrient application requires a source, a rate, a placement and a time of application; no correlation exists between total nutrient presence in the soil and availability; and plant nutrient concentration and yield are related) must be followed throughout the crop growth cycle. In certified organic production in the United States, cover crops, manure and composts may be used together with Organic Materials Review Institute–approved fertilizer products. A fertilization program usually includes (1) soil sampling and understanding the recommendation; (2) adjusting pH if necessary; (3) applying preplant fertilizer and developing a schedule for sidedressing or fertigation; (4) using foliar fertilization; (5) monitoring plant nutrient status; and (5) keeping fertilization records. The components of an irrigation schedule are (1) determining a target irrigation volume based on reference evapotranspiration and crop age; (2) adjusting this amount based on soil moisture content; (3) determining the contribution of rainfall; (4) developing a rule for splitting irrigation; and (5) keeping irrigation records. A poorly designed irrigation program can negate the benefits of a sound fertilization program. Challenges encountered in conventional and organic production include predicting nutrient release rates from organic materials, supplying enough N throughout the cropping season, identifying rescue strategies, keeping production costs low, and meeting the additional legal requirements of the food safety and best management practices programs.

Keywords: manure; compost; cover crop; soil testing

1. Introduction

Vegetable crop production today is broadly divided into conventional, sustainable, and organic methods. From an inputs standpoint, "conventional vegetable crop production" usually refers to methods of farming in which the use of synthetic fertilizers, pesticides and herbicides and genetically modified organisms is allowed. "Sustainable" does not have a standard definition, whereas "organic" refers to a system of farming that does not use synthetic chemicals and, instead, mimics natural systems. This may encompass different farm sizes, practices and philosophies that, at their core, reject the use of most synthetic chemicals.

In the United States, both conventional and organic crop production are regulated. The Federal Insecticide, Fungicide, and Rodenticide Act (FIFRA; U.S. Public Law 92-516, 21 October 1972) and the Clean Water Act (U.S. Public Law 95-217, 27 December 1977) widely regulate the agricultural use of pesticides and fertilizers, respectively. The U.S. organic standards, organic certification, and accreditation, compliance and enforcement are governed by the National Organic Program [1]. The National List of Allowed and Prohibited Substances identifies substances that may or may not be used in certified organic crop production [1].

Vegetable quality at the consumer level depends on four main factors: the genetic potential of the variety used, field production practices used (pre-harvest factors), produce physiological stage at time of harvest, and postharvest handling. This review focuses on water and nutrient management before harvest and covers (1) the principles of plant nutrition; (2) developing irrigation schedules and (3) fertility plans; (4) how to correctly make simple calculations; (5) practices, challenges and successes in the implementation of these plans; and (6) some consequences of water and nutrient deficiencies on vegetable postharvest behavior.

2. The Universal Principles of Plant Nutrition

While the methods and tools to manage fertilization and irrigation are different among all three production systems, the fundamental chemical concepts that drive plant nutrition in the soils and in the plant are universal: pH, oxido-reduction, and solubility [2]. The five fundamental principles of plant nutrition also apply equally in all production systems:

Principle 1. Plants absorb ions, not fertilizers. Essential-element containing ions (and their chemical symbols) commonly absorbed by plants are nitrate (NO_3^-), phosphate ($H_2PO_4^-$ and/or HPO_4^{-2}), sulfate (SO_4^{-2}), borate ($H_2BO_3^-$), chloride (Cl^-), molybdate (MoO_4^-), ammonium (NH_4^+), potassium (K^+), calcium (Ca^{+2}), magnesium (Mg^{+2}), copper II (Cu^{+2}), iron II (Fe^{+2}), manganese II (Mn^{+2}), and zinc II (Zn^{+2}). Chemical forms of the essential elements that are poorly taken up, toxic in small quantities or not physiologically active include nitrite (NO_2^-), nitrogen gas (N_2), phosphite ($H_2PO_3^-$), various oxides and dioxides (CaO, MgO, P_2O_5, K_2O, SO_2), ammonia gas (NH_3), copper I (Cu^+), iron III (Fe^{+3}), manganese I (Mn^+), and zinc I (Zn^+) [3]. Despite some incorrect claims, phosphites, which are registered as the active ingredient of a group of fungicides, are not a source of readily available P as a plant nutrient [4].

Principle 2. The law of limiting factors. The application of Leibeig's law of the minimum to plant nutrition states that crop growth and yield are limited by the essential element in shortest supply. Adding more of any other element (that is therefore not limiting) will not result in a plant response.

Principle 3. The four Rs of nutrient application. Fertilizer recommendations may result in correct nutrient application only when they include a Right fertilizer source, a Right rate (based on field surface or length of rows fertilized), a Right placement or application method (broadcast, modified-broadcast, banded, or injected) and a Right time of application (based on days after emergence or transplanting, or preferably based on the crop growth stage). Hence, the popular discussion of total seasonal rates alone is insufficient to implement a fertility plan.

Principle 4. No correlation exists between total nutrient amount in the soil and availability for uptake by the plant. Soil pH (that controls the oxidation state, dominant chemical species, and possibly favoring the release of aluminum [2]), cross-precipitation (possibly Ca with P or Ca with S), competition effects (Na with K, P with Al, or Fe with Mn, for example), and constituents of soil organic matter (mostly for N, P and S) may result in reduced availability. The important soil nutrient fraction is the extractable fraction, which is the one recovered by soil extractants during soil testing [5,6]. Principle 5. Leaf nutrient concentration and crop growth/yield are related. Insufficient nutrient supply and concentration in plants may result in deficiency symptoms. Several generic keys are available online to identify nutrient deficiencies in vegetable crops [7–9].

Visual symptoms need to be confirmed with a diagnostic of the crop nutritional status by using tissue analysis [10] or petiole sap testing [11]. With both methods, plant nutrient (content in % or mg/kg (ppm) on a dry weight basis for tissue analysis or in mg/L (ppm) of sap for petiole sap testing) is compared to established sufficiency thresholds. These diagnostic methods are the basis for supplemental fertilizer applications in the Florida Best Management Practices program for vegetable crops [12].

3. Irrigation Scheduling

Most vegetable crops are irrigated with seepage irrigation, overhead irrigation or drip irrigation [13,14]. Scheduling irrigation involves determining when to irrigate and how much to apply. For all irrigation methods, the steps of an irrigation schedule are: (1) determine a target irrigation volume based on reference evapotranspiration (ETo) or Class A Pan evaporation (Ep) and crop age; (2) adjust this amount based on soil moisture content; (3) determine the contribution of rainfall; (4) develop a rule for splitting irrigation; and (5) keep irrigation records [15–17].

Poorly designed irrigation systems, insufficient irrigation system maintenance plans or poor irrigation scheduling practices may negate the benefits of a sound fertility plan. Poorly designed irrigation systems may break, fail, or operate at insufficient pressure. Poorly maintained irrigation systems will result in non-uniform water applications. Excessive water (from irrigation or rainfall) may move soluble nutrients below the root zone, especially in course-textured soils [18].

4. Fertility Plans

The crop nutritional requirement (CNR) is the total amount of nutrients needed to produce a crop. The CNR is supplied by the soil and by fertilization [19]. Hence, the goal of a fertility plan is to ensure that the supply of essential elements does not limit productivity. A fertility plan starts with soil testing and determining the potential contribution of the soil to supplying nutrient needs [20,21]. The difference is typically supplied by combinations of cover crop residues, compost, animal manures, organic amendments, and fertilizers. In certified organic production, cover crops, manure and composts are used together with Organic Materials Review Institute—approved fertilizer products [22].

The components of a fertility plan are: (1) soil test, understand the recommendation and make the correct calculations; (2) lime if necessary; (3) apply organic amendments (cover crop, compost, or manure); (4) incorporate the preplant fertilizer, then sidedress or develop a daily/weekly fertigation schedule, adjusting the amount to the crop growth stage; (5) use foliar fertilization (this practice is recommended for the application of micronutrients to high-pH soils when needed); (6) assess the efficacy of the fertilizer program through leaf sampling or petiole sap analysis; (7) trap residual nutrients at the end of the season with a cover crop; and (8) record the date of application, material, placement and source of all fertilizer used. For practical purposes, the contribution of each nutrient source in the fertility plan may be tabulated to avoid insufficient or excessive nutrient applications (Table 1).

Table 1. Generic blank fertilization worksheet for vegetable crop production.

Nutrient Source [z]	Material (kg/HA [y])	N (kg/HA)	P_2O_5 (kg/HA)	K_2O (kg/HA)
Soil test recommendation	–x–			
Compost				
Cover crop				
Manure				
Soil organic matter decomposition				
Fertilizer needs [x]				
Preplant application				
Sidedress				
Fertigation				
Foliar fertilization [w]				

[z] The nutrient contribution of compost, cover crops and manures depends on the total nutrient content of the material, the method of application (banded vs. broadcast) and the mineralization rate. [y] 1 HA = number of linear meters of bed (or row) in one planted hectare at standard bed (or row) spacing. [x] Fertilizer needs = soil test recommendation – compost – cover crop – manure – soil organic decomposition; Fertilizer needs = Preplant application + Sidedress or Fertigation + Foliar fertilization. [w] Recommended to correct micronutrient deficiencies in high-pH soils only. Macronutrients may be injected in the irrigation water when overhead systems are used.

5. Correct and Simple Fertilizer Calculations

All the benefits of soil testing, plant analysis and other cultural practices may be negated by the incorrect calculation of nutrient rates and ensuing fertilizer application rates. Fertilizer calculation rates may be grouped into five types: (1) length of the row in a planted hectare at standard bed spacing [21]; (2) broadcast application rates; (3) modified broadcast or banded application rates; (4) injected rates of liquid fertilizers [23]; (5) fertilizer applications when non-standard bed (or row) spacings are used [24]; and (6) calibration of fertilizer application equipment.

6. Extension Challenges and Successes in Vegetable Production

The mission of the Florida Cooperative Extension Service (FCES) is to provide vegetable growers the latest science-based information they need to solve their water and nutrient management challenges and enable them to do so in an environmentally friendly, economically sound and consumer-safe manner. Growers benefit from the FCES in two main ways: preventive education and diagnostic assistance during crop production (Table 2).

Table 2. Common tools and educational events from the University of Florida (UF/IFAS) available to conventional and organic vegetable producers for improving water and nutrient management on their farms.

Tool	Description
Soil testing	Use representative sample; select appropriate extractant for soil type [25]
Livestock waste testing	Laboratory analysis of a representative animal waste sample includes a test for total Kjeldhal nitrogen (TKN), ammonium (NH_4^+-N), phosphorus (P) and potassium (K), moisture content (%), total solid content (%), total ash (%), and pH. Based on test results, nutrient recommendations for N, P, and K are provided for selected crops [26]
Calculations	Correct conversions of nutrient rates to material rates; adjust total nutrient rates to placement method and mineralization rates; calibrate equipment
Soil moisture measuring device	Note if device measures a soil water tension (tensiometer, gypsum block; granulometric water sensor) or a volumetric water content (Time Domain Reflectometry probes); place correctly in the bed in relation to water source; service and maintain as needed; read daily Multi-depth scanning probes connected to a computer by satellite or a cellular phone line allow for real-time, continuous and remote soil moisture monitoring Weather data and irrigation tools are available on the Florida Automated Weather Network website [27]
Sap testing/leaf analysis	Follow a weekly schedule if possible; compare results to sufficiency ranges; adjust fertilization application accordingly
In-field dye tests	Visualize the movement and distribution of water in the soil relative to the rooting zone
Water and nutrient management virtual field day	Watch short videos on how to conduct on-farm dye demonstrations and interpret the results together with UF/IFAS recommendations [28]
Florida drip irrigation school	A UF/IFAS one-day-long educational program with hands-on demonstrations of drip irrigation installation, maintenance and operation, nutrient movement in the soil and correct calculation of nutrient injection rates
Face-to-face food safety training	How to develop a farm and packing house food safety manual and prepare and respond to a food safety audit as required by the buyer
Phone apps	From irrigation suppliers (HydraWise, Crop Water, Irrigate Cost, Irrigate Pump [z]) From universities (Cotton Irrigation App; Urban Turf App [29])

[z] For information only; not recommended by the University of Florida.

The most common mistakes made by vegetable growers may be categorized as: (1) improper soil sampling; (2) incorrect calculations; (3) unnecessarily high preplant fertilizer rates; (4) relying solely on observation (or "guessing") to determine crop nutrient and water needs; and (5) believing that "more" nutrients and "more" water are good for the crop. Historically, organic growers have taken the approach to add fertilizer or compost at the beginning of the season at a rate that will provide the most N possible without risk of burning the crop, but this practice is not ideal. High preplant rates of organic fertilizers (compost or manures) in warm, sandy soils are much more vulnerable to mineralization and leach N faster than in cooler conditions [22].

On-farm demonstrations are one way Extension agents can help vegetable growers improve their practices. For example, on-farm trials have been regularly conducted in North Florida by Extension faculty for the last 30 years. By monitoring crop nutrient status with ion-specific electrodes, by measuring soil moisture content with hand-held Time Domain Reflectometry probes, and by injecting dye that shows where the water moves in their mulched beds, growers have learned that limiting the application of fertilizer in the bed row rather than broadcast over the entire field reduces the risk of nutrient leaching and reduces fertilizer expense. They also learned that soluble nutrients such as N and K move in the soil solution downward through the soil profile with excessive irrigation [18]. Consequently, they have reduced their irrigation rates rather than increasing fertilizer rates in order to avoid nutrient deficiencies. They also learned that application rates of organic-compliant fertilizers must take into account anticipated mineralization rates based on temperature and moisture which are controlled by cultural practices and environmental conditions.

Extension classes and demonstrations have resulted in broad changes by the vegetable industry in North Florida, especially by watermelon growers. Watermelon growers have reduced the N rate applied prior to bedding and mulching from 100–150 kg/ha down to 25–50 kg/ha since the beginning of these programs 20 years ago. The remaining N is now calculated and applied correctly through the drip system over the season. Therefore, a total reduction of 75 to 100 kg/ha of N per year has resulted directly from these demonstrations without affecting watermelon yield or quality. The final resulting N and water management programs on these farms are accepted Best Management Practices for growing watermelons using plasticulture [12].

In organic systems, nutrient management decisions (selection of sources and timing of application) are further complicated by the food-safety mandates of the U.S. Food Safety Modernization Act of 2011 (PL 111-353). Nutrient sources that include animal wastes are of particular concern. Farmers have to consider if the animal waste sources are composted using standardized composting procedures, or if they are untreated animal wastes. If untreated animal wastes are used, food safety audits typically require the untreated animal wastes to be applied to the crop more than 90–120 days before harvest, depending on the crop growth habit. Those crops grown with the harvested portion in or near the ground generally require a longer period (120 days) between application and harvest.

However, in some U.S. markets, very strict guidelines may completely prohibit the use of any animal wastes even in organic production, thus drastically limiting the fertilization options. Fruit and vegetable farmers in the U.S. today have to understand the much greater complexities of regulations for both water and nutrient management, and for food safety. Often food safety regulations impose longer periods required between animal waste applications and harvest, resulting in all fertilizer applied preplant. This practice increases the risk of nutrient leaching and directly contradicts the expectations of the Best Management Practice plans.

7. Effects of Insufficient Water or Nutrients on Vegetable Postharvest

Vegetable water contents are typically above 85%, which illustrates the importance of water supply during vegetable crop production. When the water supply is limited, critical periods of crop water needs and crop drought tolerance indices have been proposed to best use the available water [30]. These periods usually correspond to the development of the harvested plant part. However, as fast-growing plants, vegetable crops need a continuous adequate supply of water and nutrients.

When soil water tension remains above 25 kPa for several hours, plant growth and yields are reduced [31].

All essential elements contribute to maximizing the postharvest potential of vegetables. Specific nutrient imbalances associated with postharvest disorders include softening caused by excess N (grey wall of tomato (*Solanum lycopersicon* L.) or hollow stem of broccoli (*Brassica oleracea var.* italica), for example) and blossom-end-rot caused by an insufficient Ca supply [32,33].

Acknowledgments: This work was supported in part by the Florida Agriculture Experiment Station, the Florida Cooperative Extension Service and the Southern Region Sustainable Agricultural Research Education Program. No funds were received for covering the costs to publish in open access.

Author Contributions: Simonne and Hochmuth jointly conceived, designed and performed the on-farm trials, compiled and analyzed data and wrote the paper.

Conflicts of Interest: The authors declare no conflict of interest.

References

1. National Organic Program. Available online: http://www.ams.usda.gov/AMSv1.0/nop (accessed on 20 June 2016).
2. Tan, K.H. *Principles of Soil Chemistry*, 4th ed.; CRC Press: Boca Raton, FL, USA, 2010; p. 362p.
3. Barker, A.V.; Pilbeam, D. *Handbook of Plant Nutrition II*; CRC Press: Boca Raton, FL, USA, 2015; p. 773p.
4. Are Phosphorous and Phosphoric Acids Equal Phosphorous sources for Plant Growth? HS1010. Available online: https://edis.ifas.ufl.edu/hs254 (accessed on 2 August 2014).
5. Jones, J.B., Jr. Universal soil extractants: Their composition and use. *Commun. Soil Sci. Plan Anal.* **1990**, *21*, 1091–1101. [CrossRef]
6. Assessing Soil Quality in Organic Agriculture. Available online: https://organic-center.org/reportfiles/SoilQualityReport.pdf (accessed on 1 August 2015).
7. Key to Nutrient Deficiencies in Vegetable Crops. Available online: http://pnwhandbooks.org/plantdisease/pathogen-articles/nonpathogenic-phenomena/key-nutrient-deficiencies-vegetable-crops (accessed on 20 June 2016).
8. Guide to Symptoms of Plant Nutrient Deficiencies. Available online: http://extension.arizona.edu/sites/extension.arizona.edu/files/pubs/az1106.pdf (accessed on 20 June 2016).
9. Diagnosing Nutrient Deficiency and Toxicity Symptoms in Fruit and Vegetable Crops. Available online: http://www.extension.umn.edu/garden/fruit-vegetable/nutrient-management-for-commercial-fruit-and-vegetables-in-mn/docs/5886_31-32.pdf (accessed on 20 June 2016).
10. Mills, H.A.; Jones, J.B. *Plant Analysis Handbook II: A Practical Sampling*; Preparation, Analysis, and Interpretation Guide, Macro-Micro Publ.: Athens, GA, USA, 1997.
11. Hochmuth, G.J. Efficiency ranges for nitrate nitrogen and potassium for vegetable petiole sap quick tests. *Hort. Technology* **1994**, *4*, 218–222.
12. Water Quality/Quantity Best Management Practices for Florida Vegetable and Agronomic Crops. Available online: http://www.freshfromflorida.com/content/download/32110/789059/Bmp_VeggieAgroCrops2005.pdf (accessed on 8 August 2015).
13. Food and Agriculture Organization of the United Nations: Rome, Italy. Available online: http://www.fao.org/docrep/X0490E/x0490e00.htm (accessed on 8 January 2015).
14. Locascio, S.J. Management of irrigation for vegetables: Past, present and future. *HortTechnology* **2005**, *15*, 482–495.
15. Dukes, M.D.; Zotarelli, L.; Morgan, K.T. Use of irrigation technologies for vegetable crops in Florida. *Hort. Technology* **2010**, *20*, 133–142.
16. Principles and Practices of Irrigation Management for Vegetables. Available online: http://edis.ifas.ufl.edu/pdffiles/cv/cv10700.pdf (accessed on 1 January 2015).
17. Drip-irrigation Systems for Small Conventional and Organic Vegetable Farms. Available online: http://edis.ifas.ufl.edu/HS388 (accessed on 1 March 2015).

18. Simonne, E.; Gazula, A.; Hochmuth, R.; DeValerio, J. Water movement in drip irrigated sandy soils. In *Microirrigation*; Goyal, L.M., Ed.; Sustainable Practices in Surface and Subsurface Micro Irrigation, CRC Press: Waretown, NJ, USA, 2014; pp. 183–210.

19. Hochmuth, G.J.; Hanlon, H. Commercial Vegetable Fertilization Principles. Available online: http://edis.ifas.ufl.edu/cv009 (accessed on 2 March 2015).

20. Hochmuth, G.J. Progress in mineral nutrition and nutrient management for vegetable crops in the last 25 years. *HortScience* **2003**, *38*, 999–1003.

21. Soil and Fertilizer Management for Vegetable Production in Florida. Available online: http://edis.ifas.ufl.edu/cv101 (accessed on 5 March 2015).

22. Ozores-Hampton, M. Developing a vegetable fertility program using organic amendments and inorganic fertilizers. *HortTechnology* **2012**, *22*, 743–750.

23. Fertigation for Vegetables: A Practical Guide for Small Fields. Available online: http://edis.ifas.ufl.edu/hs1206 (accessed on 1 May 2015).

24. Calculating Recommended Fertilizer Rates for Vegetables Grown in Raised-Bed, Mulched Cultural Systems, SL303. Available online: http://edis.ifas.ufl.edu/ss516 (accessed on 1 May 2015).

25. UF/IFAS Analytical Services Laboratories. Available online: http://soilslab.ifas.ufl.edu/ESTL%20Home.asp (accessed on 20 June 2016).

26. UF/IFAS Livestock Waste Testing Laboratory. Available online: http://soilslab.ifas.ufl.edu/LWTL%20Home.asp (accessed on 20 June 2016).

27. Florida Automated Weather Network. Available online: http://fawn.ifas.ufl.edu/tools/ (accessed on 20 June 2016).

28. UF/IFAS Virtual Field Day – Water and Nutrient Management. Available online: http://vfd.ifas.ufl.edu/water-nutrient-management.shtml (accessed on 20 June 2016).

29. Smartirrigation Apps: Urban Turf. Available online: http://edis.ifas.ufl.edu/ae499 (accessed on 20 June 2016).

30. Basics of Vegetable Crop Irrigation. Pub. Available online: http://www.aces.edu/pubs/docs/A/ANR-1169/index2.tmpl (accessed on 1 May 2015).

31. Smittle, D.A.; Dickens, W.L. Water budgets to schedule irrigation for vegetables. *Hort. Technology* **1992**, *2*, 54–59.

32. Sams, C.E.; Conway, W.S. Preharvest nutritional factors affecting postharvest physiology. In *Postharvest Physiology and Pathology of Vegetables*, 2nd ed.; Bartz, J.A., Brecht, J.K., Eds.; Marcel Dekker: New York, NY, USA, 2003; pp. 161–176.

33. Shewfelt, R.L.; Prussia, S.E. *Postharvest Handling: A Systems Approach*; Academic Press Inc.: San Diego, CA, USA, 1993.

17

Improvement of Postharvest Quality of Asian Pear Fruits by Foliar Application of Boron and Calcium

Kobra Khalaj, Nima Ahmadi * and Mohammad Kazem Souri

Department of Horticultural Sciences, Tarbiat Modares University, Tehran 336-14115, Iran;
kobra.khalaj@gmail.com (K.K.); mk.souri@modares.ac.ir (M.K.S.)
* Correspondence: ahmadin@modares.ac.ir

Abstract: The aim of this study was to evaluate the effects of foliar application of boron and calcium on postharvest storage characteristics of Asian pear (*Pyrus pyrifolia*) fruit. This experiment was carried out in the experimental orchard of Asian pear at Tarbiat Modares University, Tehran, Iran. The treatments were two concentrations of boric acid (0% and 0.5%) and three concentrations of calcium chloride (0%, 0.5%, and 0.7%) which were sprayed on trees during the growing season. The data showed that foliar application of these two elements resulted in improvement in physicochemical characteristics of the pear fruit, particularly fruit firmness and polyphenol oxidase activity. At harvest, fruit firmness and total soluble solids (TSS) were significantly improved by B and Ca application, compared to control fruit. Although fruit firmness decreased during 3 months of cold storage, fruit which received higher rates of Ca and B retained more firmness. During the storage period, total phenolic content was significantly reduced in control fruit compared to fruit which were sprayed with Ca and/or B. The highest level of polyphenol oxidase activity was in B and Ca sprayed fruit and the lowest was in control fruit. In general, foliar application of Ca and B resulted in improvement of postharvest characteristics of Asian pear fruit and reduced internal browning symptoms. Thus, application of Ca and B could be recommended as treatments for increasing postharvest life of Asian pear in similar semi-arid climates.

Keywords: boric acid; calcium chloride; foliar spray; internal browning; total soluble solids

1. Introduction

Asian pear (*Pyrus pyrifolia*), originating from the eastern part of Asia, has been cultivated for centuries. Due to the high quality of the fruit, cultivation and consumption of Asian pear fruit is increasing around the world [1]. The nutrient balance is the main factor that plays a role in postharvest quality of Asian pear fruit.

However, nutrients deficiency disorders are very common in many horticultural crops, particularly under arid and calcareous soil conditions. These disorders play important roles in quality and therefore in marketability of the crops. Calcium and boron are key elements in maintaining the quality and shelf life of fruits, particularly apple and pear. Both elements generally have similar uptake, translocation and correlated physiological roles. Cell strength, cell wall thickness and tissue firmness are highly dependent upon adequate amounts of calcium, mainly forming pectin in the middle lamella [2]. Optimum levels of calcium application have led to better stability of cell membranes and cell walls, and consequently resulted in increasing fruit quality [3]. Boron is another important element in quality parameters of fruits, in particular in pome fruit such as pear. Boron participates in many physiological reactions, particularly cell wall formation and metabolism of sugars and phenolic compounds. It helps maintain cell membrane structures integrity, and is also considered an antioxidant that prevents fruit internal browning [4].

So, the aim of this study was to evaluate the foliar application of calcium and boron on quality characteristics of Asian pear in fruit under the semi-arid climatic conditions of Tehran, Iran.

2. Experimental Section

2.1. Plant Materials

This study was conducted during 2010 and 2011 in an Asian pear orchard located at the Faculty of Agriculture, Tarbiat Modares University, in Tehran, Iran. The early Asian pear cultivar "KS_{10}" was used. The effect of foliar applications of boric acid (0%, 0.5%) and calcium chloride (0%, 0.5%, and 0.7%) were evaluated on harvest and postharvest storage quality. Trees were sprayed starting 20 days after full bloom three times with boric acid and/or six times with calcium chloride solutions at 2-week intervals (Table 1) in a factorial design within randomized complete blocks, three replications of each treatment combination, one replication (one tree) per block. Fruit were harvested approximately 90 days after full bloom, and held in cold storage at 2 °C and 85% relative humidity for 3 months. Fruit from each treatment were immediately sliced at harvest or after cold storage and frozen in liquid nitrogen, and then stored at −80 °C for further use.

Table 1. Different treatments and their abbreviations used in this experiment.

Treatment	Calcium Chloride (%)	Boric Acid (%)
c0b0	0	0
c0b1	0	0.5
c1b0	0.5	0
c1b1	0.5	0.5
c2b0	0.7	0
c2b1	0.7	0.5

2.2. Physicochemical Quality Characteristics

Physicochemical quality characteristics of fruit at harvest and after three months of cold storage were characterized. The fruit length and diameter were determined using digital calipers, and weight loss during cold storage was measured by weighing each fruit individually with a precision digital balance at the beginning and end of storage. After cutting the fruit into small pieces, they were dried at 70 °C in an oven to measure fruit dry weight. Fruit firmness and total soluble solids were evaluated using a portable penetrometer and refractometer, respectively. Titratable acidity was determined by titration of fruit juice by adding 0.1 N NaOH solution to reach pH of 8.3.

2.3. The Total Phenol Content

The total phenol content of fruit at harvest and three months after storage was determined by the Folin-Ciocalteu method [5]. To prepare the sample for total phenol quantification, frozen pieces of fruit were ground to a fine powder using a mortar and pestle cooled by liquid nitrogen during grinding. For quantification of total phenol content, 6 mL of 80% methanol was added to 1 g of powdered sample. Then samples were centrifuged at 14,000 rpm for 20 min at 4 °C. Extract (20 μL) was added to 1.58 mL of distilled water containing 100 μL Folin-Ciocalteu reagent. After 8 min, 300 μL of 20% (w/v) sodium carbonate was added to the sample and placed 2 h at room temperature in darkness. The absorption of the sample was measured with a spectrophotometer at 765 nm. The results were calculated based on using garlic acid as a standard and are expressed as mg garlic acid per g fresh weight of fruit.

2.4. Measuring the Activity of Polyphenol Oxidase Enzyme

For measuring the activity of polyphenol oxidase enzyme activity after 3 months of storage, 200 mg of fruit tissue powder was mixed with 3 mL sodium phosphate buffer (25 mM, pH 6.8) at 4 °C, and the homogenate was centrifuged for 15 min at 4 °C. The supernatant was used for measurement of enzyme activity. Activity of polyphenol oxidase in the supernatant was measured according to Ghanati et al. [6] with minor changes. Sodium phosphate buffer (50 mM, pH 6.8) was used. Methyl catechol (0.15 M) was used as substrate. The reaction mixture contained 2 mL sodium phosphate buffer (50 mM, pH 6.8), 600 µL methyl catechol (0.15 M), and 400 µL enzyme extract. Changes in the oxidation of catechol were monitored at 410 nm for 1 min. Protein content of the supernatant was measured using the Bradford assay [7]. Polyphenol oxidase enzyme activity was determined as change in absorption at 410 nm for 1 min per mg protein.

2.5. Statistical Analysis

Statistical analysis was performed using SAS software (SAS Institute Inc., Cary, NC, USA) and diagrams were drawn using Excel software. Comparison of means was conducted using Duncan's New Multiple Range Test at $p = 0.05$.

3. Results

Spray applications of boric acid and calcium chloride had no significant effect on fruit length and diameter at harvest and three months after storage (Tables 2 and 3). Fruit fresh weight was affected only by Ca spray. The highest fruit fresh weight (54.72 g) was with the 0.7% calcium treatment, and the lowest (40.77 g) was the control (Figure 1). The fruit dry weight at harvest was only affected by the interaction of Ca and B. The highest fruit dry weight (17.48 g per 100 g FW) was with 7% Ca by 0.5% B. After 3 months of storage, significant differences in fruit fresh weight were observed for the effect of Ca, and for the interaction between Ca and B (Table 3). The highest fruit fresh weight (56.63 g) was at 0.7% Ca (Figure 2).

Table 2. Sums of squares in the analysis of variance for the effects of calcium and boron on physicochemical properties of pear fruit at harvest.

Source of Variation	df	Fruit Length	Fruit Diameter	Fruit Fresh Weight	Fruit Dry Weight	Fruit Firmness	Total Soluble Solids	Titratable Acidity
Block	2	9.25 ns [z]	7.68 ns	49.14 ns	0.06 ns	0.06 ns	0.09 ns	729.5 ns
Calcium	2	22.31 ns	30.09 ns	293.9 *	1.47 ns	4.5 **	7.72 **	909.6 ns
Boron	1	1.91 ns	9.7 ns	54.21 ns	0.02 ns	0.88 **	2.72 **	122.3 ns
Calcium × Boron	2	1.48 ns	2.8 ns	22.43 ns	2.99 *	0.1 ns	4.05 **	1029.1 ns
Error	10	9.02	11.55	67.9	0.4	0.06	0.14	399.6
CV(%) [z]	-	7.97	7.56	17.14	3.84	5.08	2.83	20.72

*, ** and ns indicate significant differences at $p \leq 0.05$, 0.01, and no significant difference, respectively;
[z] Coefficient of variation.

Table 3. Sums of squares in the analysis of variance for the effects of calcium and boron on physicochemical properties of pear fruit after three months of cold storage.

Source of Variation	df	Fruit Length	Fruit Diameter	Fruit Fresh Weight	Fruit Firmness	TSS	TA
Block	2	0.66 ns	0.99 ns	5.07 ns	0.07 ns	1.39 *	1.93 ns
Calcium	2	16.61 ns	7.52 ns	216.64 **	1.39 **	31.35 **	630.396 ns
Boron	1	3.78 ns	0.01 ns	21.15 ns	0.02 ns	6.12 **	360.08 ns
Calcium × Boron	2	3.59 ns	0.65 ns	112.80 **	0.02 ns	12.12 **	1109.26 *
Error	10	5.15	7.98	10.04	0.03	0.29	164.43
CV(%) [z]	-	6.80	6.99	6.57	5.73	3.36	13.99

*, ** and ns indicate significant differences at $p \leq 0.05$, 0.01, and no significant difference, respectively;
[z] Coefficient of variation.

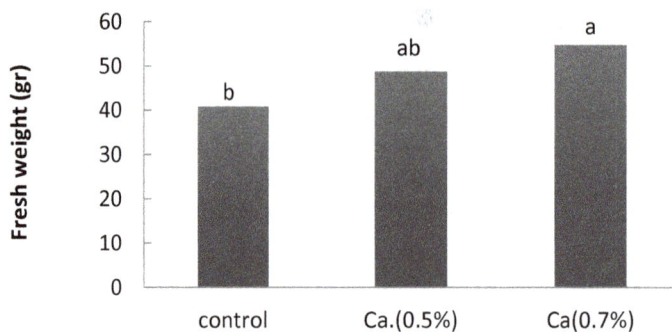

Figure 1. The effect of calcium (%) on fruit fresh weight at harvest. Means with different letters were significantly different by Duncan's New Multiple Range Test at $p = 0.05$.

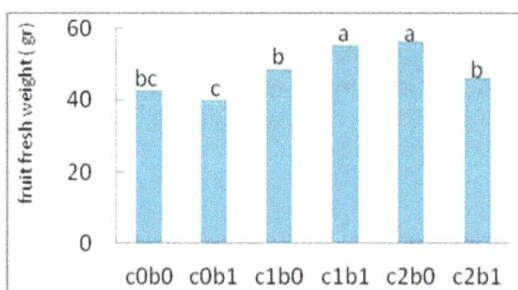

Figure 2. Interaction between calcium and boron on fruit fresh weight after three months of cold storage. See Table 1 for definitions of treatment abbreviations. Means with different letters were significantly different by Duncan's New Multiple Range Test at $p = 0.05$.

Fruit firmness at harvest was affected by both Ca and B, but not by their interaction (Table 2). The highest fruit firmness (5.83 kg/cm^2) was with 0.7% Ca, and the lowest firmness (4.1 kg/cm^2) was the control (Figure 3). Similarly, fruit firmness was significantly higher at 0.5% B (5.17 kg/cm^2) compared to the control (4.73 kg/cm^2) (data not shown). Analysis of variance showed that after three months of cold storage, firmness was only affected by Ca (Table 3). The highest fruit firmness (3.23 kg/cm^2) was at 0.7% Ca, and the lowest (2.3 kg/cm^2) was the control (Figure 3).

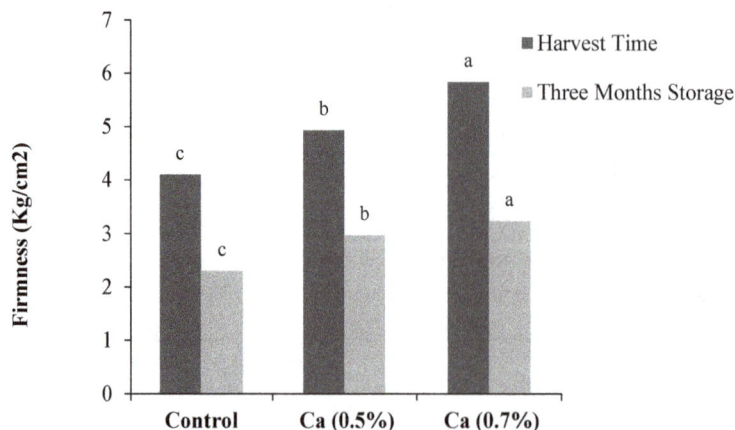

Figure 3. The effect of calcium (%) on fruit firmness at harvest and after three months of cold storage. Means with different letters within harvest or storage time were significantly different by Duncan's New Multiple Range Test at $p = 0.05$.

Total soluble solids at harvest showed significant effects of Ca, B and their interaction (Table 2). The highest TSS (16 °Brix) was the control (Figure 4). After 3 months of cold storage, TSS showed significant effects of Ca, B, and their interaction (Table 3). The highest TSS (20 °Brix) was in control fruit (Figure 4). At harvest, there were no significant differences among treatments in TA values. However, after 3 months of storage, TA was significantly affected by the interaction of B and Ca (Table 3), as the highest was measured in the 0.7% Ca treatment.

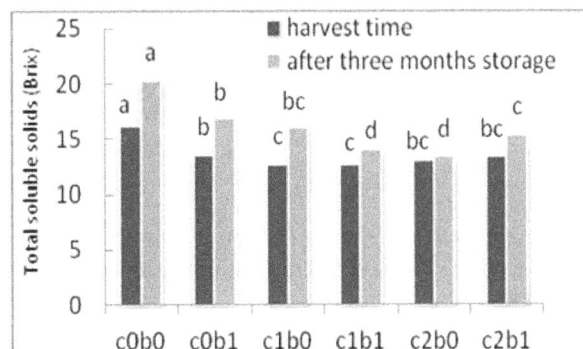

Figure 4. Interaction between calcium and boron on total soluble solids at harvest and after three months of cold storage. See Table 1 for definitions of treatment abbreviations. Means with different letters within harvest or storage time were significantly different by Duncan's New Multiple Range Test at $p = 0.05$.

At harvest, control fruit had the highest total phenol content, while application of Ca and B reduced total phenol content. Fruit treated with 0.7% Ca and 0.5% B (c_2b_1) showed the lowest phenol content (Figure 5). In addition, total phenol content showed a significant reduction during cold storage, and the reduction in the control samples was more than in B and Ca treated fruit. Control fruit showed the lowest total phenol after 3 months of cold storage. For fruit receiving 0.7% Ca and 0.5% B (c_2b_1) treatment, the total phenol was maintained (Figure 5). Ca application reduced the activity of polyphenol oxidase at 3 months (Figure 6). Thus, after storage, PPO activity was highest in control fruit, which was accompanied by a reduction in phenolic content. Thus, treating fruit with B and Ca during the growing season reduced the activity of PPO at the end of the storage period (Figure 6).

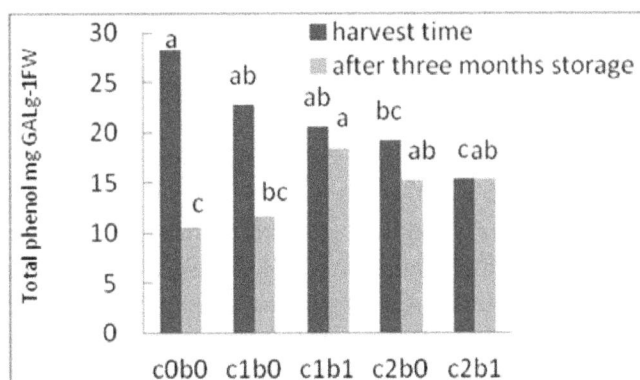

Figure 5. Total phenol at harvest and after three months of cold storage. See Table 1 for definitions of treatment abbreviations. Means with different letters within harvest or storage time were significantly different by Duncan's New Multiple Range Test at $p = 0.05$.

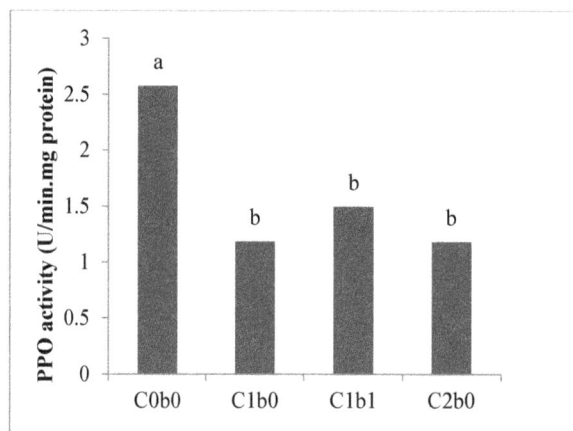

Figure 6. Activity of polyphenol oxidase after three months of cold storage. See Table 1 for definitions of treatment abbreviations. Means with different letters were significantly different by Duncan's New Multiple Range Test at $p = 0.05$.

4. Discussion

In this study, fresh weight was increased by Ca, and dry weight of fruit was increased by Ca × B applications. Increasing cell wall and pectin amounts probably were involved in these responses. Other studies also reported improvement of fruit weight with foliar sprays of Ca and B [8,9]. Fruit firmness was also significantly improved by Ca and B sprays. Increasing pectin levels in the cell wall also probably played a role in this regard [10]. Fruit firmness declined during storage. However, Ca application resulted in a higher fruit firmness compared to the control both at harvest as well as after 3 months of cold storage. Treatment of apple fruit two weeks before maturity with 8% calcium chloride has significantly increased fruit firmness [11]. Also, by application of 4% calcium chloride, Asian pear fruit firmness increased compared to a control [12]. In this study, Ca reduced fruit TSS, similar to results obtained by some others [13], although other research has had different results [8,9,14]. Betts and Bramlage [14] reported that in apple fruit, Ca did not affect TSS as application of Ca reduced respiratory rate and consequently degradation of starch to simple sugars, as well as total soluble solids. Total soluble solids increased during storage, probably due to conversion of starch to sugar. Titratable acidity of fruit decreased during storage. Previous reports indicated that acidity in pear and other fruits gradually increased until final fruit size, while it decreased during maturation, ripening and storage of apple [15]. Similar results were obtained in the present study. Other studies also reported reduction of organic acids during storage of Yali pear [16], although there are reports showing that acidity was not affected by Ca [11]. The role of calcium in cell wall strength is probably responsible for most quality-related properties of this element. Calcium was shown to reduce the production of phenolic compounds in fruit [17], and also increase the polysaccharides and non-alcohol-soluble solids in the fruit cell walls. It also maintained cell membrane stability [3,18,19]. The reduction in total phenolic levels in control fruit after three months of cold storage was likely due to the higher activity of polyphenol oxidase enzyme (Figure 6), resulting in oxidation of phenolic compounds and internal browning in fruit tissue (Figure 7). Generally, prevention of phenol oxidation by application of Ca and B has mainly been related to their effects on cell membrane integrity and membrane permeability [20,21]. Boron, on the other hand, facilitates the role of calcium in cell membrane and cell wall interactions. Nevertheless, both Ca and B have restricted transport within plants, in particular under semi-arid and calcareous soil conditions. Therefore, foliar sprays of these two nutrient elements several times during a growing season can improve Ca and B concentration in fruit tissues [21].

(a)   (b)

Figure 7. Internal browning after three months of cold storage (2 °C, 85% relative humidity). (**a**) Ca at 0.7%; (**b**) Control.

5. Conclusions

In the present study, application of both Ca and B improved the physical and chemical characteristics of fruit compared to a control. Increasing fruit firmness and TSS due to Ca and B applications could have important implications, particularly in terms of fruit quality and storage life. Foliar sprays of both Ca and B reduced the activity of polyphenol oxidase enzyme, which is mainly responsible for internal browning of fruit. The role of calcium in cell wall strength was probably responsible for most quality-related properties of this element. So, using foliar applications of Ca and B during the growing season of Asian pear trees is recommended for improving postharvest quality parameters of fruit in semi-arid regions.

Acknowledgments: We would like to thank Kazem Arzani, Department of Horticultural Sciences, Tarbiat Modares University, for providing the plant materials and samples used in this research.

Author Contributions: Nima Ahmadi and Mohammad Kazem Souri designed the study. Running experiment, sampling and evaluation of characteristics were performed by Kobra Khalaj through supervision and advising of Nima Ahmadi and Mohammad Kazem Souri respectively, by providing necessary suggestions during the research. The data analysis and interpretation was performed by Kobra Khalaj, Nima Ahmadi and Mohammad Kazem Souri. The manuscript was prepared by Kobra Khalaj, Nima Ahmadi and Mohammad Kazem Souri. All authors reviewed and approved of the final manuscript.

Conflicts of Interest: The authors declare no conflict of interest.

References

1. Chen, J.; Wang, Z.; Wu, J.; Wang, Q.; Hu, X. Chemical compositional characterization of eight pear cultivars. *J. Food Chem.* **2007**, *104*, 268–275. [CrossRef]
2. Subburamu, K.; Singaravelu, M.; Nazar, A.; Irulappan, I. Pre-harvest spray of calcium in grapes (Vitis vinifera). *South Indian Hort.* **1990**, *5*, 268–269.
3. Sharpels, R.O.; Johnson, D.S. The influence of calcium on senescence changes in apples. *Ann. Appl. Biol.* **1977**, *85*, 150–453.
4. Dzondo-Gadet, M.; Mayap-Nzietchueng, R.; Hess, K.; Nabet, P.; Belleville, F.; Dousset, B. Action of boron at the molecular level: Effects on transcription and translation in an acellular system. *Biol. Trace Elem. Res.* **2002**, *85*, 23–33. [CrossRef]
5. Singh, R.P.; Murthy, K.N.C.; Jayaprakasha, G.K. Studies on the antioxidant activity of pomegranate (*Punica granatum*) peel and seed extracts using in vitro models. *J. Agric. Food Chem.* **2002**, *50*, 81–86. [CrossRef] [PubMed]
6. Ghanati, F.; Morita, A.; Yokota, H. Induction of suberin and increase of lignin content by excess boron in tobacco cell. *Soil Sci. Plant Nutr.* **2002**, *48*, 357–364. [CrossRef]
7. Bradford, M. A rapid and sensitive method for the quantitation of protein utilizing the principle of protein-dye binding. *Anal. Biochem.* **1976**, *72*, 248–254. [CrossRef]

8. Khoshgalb, H. The Effects of Calcium, Zinc and Boron on the Chemical Composition of Fruits, Postharvest Shelf Life and Reduce Symptoms of Internal Browning of Fruits of Two Varieties of Asian Pear (*Pyrus serotina* Rehd.) in Tehran Climate. Ph.D. Thesis, Faculty of Agriculture, Tarbiat Modarres University, Tehran, Iran. (In Persian)

9. Akl, A.M.; Eid, A.F.M.; Hegab, M.Y. Effect of urea, some micronutrients and growth-regulators foliar spray on the yield, fruit quality, and some vegetative characteristics of "Washington Navel" orange trees. *Hort. Sci.* **1995**, *30*, 774–780.

10. Veltman, R.H.; Sanders, M.G.; Persijn, S.T.; Peppelenbos, H.W.; Oosterhaven, J. Decreased ascorbic acid levels and brown core development in pears (*Pyrus communis* cv. "Conference"). *Physiol. Plant.* **1999**, *107*, 39–45. [CrossRef]

11. Conway, W.S.; Sams, C.E. Effect of postharvest calcium treatment on decay of Delicious apples. *Plant Dis.* **1982**, *66*, 402–403. [CrossRef]

12. Dhatt, A.S.; Mahajan, B.V.C.; Bhatt, A.R. Effect of pre and post-harvest calcium treatments on the storage life of Asian pear. *ISHS Acta Hortic.* **2005**, *696*, 497–501. [CrossRef]

13. Dolati Baneh, H.; Hacani, A.; Majidi, A.; Zomorrodi, S.; Hacani, G.; Malakoti, M.J. Effect of calcium chloride on foliar concentration and frequency characteristics of stiffness and a Lebanese Red apples stored in Urmia region. *J. Agric. Sci.* **2002**, *12*, 47–54. (In Persian)

14. Betts, H.A.; Bramlag, W.J. Uptake of calcium by apples from postharvest dips in calcium chloride solutions. *J. Am. Soc. Hortic. Sci.* **1979**, *102*, 785–788.

15. Ackerman, J.; Fischer, M.; Amado, R. Changes in sugars, acids, and amino acids during ripening and storage of apples (cv. Glockenapfel). *J. Agric. Food Chem.* **1992**, *40*, 1131–1134. [CrossRef]

16. Chen, J.L.; Yan, S.J.; Feng, Z.; Xiao, L.; Hu, X.S. Changes in the volatile compounds and chemical and physical properties of "YaLi" pear (*Pyrus bertschneideri* Rehd) during storage. *Food Chem.* **2006**, *97*, 248–255. [CrossRef]

17. Coseteng, M.Y.; Lee, C.Y. Changes in apple polyphenoloxidase and polyphenol concentrations in relation to degree of browning. *J. Food Sci.* **1987**, *52*, 985–989. [CrossRef]

18. Malakoti, M.J.; Tehrani, M.M. *The Role of Micronutrients in Increasing Yield and Improving the Quality of Agricultural Products*, 3rd ed.; Tarbiat Modares University: Tehran, Iran, 2001.

19. Dunn, J.L.; Able, A.J. Pre-harvest calcium effects on sensory quality and calcium mobility in strawberry fruit. *Acta Hortic.* **2006**, *708*, 307–312. [CrossRef]

20. Tomas-Barberan, F.A.; Gil, M.I.; Castaner, M.; Artes, F.; Saltveit, M.E. Effect of selected browning inhibitors on phenolic metabolism in stem tissue of harvested lettuce. *J. Agric. Food Chem.* **1997**, *45*, 583–589. [CrossRef]

21. Tohidloo, G.; Souri, M.K. Uptake and translocation of boron in two different tomato genotypes. *Hortic. Environ. Biotechnol.* **2009**, *50*, 487–491.

18

Nitrogen Related Diffuse Pollution from Horticulture Production—Mitigation Practices and Assessment Strategies

Maria do Rosário Cameira [1,2,*] **and Mariana Mota** [1,2]

[1]   Department of Biosystems Engineering, Instituto Superior de Agronomia, University of Lisbon, Tapada da Ajuda, 1349-017 Lisboa, Portugal; mariana@isa.ulisboa.pt

[2]   Linking Landscape, Environment, Agriculture and Food (LEAF), Instituto Superior de Agronomia, University of Lisbon, Tapada da Ajuda, 1349-017 Lisboa, Portugal

*   Correspondence: roscameira@isa.ulisboa.pt

Abstract: Agriculture is considered one of the main nitrogen (N) pollution sources through the diffuse emissions of ammonia (NH_3) and nitrous oxide (N_2O) to the atmosphere and nitrate (NO_3^-) to water bodies. The risk is particularly high in horticultural production systems (HPS), where the use of water and fertilizers is intensive and concentrated in space and time, and more specifically, in the case of vegetable crops that have high growth rates, demanding an abundant supply of water and nitrogen forms. Therefore, to comply with the EU environmental policies aimed at reducing diffuse pollution in agriculture, there is the need for mitigation practices or strategies acting at different levels such as the source, the timing and the transport of N. HPS are often well suited for improvement practices, but efficient and specific tools capable of describing and quantifying N losses for these particular production systems are required. The most common mitigation strategies found in the literature relate to crop, irrigation and fertilization management. Nevertheless, only the success of a mitigation strategy under specific conditions will allow its implementation to be increasingly targeted and more cost effective. Assessment methods are therefore required to evaluate and to quantify the impact of mitigation strategies in HPS and to select the most promising ones.

Keywords: horticulture; diffuse pollution; N emissions; N leaching; mitigation strategies; fertigation management; crop management

1. Introduction

Diffuse or non-point source pollution refers to both water and air pollution caused by a variety of activities that have no specific point of discharge. Furthermore, the long-range transport ability and multiple sources of the pollutant contribute to the diffuse nature of the process. The management of diffuse pollution is complex and requires the careful analysis and understanding of various processes [1].

Agriculture is seen as one of the main N pollution sources through the diffuse emission of ammonia (NH_3) and nitrous oxide (N_2O) to air and nitrate (NO_3^-) to surface and ground waters [2].

NO_3^- leaching to ground waters represents a loss of soil fertility and also a threat to the wider environment and human health [3,4]. NO_3^- that enters drinking water supplies creates a risk of methemoglobinemia in infants/young children and has been linked to cancer and heart disease [5]. Half of the European population live in areas where concentrations in drinking water exceed 5.6 mg $N-NO_3 \cdot L^{-1}$, and about 20% live in areas where concentrations exceed the recommended level of 11.3 mg $N-NO_3 \cdot L^{-1}$ [6]. NO_3^- and ammonium (NH_4^+) transported with subsurface flow or surface runoff entering rivers or lakes contributes, together with phosphorous, to eutrophication, resulting in algae blooms and suffocation of aquatic life [7].

Atmospheric emissions of some nitrogen oxides, e.g., N_2O (a greenhouse gas) and NOx, and NH_3 are contributing, directly and indirectly, to negative effects on human health [8]. N_2O in the atmosphere contributes to the depletion of the ozone layer and makes a significant contribution to climate change. The NH_3 emissions contribute to acid rain causing acidification and eutrophication of the ecosystems. The latter also represent an indirect source of N_2O greenhouse gas [9].

In response to the main environmental and health threats posed by agriculture, several countries around the world have implemented various policy measures and regulation tools developed under international conventions. Figure 1 shows the example of how, for the European Union (EU), agriculture can comply with these regulations [10]. Thus, the application of fertilizers, manures and other organic materials has to comply with policy measures dealing with the emissions to air, ground water and surface water. This is a difficult task since measures to reduce losses to one compartment will often have an impact on the emissions to other compartments as a result of the nitrogen cycle (pollution swapping) [11].

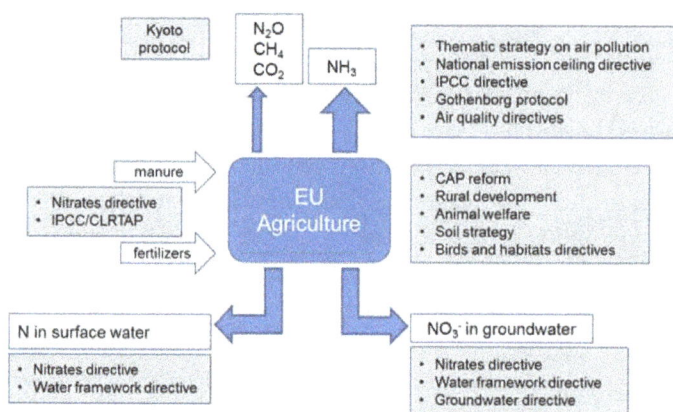

Figure 1. Overview of the EU policy instruments affecting the use and losses of nitrogen in agriculture (CLRTAP, Convention on Long Range Transport of Atmospheric Pollutants; IPPC, directive on integrated pollution, prevention and control; CAP, common agricultural policy) (adapted from [10]).

Although some improvements have led to the reduction of N inputs, diffuse pollution of agricultural origin remains a major threat for waters. The technologies and measures to reduce these emissions exist, but the diversity of the cropping systems and the complex diffuse N pathways have resulted in regulatory obligations that are not equally efficient for different types of production systems. Some authors state that the risk is higher in horticultural production systems (HPS) than in arable lands [12]. Vegetable production, in particular, requires an intensive use of resources, namely water and fertilizers, in concentrated space and time. Reported N application rates show values as high as 600–900 kg·ha^{-1} [13–15]. Vegetable crops in general present shallower root systems compared to arable plants. Furthermore, the nitrogen use efficiency (NUE) of many vegetables is often less than 50% and can be as low as 20%, [14,16]. Contrary to arable crops, vegetables are harvested in a vegetative stage when daily N uptake rates are still high, leaving soils with considerable mineral N amounts. Their residues also take a particular position relative to arable crops due to often large amounts of biomass, with a high N content (up to 200 kg·ha^{-1}) and low C:N ratios, left behind on the field [16,17]. These adverse characteristics can be exacerbated by the incorrect management of irrigation and/or by the precipitation occurring during the post-harvest season, both situations maintaining excessively high moisture conditions in the soil surface and drainage fluxes out of the root zone.

Considering fruit crops, which constitute another important part of HPS, the use of resources is not so intensive, but it follows a similar tendency. In particular, intensive and hedgerow orchards are characterized by high plant densities, contributing to a weaker and more superficial root development. This is reinforced by the application of water and nutrients exclusively to the root zone [18,19].

Moreover, the need to correct the soil reaction and/or to supply micronutrients may require the application of N forms that will increase pollution.

Horticultural production is important worldwide. For example, in the EU, it was, for the year of 2014, 30 million tons for fruits and 63 million tons for vegetables, the latter having increased 4% in relation to 2013 [20]. Water and N availability remain globally the most limiting plant growth factors, and water application is a management option in irrigated systems that strongly interacts with the efficient use of N [21]. Therefore, irrigated horticulture requires specific practices to increase water and nutrient use efficiency, which is considered to be a major challenge for the food production during this century [22]. Nitrogen losses to water bodies and the atmosphere and the correspondent mitigation strategies are thus significant concerns in HPS. Recent studies highlighted that there is the danger of pollution swapping between nitrate (NO_3^-) leaching and N_2O and NH_3 gaseous losses, which requires a holistic approach to the diffuse pollution issue, including the N and water dynamics and management in the soil-plant-atmosphere systems [5,23].

The objective of this review paper is to identify the main strategies available to mitigate the various N losses from HPS with emphasis on vegetable production and to present methods to assess their application success. The first part of the work outlines the threats to the environment and health associated with N losses from HPS. In the second part the EU policy tools concerning N diffuse pollution are presented, followed by the major pathways for N losses and its underlying processes. A description of the available strategies to mitigate these losses is then presented; followed by the revision of three methods to evaluate the success of the different types of measures.

2. Nitrogen Loss Pathways and Processes

The N present in the soil is subject to several transformations that influence its availability to plants and influence the potential for losses though the various pathways. Horticulture soils generally contain large pools of N, most of which are in organic forms. Around 1%–3% of the N bound in organic forms may become mineralized and available to crop uptake or be potentially lost within the growth period [24]. In this process, bacteria digest organic material and release NH_4^+, which is positively charged and therefore can be bound to negatively-charged soil particles and organic matter (OM). Thus, NH_4^+ does not move downward in soils and can be volatilized at the soil surface. It can also be absorbed by the plants or subject to nitrification, which is the conversion to NO_3^-. This N form is negatively charged and water soluble, and depending on the conditions, it can move below the crop rooting zone (leached), be absorbed by crops or denitrified. A temporary reduction in the available N can occur (immobilization) when the bacteria that decompose high C:N residues use the soil N to grow and build bacterial biomass. However, there is often a net gain of N during the growing season. Tracing the N path through the environmental reservoirs is thus a challenge due to the complex N cycle [25]. In HPS, all of the tree N loss pathways occur: leaching, volatilization and denitrification [9] (Figure 2).

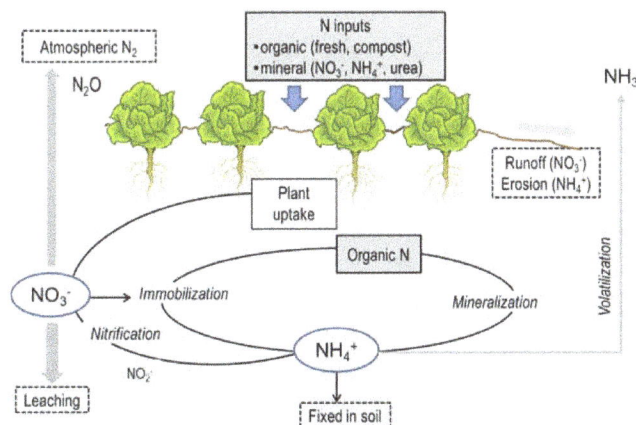

Figure 2. Simplified diagram of the nitrogen cycle in a horticultural system, showing the main transformations and pathways for loss.

2.1. Nitrate Leaching

Leaching refers to the loss of NO_3^- from the soil root zone into ground and surface waters and results from the combination of three physical processes: convection, diffusion and hydrodynamic dispersion [26]. The convective transport of nitrate in the root zone occurs due to the mass flow of water through the soil during drainage after precipitation and/or irrigation events. Diffusive transport occurs due to the concentration gradient between the soil solution with nitrate that is moving by convection and the surrounding soil with a lower concentration. During irrigation and precipitation events, the convective transport will predominate while diffusion gains importance between events, when the soil water fluxes are low. However, the large variations in soil pore sizes cause a range of pore water velocities. As a result, some portions of the flowing solution move ahead while other portions lag behind, causing the incoming solution to mix or disperse with the antecedent solution. This process is called hydrodynamic dispersion and can predominate over diffusion when the convective velocity is sufficiently high.

Factors affecting nitrate leaching have been reviewed, although more extensively for arable crops than for horticultural crops (e.g., [9,12]). The amount of NO_3^- leached from the root zone is determined by its concentration in the soil solution and the drainage flux through the soil. While the amount of NO_3^- present in the soil solution is a result of the N budget, the drainage flux depends on the soil hydraulic properties, temperature and water inputs.

Horticultural crops, vegetables in particular, are mainly produced in sandy to loamy soils, which are less retentive for water and highly to moderately conductive. Soil water contents are usually high since vegetables are irrigated, increasing the leaching risk during the crop season. When irrigation is correctly managed, leaching will be minimized, but N will accumulate in the soil profile, increasing the post-harvest leaching potential. Leaching losses for a variety of vegetable crops, climates, soils and management practices are presented by several authors [12,21,27–35]. A leaching loss of 207 kg NO_3-N ha^{-1} was reported from a cabbage crop on a sandy soil harvested in September; a broccoli crop in a loamy soil harvested in November leached 282 kg·NO_3-N ha^{-1}, while in a sandy soil planted with cauliflower harvested in November, leaching reached a value of 293 kg·NO_3-N ha^{-1}.

Nitrate losses from other horticultural systems, e.g., orchards have been less studied. Data collected in an apple orchard show concentrations up to 66 mg NO_3-N·L^{-1} in the leachate [36]. After the analysis of current fertigation practices in an intensive olive grove, it was found that nitrate leaching accounted for 75% of the fertilization inputs [37]. Other researchers found that 53%–78% of the applied nitrogen was leached in a commercial apple orchard [38]. Results suggest that if an irrigated orchard is located in a light textured and free draining soil and there is a high N input, the potential for leaching can be high, thus requiring the application of mitigation strategies [39,40].

2.2. Ammonia Volatilization

Ammonia volatilization is the transformation of NH_4^+ into NH_3 in the soil and its transport from the soil surface to the atmosphere [9]. Agriculture accounts for 50% of the NH_3 that is volatilized worldwide [41], which is undesirable since it represents a loss of N from the soil-plant system and a threat to the environment. The relative concentrations of NH_4^+ and NH_3 in the soil depend on the pH of the soil solution, being favored by high values. Other factors affecting NH_3 production are temperature, soil texture, soil cation exchange capacity and soil moisture [9]. The rate of exchange of NH_3 between soil surface or plant material and the atmosphere depends on micrometeorological circumstances (especially wind speed and temperature) and ambient NH_3 concentration [42].

This process is the most important N loss pathway from agricultural soils receiving NH_4^+-based synthetic fertilizers and some type of manures. NH_3 volatilization can also be produced by mineralization of soil OM and crop residues, especially from vegetable crops [43], which have high N levels and low C:N ratios. About 5%–16% of the N content of vegetable crop residues can be lost as NH_3 [44]. Volatilization losses from poultry manure and dairy slurry were found to be 9%–20% and 14%–35% of total N applied, respectively [44,45]. The potential risk of ammonia volatilization from urea fertilizer can represent up to 65% of the N applied, depending on soil and climatic conditions [46]. This aspect is very important in orchards, as urea application is common at the post-harvest stage, to enable a fast-nutrient uptake that confers adequate nutrient storage and a good plant performance for the following spring.

2.3. Nitrification and Denitrification

Agriculture accounts for 70% of the N_2O emissions to the atmosphere in Europe [47]. Some of the highest fluxes to date have been measured in irrigated and fertilized systems [48,49]. N_2O emissions from cultivated soils result from the biological nitrification and denitrification processes [50]. Thus, the understanding of the soil and environmental factors regulating the microbial populations is necessary to choose appropriate N_2O mitigation strategies. The major soil parameters that regulate the nitrification-denitrification processes are temperature, percentage of soil pores filled with water, oxygen level and pH.

Nitrification is an aerobic process resulting from two distinct steps: the oxidation of NH_3 or NH_4^+ to NO_2^- by ammonia-oxidizing bacteria, such as Nitrosospira and Nitrosomonas, and the oxidation of NO_2^- to NO_3^- by the nitrite-oxidizing bacteria (e.g., Nitrobacter) [51]. During the first step, a portion of the NH_4^+ may be converted to N_2O during the decomposition of nitrite [52,53].

The optimum soil temperatures for the nitrifying microbial populations range from $25\,°C$–$30\,°C$ [54]. However, lower nitrification rates have been found at temperatures below $5\,°C$ [53]. As to the optimum soil pH, it varies between 4.5 and 7.5 [54].

Nitrifying bacteria are sensitive to changes in soil water content. The maximum nitrification rate occurs when the soil moisture is near field capacity (-10 kPa) [54]. As the soil gets wetter, the nitrification rate decreases, and it is estimated that the process becomes restricted when the water-filled pore space is greater than 60%. However, as water drains and oxygen re-enters the soil, nitrification quickly resumes as the bacteria population recovers. The nitrification rate is also significantly slower when the soil is dry, although it still can occur near the wilting point (-1500 kPa) [55]. In dry environments, a surge in microbial activity (including nitrification) is commonly seen when a soil is rewetted by irrigation or rainfall following a prolonged dry period.

The whole process of nitrification is dependent on the ammonia available in the soil; however, high concentrations of ammonium/ammonia can restrict the activity of Nitrobacter [55]. Thus, the process is dependent on management practices, particularly from the type of fertilizer applied.

The other main driver for N_2O production is denitrification. It consists of a microbially-mediated process of the reduction of NO_3^- and/or nitrite (NO_2^-), which may ultimately produce N_2 [56]. In addition to the terminal product, NO and N_2O are generated as obligatory free intermediates [57]. The rate of denitrification and the relative proportions of NO, N_2O and N_2 produced depend on the outcomes of complex interactions between soil properties, soil micro-organisms, climatic factors and management practices. Increased soil N supply, decreased soil pH, C availability and water content generally increase the $N_2O:N_2$ ratio, thus increasing N_2O emissions. Mitigation approaches should focus on ways to enhance the reduction of N_2O to N_2, thus lowering the $N_2O:N_2$ product ratio.

As denitrification is favored by high NO_3^- and C availability, it is likely to occur in wet horticulture soils, in particular immediately after rainfall or an irrigation event when soil pores may become filled with water and the oxygen supply may be temporarily restricted [58]. Extensive descriptions of the denitrification processes and the regulation of the $N_2O:N_2$ ratio can be found in previous literature [56].

Available data provide evidence of the need for strategies to reduce soil N_2O emissions in HPS. Losses are expected to be derived primarily from denitrification during peak soil moisture periods, when more than 60% of the pore space is filled with water [59]. At lower soil moistures, nitrification will be the major source of N_2O. Losses are important especially for cole crop production and must account for the post-harvest period. Field emissions of N_2O following a cauliflower crop harvested in September on a silty clay loam varied from 1.1%–3.7% of the N content of the residues [60]. High N_2O

emissions have also been reported during winter after incorporation into soil of Brussels sprout and broccoli residues (up to 14% of total N), due to the high moisture content of the residues [61]. Cereal crop residues have significantly lower emissions [62].

Thus, the need for mitigation measures for leaching and gaseous losses in HPS, in particular in vegetable production, is largely justified.

3. Strategies to Mitigate Nitrogen Diffuse Pollution from Horticultural Production

N loss mitigation strategies or methods have three different targets: (a) the nutrient availability, aiming to redress the balance between nutrient supply and demand (source methods); (b) the timing for agricultural practices, aiming to synchronize N availability and plant uptake (timing methods); and (c) the nutrient delivery to the receiving environmental compartment, aiming to reduce nutrient mobility and modifying the transport pathways (transport methods) [63]. Whatever strategy is chosen, it must be well designed and adapted to local farming conditions, soils and climate.

Some reports suggest that strategies that reduce N losses after harvesting the vegetable crop may be more significant than in season strategies. In fact, over 400 $kg \cdot N \cdot ha^{-1}$ may remain in the field as soil mineral N due to uncomplete uptake by crop and/or to a low harvesting index and as readily mineralizable organic forms in vegetable crop residues [17]. However, in permanent horticultural crops, as fruit trees or vines, a significant part of the biomass remains in the permanent structure of the plant, and only a minor part is removed through the leaves (in the deciduous species) and the pruning wood [64]. The most common N loss mitigation strategies applicable to HPS are approached next.

3.1. Crop Management

Mitigation strategies based on crop management include residue management after harvest, the use of cover crops, crop rotations and intercropping and the growing of N use-efficient plants.

3.1.1. Management of Crop Residues

Vegetable crop residues often consist of a large amount of biomass with high N contents and C:N ratios ranging between 10 and 20, thus mineralizing rapidly even at low soil temperatures [4,65–68]. This means that when assessing the risk of post-harvest losses, the N content in the crop residues must be considered in addition to the potential for soil OM mineralization. Up to 198 $kg \cdot N \cdot ha^{-1}$ may result from leaving broccoli crop residues in the field after harvest, representing a significant risk [69]. This situation may occur mainly for brassica crops, which are the ones with higher N contents (Table 1). Thus, it is very important to manage these residues to conserve N for the next crop [16].

Table 1. N content of vegetable crop residues (adapted from [16]).

Crop	N Content ($kg \cdot ha^{-1}$)	Reference
Cabbage	170–200	[70]
Cauliflower	150	[70]
	98–128	[71]
	193	[72]
Brussels sprouts	140–240	[73]
Broccoli	76–304	[74]
Leek	54	[73]
Celery	25–60	[75]

The in situ management options are:

- Incorporation of the vegetable crop residues into the soil reduces NH_3 emissions [43]. However, this practice may increase the leaching losses (pollution swapping) (source);
- Leaving the crop residues intact on the soil surface following harvest in autumn will slow down mineralization [76]; incorporation can be delayed to a period where the risks of N leaching losses

are lower [77]. However, this practice may lead to an increase in gaseous emissions and also to severe phytosanitary problems, as the inocula remains in the fields from one season to the other [78] (source);

- Co-incorporation of crop residues with other residues presenting a higher C:N ratio may decrease mineral N availability through the immobilization process or even by reducing the mineralization rate of the residues; this practice was successfully tested using different types of materials as wheat straw and green waste compost (e.g., [29,67]); also, this material will be free of fungal inocula from the previous season, as spores usually lose their viability during the composting procedure [79] (source, timing and transport);

- When the harvest method leaves the root systems intact (e.g., cauliflower and broccoli), they grow and act as a catch crop during winter. This practice reduced the soil nitrate contents during winter by 39% as compared with the no catch crop situation [80] (transport);

- Removal of the crop residues from the field, which can be applied later, will reduce the potential for all kinds of losses [71]. The application time requires the synchronization of crop nutrient demand and nutrient availability from the previous crop residues. A leaching reduction of 8% when 20% of cauliflower residues were removed was reported [16] (source, timing and transport).

3.1.2. Use of Cover Crops

N losses during autumn/winter, where the conditions are less favorable to physiological activity production, can be significantly reduced by the presence of cover crops in comparison with bare ground acting as a mitigation practice at the source [80]. Cover crops must have a fast developing and deep rooting system, as well as winter hardiness [16,81]. This allows them to rapidly scavenge N from the entire soil profile. Several studies show the success of planting cover crops like oats, hairy vetch, rye, barley and ground wheat after the harvest of vegetables like endive, peas, potatoes and cabbage, in the mitigation of N gaseous and/or leaching losses (e.g., [72,82–84]). However, for some brassica crops like cauliflower or broccoli, harvested as late as November, there is no evidence of the benefits of a cover crop in reducing N losses after production [65,72]. In these cases, the cover crop must be established before the harvest of the vegetable crop (intercropping), and an adequate strategy must be implemented to avoid competition for water and nutrients.

Beside leaching and gas losses, cover crops can also mitigate nutrient losses by surface runoff and soil erosion, since water infiltration is increased and the soil surface is protected against erosive forces [85,86]. In orchards, cover crops are often used to protect the soil, enable the use of machinery with reduced soil compaction and avoid nutrient loss. However, they can constitute a reservoir of inocula, and therefore, this practice must be reviewed [87].

In a meta-analysis relative to N loss mitigation strategies, the use of cover crops ranked second best, after irrigation management [88].

3.1.3. Crop Rotations

Rotations are a technique of great importance both for the improvement of soil fertility and the mitigation of N losses at the source, timing and transport. Crop rotations, especially performed with N-fixing rotation crops, reduce the input of fertilizers and the pollution by nitrogen [89]. Legumes such as faba beans can fix large amounts of N ($100–200$ kg·N·ha^{-1}) and can thus reduce the need for N fertilizer on succeeding crops [90]. Biologically-fixed N is used most efficiently in rotations where legumes are followed by crops with high N requirements (e.g., leaves or fruits). Root depths were negatively correlated with NO_3^- leaching, and thus, rotations including crops with different root depths were good choices to improve N use efficiency and at the same time reduce nitrate leaching [91]. Deep-rooted crops (e.g., pumpkin, tomato) alternating with shallower-rooted ones (e.g., lettuce, broccoli) explore the entire profile and bring up nutrients from deeper layers that might otherwise be lost from the system [92]. Differences in plant rooting patterns including root density and root branching at different soil depths also result in more efficient extraction of nutrients from all soil layers when a series of different crops is grown.

3.1.4. Use of N Use-Efficient Plants

The use of plants associated with mycorrhizae can reduce N losses through leaching by acting at the source and transport. Mycorrhizae contribute to the formation of stable aggregates in the soil, developing a macroporous structure that favors infiltration and water retention, penetration of roots and air and prevents erosion. Furthermore, mycorrhizae promote the capacity of plants to acquire N from inorganic sources, with different studies showing that mycorrhizal plants have an increased plant N content, reducing the soil nitrate content [93]. Furthermore, an adequate choice of the rootstock is important, since fast-growing rootstocks have usually a better NUE and are less prone to N losses through leaching, when compared to slow-growing ones [39,94].

3.2. Irrigation and Fertilization Management

Vegetable crops are irrigated frequently due to their shallow rooting system and high sensitivity to moisture stress. The amount of irrigation water applied is frequently far greater than the actual crop needs. This has significant impact on soil NO_3^- level, and it is clear that controlling the amount of drainage from irrigation is critical to the protection of groundwater. Since convection is the main process of nitrate transport in the root zone, the control of the NO_3^- available for leaching must be complemented with the control of the hydraulic fluxes due to irrigation. Thus, an integrated management of both irrigation and fertilization practices should be considered.

3.2.1. Irrigation Water Management

In irrigated horticulture, excessive water applications increase N leaching, leading to a low crop N availability, which is compensated by increasing fertilization rates. As a consequence, when crops are over irrigated, it is common to observe low NUE with a negative impact on groundwater quality [21]. A meta-analysis of several strategies to control nitrate leaching led to the conclusion that the group of strategies based on improving water irrigation management had the largest effect [88]. Irrigation water management has a significant impact on N losses not only because it controls the nutrient delivery to the water bodies (transport type method), but also as a mitigation method at the source, since moisture content controls most of the biochemical reactions leading to N transformations [95]. Irrigation water management can be improved at different levels, as described next.

- Adjustment of water application to crop needs: irrigation requirements

The adjustment of water application to crop needs without decreasing yields and nutrient uptake efficiency is the irrigation management practice most effective in decreasing N losses [88]. Crop needs refer to the actual water requirements for evapotranspiration (ETc), which primarily depend on crop development and climatic factors. A suitable method for estimating ETc for vegetable crop production is the FAO method based on reference evapotranspiration (ETo) and the crop coefficients (Kc) [96]. Irrigation requirements are primarily determined by crop water requirements, but also depend on the characteristics of the irrigation system, management practices and the soil characteristics. Irrigation systems are generally rated with respect to application efficiency, which is the fraction of the water that has been applied by the irrigation system and that is available to the plant for use [97]. Applied water that is not available to the plant may have been lost from the root zone through evaporation or wind drifts of spray droplets, leaks in the pipe system, surface and subsurface runoff or deep percolation within the irrigated area.

- Irrigation scheduling

While the determination of the irrigation requirements only provides general guidelines due to the variable nature of vegetable production, irrigation scheduling provides the means to determine when and how much to irrigate according to the in situ conditions of the plants in a given field. Irrigation scheduling is generally based on experience, being generally accepted that vegetable growers over irrigate to ensure that water is not limiting the production, originating drainage fluxes that carry nitrates beyond the depth of crop roots. The use of soil moisture sensors or soil matric potential sensors is the most appropriate technique to schedule the irrigation of vegetable crops [98]. These sensors can be used as a standalone, complementing the FAO method or even complementing the grower experience. Tensiometers are an example of soil matric potential sensors whose readings are compared with the soil matric potential (SMP) at field capacity, allowing one to adjust the irrigation frequency, once the thresholds are defined. Field capacity is the moisture content or the SMP at which a soil is holding the maximum amount of water it can against the force of gravity. Table 2 contains guidelines for using SMP data to schedule irrigation events.

Table 2. Irrigation guidelines for different soil type when using tensiometers.

Soil Texture	SMP [a] (cb)	Soil Moisture Status and Irrigation Requirements
Sand, loamy sand	5–10	
Sandy loam, loam, silty loam	10–20	Soil at field capacity
Clay loam, clay	20–40	No irrigation required
Sand, loamy sand	20–40	
Sandy loam, loam, silt loam	40–60	50% available water depleted
Clay loam, clay	50–10	Irrigation required

[a] SMP, soil matrix potential.

The utility of tensiometers in fine-textured soils is limited due to the range of detection; thus, tensiometers are practical in sandy or coarse-textured soils. Details about the use of tensiometers to schedule the irrigation of vegetable crops are provided by several authors, e.g., [99–102]. SMP thresholds for several vegetables produced in a variety of soil types are also provided [98].

- Irrigation technologies

Using more efficient systems for water delivery decreases the use of water since it minimizes losses. Different types of irrigation systems have been adapted for use in horticultural crops. Hand-moved sprinklers and travelling guns are systems that present larger application efficiencies (Ea) than the surface irrigation systems; nevertheless, the uniformity of water distribution (UD) is greatly affected by wind conditions. When the UD is low, there is the tendency to over irrigate some of the areas in order to provide enough water to the others, causing considerable drainage fluxes and leaching in some parts of the field. Drip systems potentially optimize Ea and UD, and although they reduce the root development and the volume of soil used by the plants, they are very commonly used in horticultural systems, both for vegetables and for fruits trees and vineyards, as they reduce water consumption and the growth of competitor plants [103]. If fertilizer injection is combined with drip irrigation, greater yield increases are possible. The cost of installing a drip system is relatively high. However, the labor cost through the season is very low. A major advantage of drip systems is that less water is required. Drip irrigation provides the crop with a uniform supply of water through the season.

3.2.2. Fertilizer Management

Fertilizer management strategies target both N source and N timing control, since they aim to redress the balance between nutrient supply and demand to minimize the potential loss and to define the best application timing. A meta-analysis study regarding N mitigation practices ranks fertilizer management strategies after irrigation water management and the use of cover crops [86]. The largest effect was achieved by the reduction of the recommended amounts followed by improved fertilizer technologies. This group of strategies attained a leaching reduction of almost 40%.

- Fertilizer amounts and formulations

The fertilizer amount to be supplied must compensate for the exported nutrients, assuming that the remaining organs will store or restores to the soil the nutrients retained. HPS require, in some cases, high levels of N fertilizers (>200 kg·ha^{-1}) [17]. Furthermore, as horticultural crops are usually highly valued, the relative cost of this input is less important to farmers, and thus, they may apply it in excess, endangering environmental quality. The literature shows that the best relationship between nitrate leaching and yield is obtained when applying the recommended N fertilizer amount. Values below the recommended ones will reduce leaching, but also reduce yields. The focus is then the calculation of the recommended amount to meet plant demand and reducing excessive N input, considering other potential N sources, such as soil mineral N and potentially mineralizable N from the soil endogenous OM and from incorporated organic materials and the N provided by the irrigation water.

As to the formulations, the N use efficiency of foliar urea application is higher than when N is applied to the soil. In addition, it promotes fungal disease control, reducing overwintering spores [103,104]. The lowest N losses reported are associated with ammonium sulfate or diammonium phosphate fertilizers, as ammonium is stored in the soil cationic exchange sites and, therefore, is much less likely to be leached past the tree roots [9].

It is also important to consider the interaction of the N fertilizer with other nutrients, namely with Ca or S; since the need to supply these secondary macronutrients requires the use of formulations that potentially contribute to N leaching, for instance $Ca(NO_3)_2$. The same applies to the formulations used to control the soil reaction, e.g., the use of $(NH_4)_2SO_4$ to acidify the soil can contribute to N losses through volatilization [105].

- Fertilizer application method

When urea is spread at the surface, the NH_3 volatilization losses can be as high as 50% of the applied N [105]. Moreover, incorporating urea at depths higher than 7.5 cm can result in negligible NH_3 emissions and maximum N retention. Application of urea before the onset of rain can significantly reduce the amount of ammonia volatilization because it washes the urea and ammonium below the soil surface. A study reports that 10–16 mm of rainfall soon after application of urea reduced the loss by over 80% [106]. Applying irrigation water after the application of urea fertilizer can also reduce the risk of ammonia volatilization [107]. Similarly, volatilization from applications of anhydrous ammonia can be minimized if the fertilizer is injected to depths below 10 cm when soils are moist, generally after winter or rainy seasons [41]. Measurements of volatilization losses from manure account for 68% of total NH_4^+ present for surface application, 17% for surface incorporation and 2% for deep placement [42]. Foliar applications until run-off also contribute to N loss, which can be minimized by surfactants, reducing the part of the nutrient supply that drips to the soil.

- Optimized timing of fertilizer application

The date of application also affects the various pathways for the losses. For example, in the case of manure the greatest losses occur when it is applied in autumn because the soil is warm enough for mineralization. Besides, there is sufficient rainfall over the autumn and winter period to leach the resulting nitrate out of the soil profile. Woody crops, as pome or stone fruits, store N in perennial organs before leaf fall and use these reserves to enhance new metabolic activity at bud break. In this context, the delay of N supply to the active vegetative growth phase can minimize N losses. Results in young pear orchards demonstrate that the N supplied before bud break is not utilized by the trees, as they use first the stored N [108].

Applying nitrogen fertilizers and animal slurries at times when the risk of leaching is low, that is outside the rainy season, is a practice that reduces N losses associated with precipitation, which cannot be controlled in open air HPS. Splitting fertilizer applications to match plant demand will also reduce volatilization and leaching losses [106].

- Improved fertilizer technics

A reliable way to reduce NH_3 volatilization is to coat urea fertilizer with a urease inhibitor, because it reduces the conversion rate of urea into ammonium in the soil [47,109]. A reduction of 42% in the cumulative NH_3 emissions through the slower release of NH_4^+ to the soil solution from the hydrolysis of urea was reposted, which also promoted a lower exchangeable N pool [107]. Because ammonia volatilization losses from urea-based fertilizers are variable and unpredictable, the addition of a urease inhibitor can be a potentially valuable mitigation method [110].

The use of a nitrification inhibitor keeps the nitrogen in the NH_4^+ form, which the plants can use, preventing the accumulation of NO_3^- in the soil profile, thus reducing the loss potential to the water bodies and to the atmosphere as N_2O. In some cases the addition of a nitrification inhibitor reduced the N_2O emissions by 45% (in season) and 40% (winter period) during a lettuce-cauliflower rotation, as compared with conventional fertilizer [111]. When the inhibitor was applied to poultry manure and to inorganic fertilizer (broccoli, lettuce and cauliflower crops), the reductions in N_2O losses were of 64% and 32%, respectively [60]. Nevertheless, the beneficial effect of nitrification inhibitors in decreasing direct N_2O emission can be undermined or even outweighed by an increase in NH_3 volatilization [112]. The use of fertilizers with N in readably usable forms, as amino acids or algae extracts, can also contribute to reduce N losses, as it promotes a faster N uptake [113]. However, this option is still quite expensive and not overall established.

The use of fertilizer techniques can reduce nitrate leaching by 20%–30% compared with standard fertilizers. Nevertheless, it may incur an additional cost for the farmer [88].

- Organic manures and composts

Agricultural wastes, industrial wastes and sewage sludge (biosolids) usually contain large amounts of nitrogen. Much of the N in these manures and wastes can be released through mineralization processes in the soil, which ultimately results in a risk of nitrate leaching [53,114]. However, it is very difficult to predict the mineralization rate of these organic wastes and manures because of the variability of the constituent materials [115]. Nitrate leaching losses were reported to be significantly higher from poultry manure (low C:N) or slurry than from straw-based (higher C:N) farm yard manure when applied to an arable free-draining soil in the U.K. [116]. In fact, the addition of organic carbon in these wastes may also increase the rate of mineralization/immobilization turnover of soil N, as well as the applied N [117]. Well matured compost should substitute for fresh manure since it presents higher C:N, hence lower mineralization rates. The calculation of the crop needs should consider the soil mineral N at transplanting, which must be always assessed, and also, the N from organic sources by estimating a mineralization coefficient [118]. In the EU, very large manure applications are not allowed due to the EU limit of 170 $kg \cdot ha^{-1}$ for the nitrate vulnerable zones (Nitrates Directive), and the allowed application periods should be respected according to each country's action program and rainfall pattern.

- Monitoring of plant and/or soil N status

The synchronization between N fertilization and crop N demand is essential in order to minimize the amount of N that over the season is in the soil constituting a potential loss. This may be improved by regular analysis of soil solution, plant sap analysis, leaf chlorophyll measurements or plant tissue testing. This management practice is particularly important for vegetable crops, since for woody species, the buffering capacity conferred by the perennial structures reduces the impact of the monitoring. For these cases, the N correction of fertilization management will have an effect predominantly in the coming season. Tissue testing involves taking samples from the plant (most commonly leaves) at various times during the growth period and sending them to a laboratory for mineral nutrient analysis [15]. Petiole sap testing involves taking leaf petioles and collecting the sap, which is then tested for nitrate and/or potassium using portable meters [119]. Chlorophyll meters are used to measure the "greenness" of individual leaves and to monitor nitrogen status after proper calibration against the crop with proper fertilization [120,121]. Monitoring the soil N status by collecting soil samples, or placing suction cups beyond the root system to collect soil solution, also constitute efficient mitigation practices, as they provide information regarding the N stored in the soil and/or moving beyond the roots' depth.

4. Assessment of the Effectiveness of Mitigation Measures

The success of a mitigation strategy under specific conditions will allow its implementation to be increasingly targeted and more cost effective. Assessment methods are therefore required to evaluate and to quantify the impact of mitigation strategies. These methods provide evidence of a reduction in nutrient loadings or concentration in the receiving compartments (water body or atmosphere) or an improvement in ecological response [73]. Assessment methods must also recognize responses to mitigation that are often site-specific, since a positive response in one area may not be repeated elsewhere. They must be both practical and suited to their end users. The most commonly-used assessment methods are described below.

4.1. Measurements

Long time series of N concentrations in the environmental compartments provide the best analysis of mitigation success, since they describe actual changes resulting from the implementation of a mitigation practice either at the source, timing or transport (Table 3). However, it is not always clear why any response to mitigation has been achieved because of the complex environmental processes involved and our incomplete knowledge of nutrient dynamics. Variations in weather between years add another layer of complexity and make it difficult to distinguish the effect of the mitigation method from environmental noise [122].

Table 3. Assessment method features applicable at field and farm scales (from [73]).

Mitigation Type	Assessment Method		
	Measurement	Budget	Modelling
Source	Y	Y	Y
Timing	Y	N	Y
Transport	Y	N	Y
Data requirements	medium	low	medium/high
Uncertainty	low	medium	medium

Y, sensitive; N, not sensitive.

In the medium-term, the continued development and installation of automated in situ sampling and analytical equipment facilitating high frequency sampling will help improve our understanding and provide more representative assessments [123]. Moreover, sometimes, long time series are necessary to reveal the impact of a mitigation strategy. In this case, or where the measurement is not cost effective, alternatives are required, e.g., nutrient budgets and modelling [73].

4.2. Farm Surveys and Nutrient Budgets

Nutrient budgets are commonly used to assess nutrient management by evaluating inputs and outputs over a period of time. First, there is a need for a survey to collect enough information to enable a thorough system analysis of the crop production practices [124]. The surveyed information is then used to perform the particular water and N budgets. The overall objective is to compare the total amount of water applied with the crop evapotranspiration and the total amount of N fertilizer applied with crop N uptake to identify and quantify possible N surplus and its origin. As the nutrient budget is significantly influenced by horticulture practices, the balances assessment provides an overview of current practices and allows the improvement of NUE. The nutrient budget can be calculated using simple tools as spreadsheets and user-friendly interfaces. A positive balance indicates a potential loss of nutrients to the environment or a nutrient accumulation in the soil, usually designated as surplus. A negative balance signifies soil nutrient depletion. Different budget levels can be considered in association with the system boundaries according to the available data and the purposes of the study (Figure 3). Farm gate balances, for example, demand minimal and routinely available data in farm records compared with soil surface or soil system balances. Thus, the budgets quantify water and nutrients that enter and leave the farm gate with no consideration of internal transfers or loss processes [63]. The key information to be collected in the horticultural farm consists of the area occupied by each crop, the amounts of N applied in fertilizers (mineral and organic) and soil amendments, the N fixation by the crops, the irrigation volumes and yields per crop.

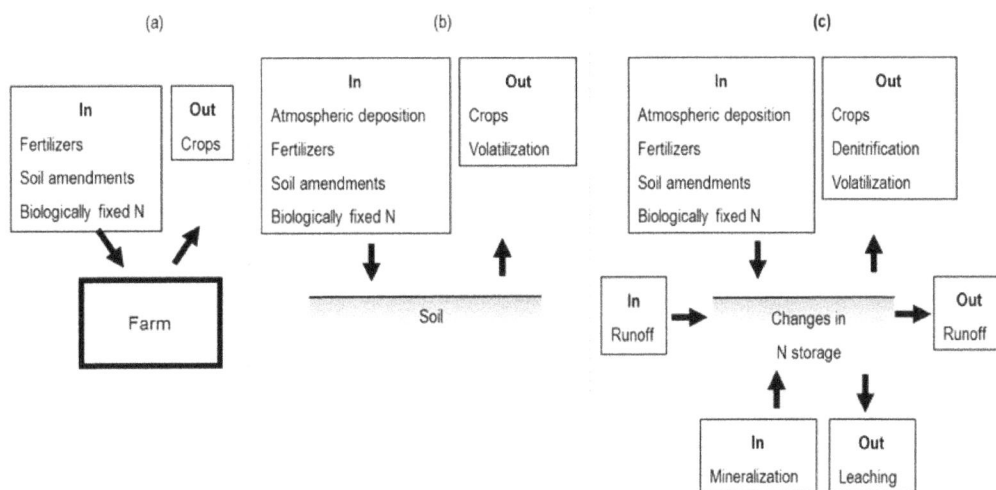

Figure 3. Schematization of the nitrogen budget for horticultural crops, with examples of inputs and outputs considered at three levels: (**a**) farm gate; (**b**) soil surface; (**c**) soil system. Adapted from [63].

The soil surface budgets account for the water and N fluxes through the soil surface boundary while the soil system budgets also consider all of the inputs and outputs resulting from biochemical processes occurring with the soil profile. These two approaches correspond to a more scientific approach often requiring field experiments and the use of mathematical models.

Some studies have calculated the N balances for different vegetable production systems and have estimated that between 9% and 90% of the N applied is not used by the crops and, thus, is potentially lost [125]. These surpluses should be used to estimate potential nutrient loss. To improve the relationship between surpluses and losses, some researchers propose that the budgets should be averaged over a significant number of years to eliminate the effect of temporal variation in climate and farming practices [126,127]. Although the variety of existent accounting systems and the extent of their adoption is encouraging, a uniform and coherent concept for budget calculations at the field and farm scale is required [126,128]. Nonetheless, the more simple budgeting approaches like the farm gate budget fail to consider the timing and transport aspects of mitigation (Table 4) and assume a direct causal relationship between potential and actual nutrient loss [73].

Table 4. Main features of selected models for the simulation of the water and N-related process in horticultural systems.

Model	Type	N Loss Mitigation Measures	Pros	Cons	Application Examples
Export-coefficient	E	fertilizer management	simplicity and minimum data requirements	does not allow extrapolation beyond the range of available information	[129]
Sticks	C	irrigation and fertilizer management	adaptability to various crops; reasonable amount of input data	so far, it has not been used much for vegetables	[130]
DNDC	PB	fertilizer and manure management; crop management (cover crops, rotation, tillage); nitrification inhibitors and slow release fertilizers; irrigation management	big detail in simulating the soil biogeochemical processes; holistic in relation to the different N path losses	big amount of crop physiological parameters as input data; needs more evaluation for vegetable production systems	[131,132]
EU-Rotate	PB	fertilizer management	database with parameters for most vegetable crops; big detail in the N transformations; simple calibration; economic assessment	research model that needs more work; additional studies are necessary to calibrate the mineralization factor of this model for Mediterranean conditions	[133,134]
RZWQM2	PB	fertilizer and manure management; crop management (rotation, mulching, tillage); cover crops; irrigation management; nitrification inhibitors and slow-release fertilizers	database with parameters for most vegetable crops and fruit trees; holistic for the soil-crop-atmosphere system and for the different N path losses; extensive database	considerable amount of soil, crop parameters; average to high difficulty in parameterization and calibration	[36,135–137]
SWAT	PB	reduced fertilization; cover crops, filter strips; crop management (rotation and tillage); fertilization strategies	holistic in relation to the different N path losses (with a modified version)	requires a big amount of soil, crop parameters; difficult parameterization and calibration	[138–140]
VegSyst	C	dry matter production and crop uptake as a result of fertilization management; when incorporated in a DSS predicts N fertilization requirements	specific for vegetable production	it does not predict N losses, but as it predicts crop uptake, it can be used to optimize crop uptake efficiency	[141–143]

E, empirical; C, conceptual; PB, process based.

4.3. Modelling Water and N-Related Processes in Horticultural Systems

The increasing concern about the quality of the environment, together with the development of hardware, software and user-friendly interfaces led to an increase in the use of modelling in agricultural systems. Lately, the evaluation of the implementation of the environmental policies has also been a driver for modelling applications. There is a range in the complexity and variety of models for assessing the environmental impact of the water and N cycles in agricultural production systems. These models provide mechanisms for comparing the relative effects of differing mitigation practices on yields and N losses for different soil-crop-climate systems. The running of a number of scenarios has already demonstrated that nitrogen management in field vegetable rotations can be improved in Europe by following at least good agricultural practices.

Considering the approach by which the various models simulate water and nitrogen dynamics in the soil, it is possible to distinguish from simple empirical applications to comprehensive, fully process-driven models. Examples of these types of models applicable at plot and farm scales are given in Table 4, including application case references. Empirical models evaluate inputs and outputs, but provide no consideration of internal dynamics, adopting instead a 'black box' approach. They are built on quantitative relationships rather than process understanding and generally have low data requirements. These features limit the evaluation of some mitigation methods. Conceptual models occupy an intermediate position in terms of complexity, being process informed and requiring greater empirical evidence to support the selection of coefficients, but not yet offering a full process representation. Process models attempt to simulate the complete systems and quantify all processes that constitute them. They are therefore computationally intense, require expertise and have large data requirements. The wide range of inputs and parameters, the inclusion of transport processes, short time steps and high spatial resolution allow the simulation of source, time and transport mitigation methods.

Simulation models need preparatory work of calibration and validation before being used with different crops and soils, which is a rather expensive and time-consuming task. Some of the models that simulate the N dynamics in the soil-plant system are too complex to be used and/or understood by farmers or are still in the state of research at the moment. They require a large amount of input data, and in some cases, research is necessary to obtain calibration factors for specific horticulture crops. The more complex models are useful to test new strategies at a regional level and for legislation purposes. In most cases, advisors should support farmers when they want to use prediction models, and the latter should be involved in the demonstration of models. Some models have been simplified to make them easier to be used by farmers and consultants, reducing the number of required inputs and assuming the possible risk of loss of accuracy. These simplified models work well and provide simple and useful information for farmers to design appropriate and to compare different fertilization and irrigation strategies.

5. Conclusions

The reduction of N emissions to water bodies and the atmosphere by horticultural production systems is required. Current directives and international conventions regarding agriculture consider the threats from NO_3^- leaching, NH_3 and N_2O emissions separately. However, when not combined with an integrated approach to N and irrigation management, the policy measures may induce mitigation strategies/practices with antagonistic effects (swapping pollution).

The combined use of fertigation and drip irrigation to frequently apply small amounts of water and N throughout a crop cycle provides the technical possibility for precise N and irrigation management. A combination of optimal water management and applying recommended fertilization rates should be the most profitable choice for the farmer. The recommendations must be corrected in situ by using soil and plant monitoring devices. In addition to the in-crop season good practices, the use of cover crops to catch the N released from the horticultural crop residues during the winter period also seems indispensable. The use of fast-growing rootstocks, mycorrhizal plants or more ready-to-assimilate fertilizers, as amino acids, are other examples of mitigation strategies that can also contribute to reduce N losses.

Mitigation strategies need to be assessed and the best ones selected and broadly applied at low cost. Simulation models can make long-term assessments as opposed to an expensive and time-consuming pure classical field research. Once models have been calibrated and validated against selected field data, they can be used before and during the cropping season considering the crop type, management practices and environmental conditions for better fertilization and irrigation strategies. A range of potential mitigation practices can be evaluated using models, and the most promising ones can be field tested. Additional studies are still necessary to create complete reference databases (climatic data, physical and chemical soil parameters and crop calibration factors) for different European areas to promote the use of prediction models. The combined use of soil and plant monitoring sensors and simulation models is a useful group of tools for designing optimal irrigation and N management practices that mitigate N losses. The combination of these techniques with crop rotation and intercropping systems will undoubtedly improve the NUE of horticulture crops and minimize N losses as a whole.

Author Contributions: Maria do Rosário Cameira outlined the paper; Maria do Rosário Cameira and Mariana Mota wrote the paper.

Conflicts of Interest: The authors declare no conflict of interest.

References

1. World Meteorological Organization. *Planning of Water Quality Monitoring Systems*; Technical report series 3; WMO: Geneva, Switzerland, 2013; p. 117.
2. Oenema, O.; Bleeker, A.; Braathen, N.A.; Budnakova, M.; Bull, K.; Cermak, P.; Geupel, M.; Hicks, K.; Hoft, R.; Kozlova, N.; et al. Nitrogen in current European policies. In *The European Nitrogen Assessment*; Sutton, M., Howard, C., Erisman, J., Billen, G., Bleeker, A., Grennfelt, P., Grinsven, H., Grizzetti, B., Eds.; Cambridge University Press: Cambridge, UK, 2011; pp. 62–81.
3. Addiscott, T.M. Fertilizers and nitrate leaching. In *Agricultural Chemicals and the Environment Issues*; Hester, R.E., Harrison, R.M., Eds.; CABI: Oxfordshire, UK, 1996; pp. 1–26.
4. Di, H.J.; Cameron, K.C. Nitrate leaching in temperate agroecosystems: Sources, factors and mitigating strategies. *Nutr. Cycl. Agroecosyst.* **2002**, *64*, 237–256. [CrossRef]
5. Ward, M.H.; DeKok, T.M.; Levallois, P.; Brender, J.; Gulis, G.; Nolan, B.T.; VanDerslice, J. Workgroup report: Drinking-water nitrate and health-recent findings and research needs. *Environ. Health Perspect.* **2005**, *113*, 1607–1614. [CrossRef] [PubMed]
6. Grizzetti, B.; Bouraoui, F.; Billen, G.; van Grinsven, H.; Cardoso, A.C.; Thieu, V.; Garnier, J.; Curtis, C.; Howarth, R.W.; Johnes, P. Nitrogen as a threat to European water quality. In *The European Nitrogen Assessment*; Sutton, M., Howard, C., Erisman, J., Billen, G., Bleeker, A., Grennfelt, P., Grinsven, H., Grizzetti, B., Eds.; Cambridge University Press: Cambridge, UK, 2011; pp. 379–404.
7. Smith, V.H.; Tilman, G.D.; Nekola, J.C. Eutrophication: Impacts of excess nutrient inputs on freshwater, marine, and terrestrial ecosystems. *Environ. Pollut.* **1999**, *100*, 179–196. [CrossRef]
8. Moldanováá, J.; Grennfelt, P.; Jonsson, Å.; Simpson, D.; Spranger, T.; Aas, W.; Munthe, J.; Rabl, A. Nitrogen as a threat to European air quality. In *The European Nitrogen Assessment*; Sutton, M., Howard, C., Erisman, J., Billen, G., Bleeker, A., Grennfelt, P., Grinsven, H., Grizzetti, B., Eds.; Cambridge University Press: Cambridge, UK, 2011; pp. 405–433.
9. Cameron, K.C.; Rate, A.W.; Noonan, M.J.; Moore, S.; Smith, N.P.; Kerr, L.E. Lysimeter study of the fate of nutrients following subsurface injection and surface application of dairy pond sludge to pasture. *Agric. Ecosyst. Environ.* **1996**, *58*, 187–197. [CrossRef]
10. Oenema, O.; Oudendag, D.A.; Witzke, H.P.; Monteny, G.J.; Velthof, G.L.; Pietrzak, S.; Pinto, M.; Britz, W.; Schwaiger, E.; Erisman, J.W.; et al. *Integrated Measures in Agriculture to Reduce Ammonia Emissions*; Final Summary Report; Alterra: Wageningen, The Netherlands, 2007.
11. Stevens, C.J.; Quinton, J.N. Diffuse pollution swapping in arable agricultural systems. *Crit. Rev. Environ. Sci. Technol.* **2009**, *39*, 478–520. [CrossRef]

12. Goulding, K. Nitrate leaching from arable and horticultural land. *Soil Use Manag.* **2000**, *16*, 145–151. [CrossRef]

13. Pionke, H.B.; Sharma, M.L.; Hirschberg, K.J. Impact of irrigated horticulture on nitrate concentrations in groundwater. *Agric. Ecosys. Environ.* **1990**, *32*, 119–132. [CrossRef]

14. Zhu, J.H.; Li, X.L.; Christie, P.; Li, J.L. Environmental implications of low nitrogen use efficiency in excessively fertilized hot pepper (*Capsicum frutescens* L.) cropping systems. *Agric. Ecosys. Environ.* **2005**, *111*, 70–80. [CrossRef]

15. Peña-Fleitas, M.T.; Gallardo, M.; Thompson, R.B.; Farneselli, M.; Padilla, F.M. Assessing crop N status of fertigated vegetable crops using plant and soil monitoring techniques. *Ann. Appl. Biol.* **2015**, *167*, 387–405. [CrossRef] [PubMed]

16. Agneessens, L.; De Waele, J.; De Neve, S. Review of alternative management options of vegetable crop residues to reduce nitrate leaching in intensive vegetable rotations. *Agronomy* **2014**, *4*, 529–555. [CrossRef]

17. Congreves, K.A.; Van Eerd, L.L. Nitrogen cycling and management in intensive horticultural systems. *Nutr. Cycl. Agroecosyst.* **2015**, *102*, 299–318. [CrossRef]

18. Connor, D.J.; Gómez-del-Campo, M.; Rousseaux, M.C.; Searles, P.S. Structure, management and productivity of hedgerow olive orchards: A review. *Sci. Hortic.* **2014**, *169*, 71–93. [CrossRef]

19. Rufat, J.; Villar, J.M.; Pasqual, M.; Falguera, V.; Arbonés, A. Productive and vegetative response to different irrigation and fertilization strategies of an Arbequina olive orchard grown undersuper-intensive conditions. *Agric. Water Manag.* **2014**, *144*, 33–41. [CrossRef]

20. FAO. *Statistical Year Book, Europe and Central Asia Food and Agriculture*; FAO: Budapest, Hungary, 2014.

21. Vázquez, N.; Pardo, A.; Suso, M.L.; Quemada, M. Drainage and nitrate leaching under processing tomato growth with drip irrigation and plastic mulching. *Agric. Ecosyst. Environ.* **2006**, *112*, 313–323.

22. Tilman, D.; Cassman, K.G.; Matson, P.A.; Naylor, R.; Polasky, S. Agricultural sustainability and intensive production practices. *Nature* **2002**, *418*, 671–677. [CrossRef] [PubMed]

23. Agostini, F.; Tei, F.; Silgram, M.; Farneselli, M.; Benincasa, P.; Aller, M.F. Decreasing nitrate leaching in vegetable crops with better N management. In *Genetic Engineering, Biofertilisation, Soil Quality and Organic Farming*; Sustainable Agriculture Reviews; Lichtfouse, E., Ed.; Springer: Heidelberg, Germany, 2010; Volume 4, pp. 147–200.

24. Christensen, B.T. Tightening the Nitrogen Cycle. In *Managing Soil Quality—Challenges in Modern Agriculture*; CAB International: Wallingford, UK, 2004; pp. 47–67.

25. Galloway, J.N.; Dentener, F.J.; Capone, D.G.; Boyer, E.W.; Howarth, R.W.; Seitzinger, S.P.; Asner, G.P.; Cleveland, C.C.; Green, P.A.; Holland, E.A.; et al. Nitrogen cycles: Past, present, and future. *Biogeochemistry* **2004**, *70*, 153–226. [CrossRef]

26. Hillel, D. *Environmental Soil Physics: Fundamentals, Applications, and Environmental Considerations*; Academic Press: London, UK, 1998; p. 771.

27. Mitchell, R.; Webb, J.; Harrison, R. Crop residues can affect N leaching over at least two winters. *Eur. J. Agron.* **2001**, *15*, 17–29. [CrossRef]

28. Wyland, L.J.; Jackson, L.E.; Chaney, W.E.; Klonsky, K.; Koike, S.T.; Kimple, B. Winter cover crops in a vegetable cropping system: Impacts on nitrate leaching, soil water, crop yield, pests and management costs. *Agric. Ecosyst. Environ.* **1996**, *59*, 1–17. [CrossRef]

29. De Neve, S.; Sáez, S.G.; Daguilar, B.C.; Sleutel, S.; Hofman, G. Manipulating N mineralization from high N crop residues using on-and off-farm organic materials. *Soil Biol. Biochem.* **2004**, *36*, 127–134. [CrossRef]

30. Feaga, J.B.; Selker, J.S.; Dick, R.P.; Hemphill, D.D. Long-term nitrate leaching under vegetable production with cover crops in the Pacific Northwest. *Soil Sci. Soc. Am. J.* **2010**, *74*, 186–195. [CrossRef]

31. Campiglia, E.; Mancinelli, R.; Radicetti, E.; Marinari, S. Legume cover crops and mulches: Effects on nitrate leaching and nitrogen input in a pepper crop (*Capsicum annuum* L.). *Nutr. Cycl. Agroecosyst.* **2011**, *89*, 399–412. [CrossRef]

32. Bruin, A.J.; Ball Coelho, B.R.; Beyaert, R.P.; Reeleder, R.D.; Roy, R.C.; Capell, B. High value crops in coarse-textured soil and nitrate leaching—How risky is it? *Can. J. Plant Sci.* **2010**, *90*, 515–528. [CrossRef]

33. Sharma, P.; Shukla, M.K.; Sammis, T.W.; Adhikari, P. Nitrate-nitrogen leaching from onion bed under furrow and drip irrigation systems. *Appl. Environ. Soil Sci.* **2012**, *2012*, 650206. [CrossRef]

34. Chotangui, A.H.; Sugahara, K.; Okabe, M.; Kasuga, S.; Isobe, K.; Higo, M.; Torigoe, Y. Evaluation of NO_3-N Leaching in Commercial Fields of Leafy Vegetables by the Soil Nitrogen Balance Estimation System. *Environ. Control Biol.* **2015**, *53*, 145–157. [CrossRef]

35. Woli, P.; Hoogenboom, G.; Alva, A. Simulation of potato yield, nitrate leaching, and profit margins as influenced by irrigation and nitrogen management in different soils and production regions. *Agric. Water Manag.* **2016**, *171*, 120–130. [CrossRef]

36. Merwin, I.A.; Ray, J.A.; Steenhuis, T.S.; Boll, J. Groundcover management systems influence fungicide and nitrate-N concentrations in leachate and runoff from a New York apple orchard. *J. Am. Soc. Hortic. Sci.* **1996**, *121*, 249–257.

37. Cameira, M.R.; Pereira, A.; Ahuja, L.; Ma, L. Sustainability and environmental assessment of fertigation in an intensive olive grove under Mediterranean conditions. *Agric. Water Manag.* **2014**, *146*, 346–360. [CrossRef]

38. Hardie, M.A.; Oliver, G.; Clothier, B.E.; Bound, S.A.; Green, S.A.; Close, D.C. Effect of biochar on nutrient leaching in a young apple orchard. *J. Environ. Qual.* **2015**, *44*, 1273–1282. [CrossRef] [PubMed]

39. Lea-Cox, J.D.; Syvertsen, J.P.; Graetz, D.A. Springtime [15]Nitrogen uptake, partitioning, and leaching losses from young bearing citrus trees of differing nitrogen status. *J. Am. Soc. Hort. Sci.* **2001**, *126*, 242–251.

40. Lehmann, J.; Schroth, G. Nutrient leaching. In *Trees, Crops and Soil Fertility*; Schroth, G., Sinclair, F.L., Eds.; CABI: Oxfordshire, UK, 2003; pp. 151–166.

41. Sommer, S.G.; Schjoerring, J.K.; Denmead, O.T. Ammonia emission from mineral fertilizers and fertilized crops. *Adv. Agron.* **2004**, *82*, 557–622.

42. Huijsmans, J.F.M.; Hol, J.M.G.; Vermeulen, G.D. Effect of application method, manure characteristics, weather and field conditions on ammonia volatilization from manure applied to arable land. *Atmos. Environ.* **2003**, *37*, 3669–3680. [CrossRef]

43. Miola, E.C.; Rochette, P.; Chantigny, M.H.; Angers, D.A.; Aita, C.; Gasser, M.O.; Pelster, D.E.; Bertrand, N. Ammonia volatilization after surface application of laying-hen and broiler-chicken manures. *J. Environ. Qual.* **2014**, *43*, 1864–1872. [CrossRef] [PubMed]

44. De Ruijter, F.J.; Huijsmans, J.F.M.; Rutgers, B. Ammonia volatilization from crop residues and frozen green manure crops. *Atmos. Environ.* **2010**, *44*, 3362–3368. [CrossRef]

45. Sadeghpour, A.; Hashemi, M.; Weis, S.A.; Spargo, J.T.; Mehrvarz, S.; Herbert, S.J. Assessing tillage systems for reducing ammonia volatilization from spring-applied slurry manure. *Commun. Soil Sci. Plant Anal.* **2015**, *46*, 724–735. [CrossRef]

46. Sutton, M.A.; Grinsven, H. Summary for policy makers. In *The European Nitrogen Assessment*; Sutton, M.A., Howard, C.M., Erisman, J.W., Billen, G., Grennfelt, P., van Grinsven, H., Grizzetti, B., Eds.; Cambridge Univerity Press: Cambridge, UK, 2011; pp. 2–18.

47. Bishop, P.; Manning, M. *Urea Volatilisation: The Risk Management and Mitigation Strategies*; Occasional Report no. 24; Massey University: Palmerston North, New Zealand, 2010.

48. Rochette, P.; Tremblay, N.; Fallon, E.; Angers, D.A.; Chantigny, M.H.; MacDonald, J.D.; Bertrand, N.; Parent, L.É. N_2O emissions from an irrigated and non-irrigated organic soil in eastern Canada as influenced by N fertilizer addition. *Europ. J. Soil Sci.* **2010**, *61*, 186–196. [CrossRef]

49. Trost, B.; Prochnow, A.; Drastig, K.; Meyer-Aurich, A.; Ellmer, F.; Baumecker, M. Irrigation, soil organic carbon and N_2O emissions. A review. *Agron. Sustain. Dev.* **2013**, *33*, 733–749. [CrossRef]

50. Benckiser, G.; Schartel, T.; Weiske, A. Control of NO_3^- and N_2O emissions in agroecosystems: A review. *Agron. Sustain. Dev.* **2015**, *35*, 1059–1074. [CrossRef]

51. Prosser, J.I. The ecology of nitrifying bacteria. In *Biology of the Nitrogen Cycle*; Bolhe, H., Fergusoin, S.J., Newton, W.E., Eds.; Elsevier: Oxford, UK, 2006; pp. 223–245.

52. Kool, D.M.; Müller, C.; Wrage, N.; Oenema, O.; Van Groenigen, J.W. Oxygen exchange between nitrogen oxides and H_2O can occur during nitrifier pathways. *Soil Biol. Biochem.* **2009**, *41*, 1632–1641. [CrossRef]

53. Cameron, K.C.; Di, H.J.; Moir, J.L. Nitrogen losses from the soil/plant system: A review. *Ann. Appl. Biol.* **2013**, *162*, 145–173. [CrossRef]

54. Haynes, R. *Mineral Nitrogen in the Plant-Soil System*; Elsevier: Amsterdam, The Netherlands, 2012.

55. Monaghan, R.M.; Barraclough, D. Some chemical and physical factors affecting the rate and dynamics of nitrification in urine-affected soil. *Plant Soil* **1992**, *143*, 11–18. [CrossRef]

56. Saggar, S.; Jha, N.; Deslippe, J.; Bolan, N.S.; Luo, J.; Giltrap, D.L.; Kim, D.-G.; Zaman, M.; Tillman, R.W. Denitrification and N_2O: N_2 production in temperate grasslands: Processes, measurements, modelling and mitigating negative impacts. *Sci. Total Environ.* **2013**, *465*, 173–195. [CrossRef] [PubMed]

57. Ye, R.B.; Averill, B.A.; Tiedje, J.M. Denitrification: Production and consumption of nitric oxide. *Appl. Environ. Microbiol.* **1994**, *60*, 1053–1058. [PubMed]

58. Ussiri, D.; Lal, R. The role of nitrous oxide on climate change. In *Soil Emission of Nitrous Oxide and Its Mitigation*; Springer: Dordrechet, The Netherlands, 2013; pp. 1–28.

59. Davidson, E.A. Soil water content and the ratio of nitrous oxide to nitric oxide emitted from soil. In *Biogeochemistry of Global Change*; Springer: Berlin, Germany, 1993; pp. 369–386.

60. Pfab, H.; Palmer, I.; Buegger, F.; Fiedler, S.; Müller, T.; Ruser, R. Influence of a nitrification inhibitor and of placed N-fertilization on N_2O fluxes from a vegetable cropped loamy soil. *Agric. Ecosyst. Environ.* **2012**, *150*, 91–101. [CrossRef]

61. Zhu, T.; Zhang, J.; Yang, W.; Cai, Z. Effects of organic material amendment and water content on NO, N_2O, and N_2 emissions in a nitrate-rich vegetable soil. *Biol. Fertil. Soils* **2013**, *49*, 153–163. [CrossRef]

62. Velthof, G.L.; Kuikman, P.J.; Oenema, O. Nitrous oxide emission from soils amended with crop residues. *Nutr. Cycl. Agroecosyst.* **2002**, *62*, 249–261. [CrossRef]

63. Cherry, K.A.; Shepherd, M.; Withers, P.J.A.; Mooney, S.J. Assessing the effectiveness of actions to mitigate nutrient loss from agriculture: A review of methods. *Sci. Total Environ.* **2008**, *406*, 1–23. [CrossRef] [PubMed]

64. Palmer, J.W.; Dryden, G. Fruit mineral removal rates from New Zealand apple (*Malus domestica*) orchards in the Nelson region. *N. Z. J. Crop Hortic. Sci.* **2006**, *34*, 27–32. [CrossRef]

65. De Neve, S.; Hofman, G. N mineralization and nitrate leaching from vegetable crop residues under field conditions: A model evaluation. *Soil Biol. Biochem.* **1998**, *30*, 2067–2075. [CrossRef]

66. Chaves, B.; De Neve, S.; Cabrera, M.D.C.L.; Boeckx, P.; Van Cleemput, O.; Hofman, G. The effect of mixing organic biological waste materials and high-N crop residues on the short-time N_2O emission from horticultural soil in model experiments. *Biol. Fertil. Soils* **2005**, *41*, 411–418. [CrossRef]

67. De Neve, S.; Pannier, J.; Hofman, G. Temperature effects on C-and N-mineralization from vegetable crop residues. In Proceedings of the 8th Nitrogen Workshop, Ghent, Belgium, 5–8 September 1996; pp. 41–46.

68. Trinsoutrot, I.; Nicolardot, B.; Justes, E.; Recous, S. Decomposition in the field of residues of oilseed rape grown at two levels of nitrogen fertilisation. Effects on the dynamics of soil mineral nitrogen between successive crops. *Nutr. Cycl. Agroecosyst.* **2000**, *56*, 125–137. [CrossRef]

69. Congreves, K.A.; Vyn, R.J.; Van Eerd, L.L. Evaluation of post-harvest organic carbon amendments as a strategy to minimize nitrogen losses in cole crop production. *Agronomy* **2013**, *3*, 181–199. [CrossRef]

70. Rahn, C.R.; Vaidyanathan, L.V.; Paterson, C.D. Nitrogen residues from brassica crops. *Asp. Appl. Biol.* **1992**, *30*, 263–270.

71. Everaarts, A.P.; De Moel, C.P.; Van Noordwijk, M. The effect of nitrogen and the method of application on nitrogen uptake of cauliflower and on nitrogen in crop residues and soil at harvest. *NJAS Wagening. J. Life Sci.* **1996**, *44*, 43–55.

72. Nett, L.; Feller, C.; George, E.; Fink, M. Effect of winter catch crops on nitrogen surplus in intensive vegetable crop rotations. *Nutr. Cycl. Agroecosyst.* **2011**, *91*, 327–337. [CrossRef]

73. Whitmore, A.P. Modelling the release and loss of nitrogen after vegetable crops. *NJAS Wagening. J. Life Sci.* **1996**, *44*, 73–86.

74. Bakker, C.J.; Swanton, C.J.; McKeown, A.W. Broccoli growth in response to increasing rates of pre-plant nitrogen. I. Yield and quality. *Can. J. Plant Sci.* **2009**, *89*, 527–537. [CrossRef]

75. Wehrmann, J.; Scharpf, H.C. Reduction of nitrate leaching in a vegetable farm: Fertilization, crop rotation, plant residues. In *Management Systems to Reduce Impact of Nitrates*; Elsevier Applied Sciences: London, UK, 1989.

76. Coppens, F.; Garnier, P.; Findeling, A.; Merckx, R.; Recous, S. Decomposition of mulched versus incorporated crop residues: Modelling with PASTIS clarifies interactions between residue quality and location. *Soil Biol. Biochem.* **2007**, *39*, 2339–2350. [CrossRef]

77. Mitchell, R.D.J.; Harrison, R.; Russell, K.J.; Webb, J. The effect of crop residue incorporation date on soil inorganic nitrogen, nitrate leaching and nitrogen mineralization. *Biol. Fertil. Soils* **2000**, *32*, 294–301. [CrossRef]

78. Smit, A.L.; de Ruijter, F.J.; ten Berge, H.F.M. The fate of nitrogen from crop residues of broccoli, leek and sugar beet. In Proceedings of the IV International Symposium on Ecologically Sound Fertilization Strategies for Field Vegetable Production, Alnarp, Sweden, 22–29 September 2008; Volume 852, pp. 157–162.

79. Nutter, F., Jr. The role of plant disease epidemiology in developing successful integrated disease management programs. In *General Concepts in Integrated Pest and Disease Management*; Ciancio, A., Mukerji, K.J., Eds.; Springer: Berlin, Germany, 2007; pp. 45–90.

80. Thorup-Kristensen, K.; Dresbøll, D.B.; Kristensen, H.L. Crop yield, root growth, and nutrient dynamics in a conventional and three organic cropping systems with different levels of external inputs and N re-cycling through fertility building crops. *Eur. J. Agron.* **2012**, *37*, 66–82. [CrossRef]

81. Thorup-Kristensen, K. Effect of deep and shallow root systems on the dynamics of soil inorganic N during 3-year crop rotations. *Plant Soil.* **2006**, *288*, 233–248. [CrossRef]

82. Gabriel, J.L.; Muñoz-Carpena, R.; Quemada, M. The role of cover crops in irrigated systems: Water balance, nitrate leaching and soil mineral nitrogen accumulation. *Agric. Ecosyst. Environ.* **2012**, *155*, 50–61. [CrossRef]

83. Campiglia, E.; Mancinelli, R.; Di Felice, V.; Radicetti, E. Long-term residual effects of the management of cover crop biomass on soil nitrogen and yield of endive (*Cichorium endivia* L.) and savoy cabbage (*Brassica oleracea* var. sabauda). *Soil Tillage Res.* **2014**, *139*, 1–7. [CrossRef]

84. Bliss, C.; Andersen, P.; Brodbeck, B.; Wright, D.; Olson, S.; Marois, J. The Influence of Bahiagrass, Tillage, and Cover Crops on Organic Vegetable Production and Soil Quality in the Southern Coastal Plain. *Sustain. Agric. Res.* **2016**, *5*, 65. [CrossRef]

85. Morgan, R.P. *Soil Erosion and Conservation*; Longman Science & Technical: Hoboken, NJ, USA, 1995.

86. Govers, G.; Poesen, J.; Goossens, D.; Christensen, B.T. Soil erosion-processes, damages and countermeasures. In *Managing Soil Quality, Challenges in Modern Agriculture*; Schjønning, P., Elmholt, S., Christensen, B.T., Eds.; Institute of Agricultural Sciences: Tjele, Denmark, 2004; pp. 199–217.

87. Wiman, M.R.; Kirby, E.M.; Granatstein, D.M.; Sullivan, T. Cover crops influence meadow vole presence in organic orchards. *HortTechnology* **2009**, *19*, 558–562.

88. Quemada, M.; Baranski, M.; Nobel-de Lange, M.N.J.; Vallejo, A.; Cooper, J.M. Meta-analysis of strategies to control nitrate leaching in irrigated agricultural systems and their effects on crop yield. *Agric. Ecosyst. Environ.* **2013**, *174*, 1–10. [CrossRef]

89. Khan, S.A.; Mulvaney, R.L.; Ellsworth, T.R.; Boast, C.W. The myth of nitrogen fertilization for soil carbon sequestration. *J. Environ. Qual.* **2007**, *36*, 1821–1832. [CrossRef] [PubMed]

90. Jensen, E.S.; Peoples, M.B.; Hauggaard-Nielsen, H. Faba bean in cropping systems. *Field Crops Res.* **2010**, *115*, 203–216. [CrossRef]

91. Li, X.; Hu, C.; Delgado, J.A.; Zhang, Y.; Ouyang, Z. Increased nitrogen use efficiencies as a key mitigation alternative to reduce nitrate leaching in north china plain. *Agric. Water Manag.* **2007**, *89*, 137–147. [CrossRef]

92. Thorup-Kristensen, K. Utilising differences in rooting depth to design vegetable crop rotations with high nitrogen use efficiency (NUE). In Proceedings of the Workshop Towards and Ecologically Sound Fertilisation in Field Vegetable Production, Wageningen, The Netherlands, 11–14 September 2000; Volume 571, pp. 249–254.

93. Asghari, H.R.; Cavagnaro, T.R. Arbuscular mycorrhizas reduce nitrogen loss via leaching. *PLoS ONE* **2012**, *7*, e29825. [CrossRef] [PubMed]

94. Syvertsen, J.P.; Smith, L.M. Nitrogen leaching, N uptake efficiency and water use from citrus trees fertilized at three N. *Proc. Fla. State Hort. Soc.* **1995**, *108*, 151–155.

95. Shang, F.; Ren, S.; Yang, P.; Chi, Y.; Xue, Y. Effects of Different Irrigation Water Types, N Fertilizer Types, and Soil Moisture Contents on N_2O Emissions and N Fertilizer Transformations in Soils. *Water Air Soil Pollut.* **2016**, *227*, 225. [CrossRef]

96. Allen, R.G.; Pereira, L.S.; Raes, D.; Smith, M. *Crop Evapotranspiration-Guidelines for Computing Crop Water Requirements*; FAO Irrigation and Drainage Paper 56; FAO: Rome, Italy, 1998.

97. Keller, J.; Bliesner, R.D. *Sprinkle and Trickle Irrigation*; The Blackburn Press: Caldwel, NJ, USA, 1990; p. 652.

98. Thompson, R.B.; Gallardo, M.; Valdez, L.C.; Fernández, M.D. Using plant water status to define threshold values for irrigation management of vegetable crops using soil moisture sensors. *Agric. Water Manag.* **2007**, *88*, 147–158. [CrossRef]

99. Granados, M.R.; Thompson, R.B.; Fernández, M.D.; Martínez-Gaitán, C.; Gallardo, M. Prescriptive–corrective nitrogen and irrigation management of fertigated and drip-irrigated vegetable crops using modeling and monitoring approaches. *Agric. Water Manag.* **2013**, *119*, 121–134. [CrossRef]

100. Rekika, D.; Caron, J.; Rancourt, G.T.; Lafond, J.A.; Gumiere, S.J.; Jenni, S.; Gosselin, A. Optimal irrigation for onion and celery production and spinach seed germination in Histosols. *Agron. J.* **2014**, *106*, 981–994. [CrossRef]

101. Létourneau, G.; Caron, J.; Anderson, L.; Cormier, J. Matric potential-based irrigation management of field-grown strawberry: Effects on yield and water use efficiency. *Agric. Water Manag.* **2015**, *161*, 102–113. [CrossRef]

102. Rahil, M.H.; Qanadillo, A. Effects of different irrigation regimes on yield and water use efficiency of cucumber crop. *Agric. Water Manag.* **2015**, *148*, 10–15. [CrossRef]

103. Berrie, A.; Ellerker, B.; Lower, K. Evaluation of alternative treatments to urea to eliminate leaf litter in organic apple production. *IOBC/WPRS Bull.* **2006**, *29*, 139.

104. Mitre, V.; Mitre, J.; Sestras, A.; Petrisor, P.; Sestras, R.E. Reducing primary inoculum of apple scab using foliar application of urea in autumn. *Bull. UASVM Hortic.* **2012**, *69*, 415–416.

105. Rochette, P.; Angers, D.A.; Chantigny, M.H.; Gasser, M.O.; MacDonald, J.D.; Pelster, D.E.; Bertrand, N. Ammonia volatilization and nitrogen retention: How deep to incorporate urea? *J. Environ. Qual.* **2013**, *42*, 1635–1642. [CrossRef] [PubMed]

106. Black, A.S.; Sherlock, R.R.; Smith, N.P.; Cameron, K.C.; Goh, K.M. Effects of form of nitrogen, season, and urea application rate on ammonia volatilisation from pastures. *N. Z. J. Agric. Res.* **1985**, *28*, 469–474. [CrossRef]

107. Sanz-Cobena, A.; Misselbrook, T.H.; Arce, A.; Mingot, J.I.; Diez, J.A.; Vallejo, A. An inhibitor of urease activity effectively reduces ammonia emissions from soil treated with urea under Mediterranean conditions. *Agric. Ecosyst. Environ.* **2008**, *126*, 243–249. [CrossRef]

108. Neto, C.; Carranca, C.; Clemente, J.; de Varennes, A. Nitrogen distribution, remobilization and re-cycling in young orchard of non-bearing 'Rocha' pear trees. *Sci. Hortic.* **2008**, *118*, 299–307. [CrossRef]

109. Shan, L.; He, Y.; Chen, J.; Huang, Q.; Wang, H. Ammonia volatilization from a Chinese cabbage field under different nitrogen treatments in the Taihu Lake Basin, China. *J. Environ. Sci.* **2015**, *38*, 14–23. [CrossRef] [PubMed]

110. Chambers, B.; Dampney, P. *Nitrogen Efficiency and Ammonia Emissions from Urea-Based and Ammonium Nitrate Fertilisers*; The International Fertiliser Society: Colchester, UK, 2009; p. 657.

111. Riches, D.; Mattner, S.; Davies, R.; Porter, I. Mitigation of nitrous oxide emissions with nitrification inhibitors in temperate vegetable cropping in southern Australia. *Soil Res.* **2016**, *54*, 533–543. [CrossRef]

112. Lam, S.K.; Suter, H.; Mosier, A.R.; Chen, D. Using nitrification inhibitors to mitigate agricultural N_2O emission: A double-edged sword? *Glob. Chang. Biol.* **2016**, *23*, 485–489. [CrossRef] [PubMed]

113. Shehata, S.M.; Abdel-Azem, H.S.; El-Yazied, A.A.; Abou, A.; El-Gizawy, A.M. Effect of foliar spraying with amino acids and seaweed extract on growth chemical constitutes, yield and its quality of celeriac plant. *Eur. J. Sci. Res.* **2011**, *58*, 257–265.

114. Elwell, D.L.; Keener, H.M.; Carey, D.S.; Schlak, P.P. Composting unamended chicken manure. *Compost. Sci. Util.* **1998**, *6*, 22–35. [CrossRef]

115. Van Kessel, J.S.; Reeves, J. Nitrogen mineralization potential of dairy manures and its relationship to composition. *Biol. Fertil. Soils* **2002**, *36*, 118–123. [CrossRef]

116. Chambers, B.J.; Smith, K.A.; Pain, B.F. Strategies to encourage better use of nitrogen in animal manures. *Soil Use Manag.* **2000**, *16*, 157–166. [CrossRef]

117. Heinrich, A.L.; Pettygrove, G.S. Influence of dissolved carbon and nitrogen on mineralization of dilute liquid dairy manure. *Soil Sci. Soc. Am. J.* **2012**, *76*, 700–709. [CrossRef]

118. Schepers, J.S.; Mosier, A.R. Accounting for nitrogen in nonequilibrium soil-crop systems. In *Managing Nitrogen for Groundwater Quality and Farm Profitability*; Soil Science Society of America: Fitchburg, WI, USA, 1991; pp. 125–138.

119. Farneselli, M.; Tei, F.; Simonne, E. Reliability of petiole sap test for N nutritional status assessing in processing tomato. *J. Plant Nutr.* **2014**, *37*, 270–278. [CrossRef]

120. Parry, C.; Blonquist, J.; Bugbee, B. In situ measurement of leaf chlorophyll concentration: Analysis of the optical/absolute relationship. *Plant Cell Environ.* **2014**, *37*, 2508–2520. [CrossRef] [PubMed]

121. Padilla, F.M.; Peña-Fleitas, M.T.; Gallardo, M.; Thompson, R.B. Threshold values of canopy reflectance indices and chlorophyll meter readings for optimal nitrogen nutrition of tomato. *Ann. Appl. Biol.* **2015**, *166*, 271–285. [CrossRef]

122. Lord, E.I.; Anthony, S.G.; Goodlass, G. Agricultural nitrogen balance and water quality in the UK. *Soil Use Manag.* **2002**, *18*, 363–369. [CrossRef]

123. Harris, G.; Heathwaite, A.L. Inadmissible evidence: Knowledge and prediction in land and riverscapes. *J. Hydrol.* **2005**, *304*, 3–19. [CrossRef]

124. Thompson, R.B.; Martínez-Gaitan, C.; Gallardo, M.; Giménez, C.; Fernández, M.D. Identification of irrigation and N management practices that contribute to nitrate leaching loss from an intensive vegetable production system by use of a comprehensive survey. *Agric. Water Manag.* **2007**, *89*, 261–274. [CrossRef]

125. HongYeng, L.; Agamuthu, P. Nitrogen flow in organic and conventional vegetable farming systems: A case study of Malaysia. *Nutr. Cycl. Agroecosyst.* **2015**, *103*, 131–151. [CrossRef]

126. Oborn, I.; Edwards, A.C.; Witter, E.; Oenema, O.; Ivarsson, K.; Withers, P.J.A.L.; Nilsson, S.I.; Stinzing, A.R. Element balances as a tool for sustainable nutrient management: A critical appraisal of their merits and limitations within an agronomic and environmental context. *Eur. J. Agron.* **2003**, *20*, 211–225. [CrossRef]

127. Bechmann, M.; Eggestad, H.O.; Vagstad, N. Nitrogen balances and leaching in four agricultural catchments in southeastern Norway. *Environ. Pollut.* **1998**, *102*, 493–499. [CrossRef]

128. Oenema, O.; Kros, H.; de Vries, W. Approaches and uncertainties in nutrient budgets: Implications for nutrient management and environmental policies. *Eur. J. Agron.* **2003**, *20*, 3–16. [CrossRef]

129. Ding, X.; Shen, Z.; Hong, Q.; Yang, Z.; Wu, X.; Liu, R. Development and test of the export coefficient model in the upper reach of the Yangtze River. *J. Hydrol.* **2010**, *383*, 233–244. [CrossRef]

130. Brisson, N.; Gary, C.; Justes, E.; Roche, R.; Mary, B.; Ripoche, D.; Zimmer, D.; Sierra, J.; Bertuzzi, P.; Burger, P.; et al. An overview of the crop model STICS. *Eur. J. Agron.* **2003**, *18*, 309–332.

131. Zhang, Y.; Niu, H.; Wang, S.; Xu, K.; Wang, R. Application of the DNDC model to estimate N_2O emissions under different types of irrigation in vineyards in Ningxia, China. *Agric. Water Manag.* **2016**, *163*, 295–304. [CrossRef]

132. Uzoma, K.C.; Smith, W.; Grant, B.; Desjardins, R.L.; Gao, X.; Hanis, K.; Tenuta, M.; Goglio, P.; Li, C. Assessing the effects of agricultural management on nitrous oxide emissions using flux measurements and the DNDC model. *Agric. Ecosyst. Environ.* **2015**, *206*, 71–83. [CrossRef]

133. Rahn, C.R.; Zang, K.; Lillywhite, R.D.; Ramos, C.; De Paz, J.M.; Doltra, J.; Riley, H.; Fink, M.; Nendel, C.; Thorup-Kristensen, K.; et al. *Development of a Model Based Decision Support System to Optimise Nitrogen Use in Horticultural Crop Rotations Across Europe—EU-ROTATE N*; Final Scientific Report QLK5–2002–01100; Horticulture Research International: Wellesbourne, UK, 2007.

134. Nendel, C.; Venezia, A.; Piro, F.; Ren, T.; Lillywhite, R.D.; Rahn, C.R. The performance of the EU-Rotate-N model in predicting the growth and nitrogen uptake of rotations of field vegetable crops in a Mediterranean environment. *J. Agric. Sci.* **2013**, *151*, 538–555. [CrossRef]

135. Bakhsh, A.; Kanwar, R.S.; Jaynes, D.B.; Colvin, T.S.; Ahuja, L.R. Simulating effects of variable nitrogen application rates on corn yields and NO_3–N losses in subsurface drain water. *Trans. ASAE.* **2001**, *44*, 269. [CrossRef]

136. Fang, Q.X.; Ma, L.; Halvorson, A.D.; Malone, R.W.; Ahuja, L.R.; Del Grosso, S.J.; Hatfield, J.L. Evaluating four nitrous oxide emission algorithms in response to N rate on an irrigated corn field. *Environ. Model. Softw.* **2015**, *72*, 56–70. [CrossRef]

137. Cameira, M.R.; Tedesco, S.; Leitão, T.E. Water and nitrogen budgets under different production systems in Lisbon urban farming. *Biosys. Eng.* **2014**, *125*, 65–79. [CrossRef]

138. Chaplot, V.; Saleh, A.; Jaynes, D.B.; Arnold, J. Predicting water, sediment and NO_3-N loads under scenarios of land-use and management practices in a flat watershed. *Water Air Soil Pollut.* **2004**, *154*, 271–293. [CrossRef]

139. Santhi, C.; Srinivasan, R.; Arnold, J.G.; Williams, J.R. A modeling approach to evaluate the impacts of water quality management plans implemented in a watershed in Texas. *Environ. Model. Softw.* **2006**, *21*, 1141–1157. [CrossRef]

140. Arabi, M.; Frankenberger, J.R.; Engel, B.A.; Arnold, J. Representation of agricultural conservation practices with SWAT. *Hydrol. Process.* **2008**, *22*, 3042–3055. [CrossRef]

141. Giménez, C.; Gallardo, M.; Martínez-Gaitán, C.; Stöckle, C.O.; Thompson, R.B.; Granados, M.R. VegSyst, a simulation model of daily crop growth, nitrogen uptake and evapotranspiration for pepper crops for use in an on-farm decision support system. *Irrig. Sci.* **2013**, *31*, 465–477. [CrossRef]
142. Gallardo, M.; Thompson, R.B.; Giménez, C.; Padilla, F.M.; Stöckle, C.O. Prototype decision support system based on the VegSyst simulation model to calculate crop N and water requirements for tomato under plastic cover. *Irrig. Sci.* **2014**, *32*, 237–253. [CrossRef]
143. Gallardo, M.; Fernández, M.D.; Giménez, C.; Padilla, F.M.; Thompson, R.B. Revised VegSyst model to calculate dry matter production, critical N uptake and ET of several vegetable species grown in Mediterranean greenhouses. *Agric. Syst.* **2016**, *146*, 30–43. [CrossRef]

19

Production of Konjac Glucomannan Antimicrobial Film for Extending Shelf Life of Fresh-Cut Vegetables

Panya Saeheng [1], Panuwat Eamsakulrat [1], Orachorn Mekkerdchoo [2] and Chaleeda Borompichaichartkul [1,*]

[1] Department of Food Technology, Faculty of Science, Chulalongkorn University, Bangkok 10330, Thailand; saeheng.panya@gmail.com (P.S.); bank.banks@gmail.com (P.E.)

[2] Division of Fermentation Technology, Faculty of Agro-Industry, King Mongkut's Institute of Technology Ladkrabang, Bangkok 10520, Thailand; orachorn.me@kmitl.ac.th

* Correspondence: chaleedab@hotmail.com

Abstract: The aim of this research was to produce konjac glucomannan (KGM) antimicrobial film with added sweet basil oil (SB) (*Ocimum basilicum*) as an antimicrobial agent for inhibiting coliform bacteria which is the type most often found in fresh-cut vegetables. The concentrations of SB oil in the emulsion that inhibited the most antimicrobial growth were 4% and 6% (*v/v*). Film-forming conditions were evaluated by varying the volume of KGM solution per area (0.325, 0.455 and 0.585 mL·cm^{-2}) and the concentration of SB oil (4% and 6%). After mixing the film emulsions, the emulsions were dried on a tray dryer at 50 °C for 10 h. After drying, the results showed that KGM film made at 0.325 mL·cm^{-2} with SB oil at 4% resulted in the smoothest surface. When the film was tested against *Escherichia coli*, KGM at 0.325 mL·cm^{-2} with 4% SB oil and at 455 mL·cm^{-2} with 6% SB oil produced the greatest inhibition. Film with SB oil at 4% was used to study film properties. Physical properties of the film such as tensile strength (68.08 MPa) and % elongation (33.56%) as well as water vapor transmission rate (4.44 × 10^{-3} g·cm^{-2}·h^{-1}) were determined. The KGM/SB film did not show an antimicrobial effect on packages of fresh-cut baby cos lettuce or spring onion under the experimental conditions. Further work will be carried out to study more closely the controlled release properties of KGM/SB film to enhance its antimicrobial effects. These improvements could help to develop a more successful application for its use as a natural biopreservative in minimally-processed products like fresh-cut vegetables.

Keywords: konjac glucomannan; antimicrobial films; fresh-cut vegetables

1. Introduction

Nowadays, consumers are aware of their health and the quality of food. Fresh and less processed foods are the first option for a healthy meal. Freshly-made salads from fresh-cut vegetables are increasing in popularity. However, there is often news about disease outbreaks from consuming fresh-cut vegetables.

Essential oils are aroma compounds found in plants that contain volatile and non-volatile components. The qualitative and quantitative composition of essential oils determines their characteristic. Sweet basil (*Ocimum basilicum*) is a common herb in Thai cuisine. Its essential oil contains bioactive volatile compounds such as linalool and methyl chavicol. These bioactive volatile compounds in sweet basil have antimicrobial properties against gram-positive bacteria [1].

Konjac glucomannan is a component derived from tubers of *Amorphophallus konjac* plants which grow in subtropical and tropical regions of China, Japan, and Thailand. Konjac glucomannan is composed of (1→4) linked β-D-mannose and β-D-glucose. It has a high molecular weight, is water soluble, has properties of film-formation, is edible, biodegrades, has a tendency to form a fine dense network upon drying, and is used in pharmaceutical and food applications [2–5]. Therefore, Konjac glucomannan could be a good option for entrapping target molecules. The objectives of this study were to develop natural and edible antimicrobial film by combining Konjac glucomannan with sweet basil oil and determine its mechanical and antimicrobial properties, as well as the performance of the film on the fresh cut vegetables baby cos (*Lactuca sativa* var. longifolia) and spring onion (*Allium wakegi*) during storage.

2. Materials and Methods

2.1. Materials

Konjac glucomannan (food grade) was obtained from the Yunnan Genyun Konjac Resource Corp., Kunming (Yunnan, China). Sweet basil (SB) oil was purchased from Thai-China Flavours and Fragrances Industry Co., Ltd. (Nonthaburi, Thailand). All fresh vegetables including baby cos and spring onion were purchased from the Samyan wet market, Bangkok, Thailand, on the day of the experiment. All vegetables were washed under running water for 30 s and soaked in a 50 ppm available chlorine solution for 2 min then drained and put aside for drying on a perforated tray before use. The bacterial strain *Escherichia coli* ATCC 25922 that was used for testing antimicrobial activity was obtained from standardized culture at the Department of Food Technology, Faculty of Science, Chulalongkorn University, Bangkok, Thailand.

2.2. Methods

2.2.1. Emulsion Preparation and Antimicrobial Assessment

Konjac glucomannan (KGM) solution at 1% (w/v) was prepared by slowly adding KGM powder into water and stirring with a magnetic stirrer for 3 h at ambient temperature, followed by adding 0.5 M KOH to reach 0.14% (v/v) KOH and stirring for another 15 min. The solution was left for 1 h, then glycerol was added to reach 0.3% (w/v), and the solution stirred for 20 min. SB oil was then added to the KGM solution to achieve concentrations of 1%, 2%, 3%, 4%, 5%, 6%, 7%, and 8% (v/v), and they were stirred at 20,000 rpm for 10 min at ambient temperature.

Asuitable concentration of SB oil for further work was selected by assessing antimicrobial activity using the paper disc diffusion method against *E. coli*. The method described by Mekkerdchoo et al. [6] was followed. Bacterial cultures were grown and inoculated in fresh nutrient broth (NB) medium at 37 °C for 24 h. The microorganism was inoculated on the surface of a nutrient agar plate by sterile cotton swab (initial concentration of *E. coli* was 10^7 log colony forming units (CFU)/mL). Subsequently, sterile paper discs (Antibiotica-Testblattchen paper disc, Duran, West Chester, PA, USA) (6 mm in diameter) saturated with SB oil at different concentrations (4% or 6% (w/w)) were placed on the surface of each inoculated plate. A standard antibiotic (tetracycline (Becton, Dickson and Company, Franklin Lakes, NJ, USA)) was simultaneously used as a positive control. The inoculated plates were incubated at 37 °C for 24 h under aerobic conditions. After this period, it was possible to observe a clear zone, free of visible bacterial growth. The antibacterial activity was evaluated by measuring the clear zone diameter. The sensitivity of bacterial growth to the antimicrobial solutions was determined by the diameter of the inhibition zone as: not sensitive = diameters less than 0.8 cm; sensitive = diameters 0.9–1.4 cm; very sensitive = diameters 1.5–1.9 cm; and extremely sensitive = diameters larger than 2.0 cm [7]. All treatments were performed in triplicate.

2.2.2. Film Preparation and Antimicrobial Assessment

The KGM/SB emulsion selected above was cast on hard adonized steel trays (14 cm diameter) at 0.325, 0.445 and 0.585 mL·cm^{-2}. The film solutions were dried at 50 °C by using a hot air dryer for 10 h.

After drying, antimicrobial films were conditioned at 67% RH before further testing. Antimicrobial activity of the films against *E. coli* were assessed by measuring the clear zone (cm) [6]. The bacteria culture was grown and inoculated in NB medium at 37 °C. After 24 h of growth, the microorganism was inoculated on the surface of an agar plate (initial concentration of *E. coli* was 10^7 log CFU/mL). Film samples were cut into circles (6 mm diameter) from different positions of the film and placed on the culture plate. Standard antibiotic (tetracycline) was used as positive control. The culture plate was incubated at 37 °C for 24 h under aerobic conditions. After this period, it was possible to observe a clear zone. All measurements were done in triplicate.

2.2.3. Evaluation of Mechanical Property of Antimicrobial Film

The mechanical properties of antimicrobial films that had good antimicrobial activity from above were assessed. The properties included tensile strength (TS), percent of elongation (%E) (ASTM standard test method D882 [8], film thickness (digital micrometer), and water vapor transmission rate (WVTR) [1]. All measurements were performed in triplicate.

2.2.4. Antimicrobial Assessment of Fresh-Cut Vegetables during Storage

Baby cos lettuce (95 g) was packed into plastic boxes along with a KGM/SB film cut into a 3 cm × 3 cm piece. It was stored at 4 ± 1 °C, and was sampled every day for 7 days to assess total plate count (TPC) and bacterial count of *E. coli* and coliform. Spring onion (40 g) was packed as above and stored at ambient temperature for 24 h, and then TPC and bacterial count of *E. coli* and coliform were assessed. The TPC analyses were as follows: 25 g of each vegetable, from 3 different packages, was collected and placed in a sterile plastic bag (Sterilin, Staffordshire, UK) with 90 mL of buffered 0.1% peptone water (Merck, Darmstadt, Germany). After 1 min in a Stomacher blender (Model 400 circulator, West Sussex, UK), appropriate dilutions (10^{-1} to 10^{-7}) were prepared for TPC determination on plates with Plate Count Agar and incubated at 35 °C for 24 h [9]. The number of *E. coli* and coliform were assessed using 3 M Petrifilm^{TM} *E. coli*/Coliform Count Plates (EC) by taking 25 g of sample by aseptic technique, placing the sample in 225 mL sterile peptone water in a bag, and mixing using a Stomacher mixer. Then, a serial dilution from the mixed solution was prepared at 10^{-1}, 10^{-2}, 10^{-3}, 10^{-4}, 10^{-5}, 10^{-6}, and 10^{-7}. A 1 mL solution of each dilution was pipetted onto 3M Petrifilm^{TM}, pressed with a spreader, and left for 1 min for setting the gel. The sample was incubated at 37 °C for 48 h. The number of coliform and *E. coli* colonies were measured following the 3M Petrifilm^{TM} manual, counting gas bubble-producing red coliform colonies and blue *E. coli* colonies. All measurements were performed in triplicate [9].

2.2.5. Statistical Analyses

All data were statistically analyzed (SPSS 17.0 software; IBM SPSS, Chicago, IL, USA). Data are shown as mean values with the standard deviation. Differences between mean values were established using Duncan's multiple range test with a level of significance of $p \leq 0.05$.

3. Results and Discussion

3.1. Antimicrobial Assessment of KGM/SB Emulsion and Film

Microbial assessment of the emulsion was evaluated by the diameter of the clear inhibition zone against *E. coli* (Table 1). Contact area was used to evaluate growth inhibition underneath film discs in direct contact with target microorganisms in agar [10]. Pure sweet basil oil gave a clear zone of 13.2 ± 0.8 mm, lower than tetracycline at 23.7 ± 1.6 mm but indicating inhibition of *E. coli*. According to the sensitivity levels, the response to SB oil was sensitive, a diameter of 9–14 mm [7]. When adding SB oil to KGM solution from 1%–8%, antimicrobial efficiency was reduced due to the lower oil concentration in the emulsion resulting in a decrease in active compound in the aqueous phase of the agar layer where microbial proliferation takes place [11]. Even though there were no statistical differences in the inhibition zone between 2%, 3%, 5%, 7% and 8% KGM/SB emulsion, 4% and 6% emulsion were selected for further study due to giving the largest clear zones (7.8 and 8.2 mm, respectively). According to Ponce et al. [7], at this range of the clear zone, it would be considered not sensitive for use as an antimicrobial solution.

Table 1. Antimicrobial activity of a KGM/SB emulsion film assessed by the clear zone (mm) against *E. coli*.

KGM/SB Emulsion (%, *v/v*)	Diameter of Clear Zone (mm)
Control	No clear zone
1	7.1 ± 0.5 [f]
2	7.6 ± 0.6 [d,e]
3	7.5 ± 0.6 [d,e,f]
4	7.8 ± 1.1 [c,d]
5	7.3 ± 1.1 [e,f]
6	8.2 ± 1.2 [c]
7	7.6 ± 1.0 [d,e]
8	7.7 ± 0.6 [d,e]
Sweet basil oil 100	13.2 ± 0.8 [b]
Tetracycline	23.7 ± 1.6 [a]

Different superscript letters (a–f) indicate mean values in the same column differ significantly ($p \leq 0.05$).

Emulsions of KGM/SB with 4% or 6% SB oil were cast into films at different ratios of volume per area. The concentrations of 4% SB oil in a 0.325 mL·cm^{-2} film and 6% SB oil in a0.455 mL·cm^{-2} film gave the highest diameter clear zones (Table 2). A dilution effect was only seen in the 4% SB oil samples. The physical appearance of the different KGM/SB films is shown in Figure 1. The 4% SB oil was more transparent than the 6% film. During drying, when water evaporated, the SB oil was likely becoming more concentrated and phase separation may have occurred. This was more apparent in the 6% KGM/SB film when volume-to-area ratio increased, and the phase separation occurred creating the opaque appearance. Thus, SB oil used at a high concentration, which is composed mostly of nonpolar compounds, may not mix well with KGM solution which contains a range of polarities.

Table 2. Antimicrobial activity of a KGM/SB emulsion film assessed by the clear zone (mm) against *E. coli* at different volumes per area.

Concentration (%, *v/v*)	Volume per Area (mL·cm^{-2})	Diameter of Clear Zone (mm)
6	0.325	9.6 ± 0.32 [b]
6	0.455	10.1 ± 0.41 [a]
4	0.325	10.1 ± 0.34 [a]
4	0.455	9.8 ± 0.44 [a,b]
4	0.585	9.5 ± 0.49 [b]

Different superscript letters (a,b) indicate mean values in the same column differ significantly ($p \leq 0.05$).

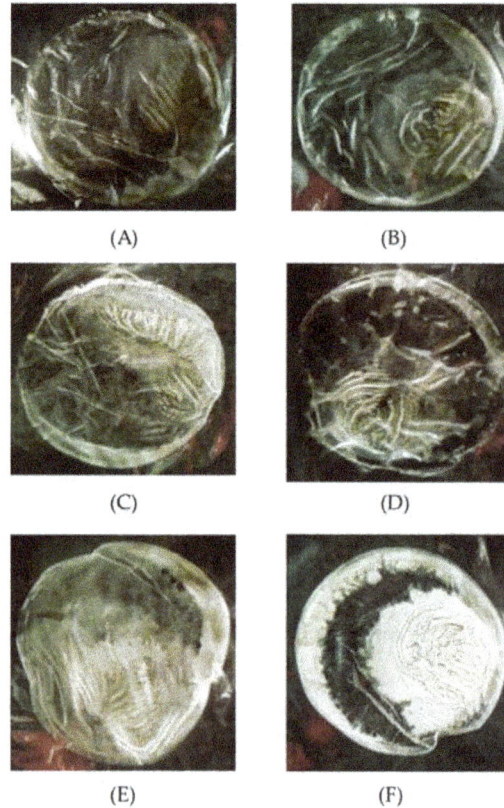

Figure 1. Appearance of antimicrobial film after drying at different KGM-to-SB oil ratios. SB oil at 4% in 0.325 (**A**); 0.445 (**B**) and 0.585 (**C**) mL·cm^{-2} KGM solution; and at 6% in 0.325 (**D**); 0.445 (**E**) and 0.585 (**F**) mL·cm^{-2}.

The KGM/SB film sample at 4% SB oil in 0.325 mL·cm^{-2} KGM was selected to determine mechanical properties (Table 3). Mechanical property determination is necessary for material that may be used as a food packaging [12]. The mechanical properties of antimicrobial film from a KGM/SB film sample at 4% SB oil in 0.325 mL·cm^{-2} KGM showed a lower tensile strength when compared to a previous study of mechanical properties of a 1% w/w KGM film (tensile strength = 88.1 ± 1.2 MPa) [10]. Normally, incorporation of additives into a film results in a lower tensile strength value [13]. On the other hand, %E and WVTR of this film were increased when compared to 1% KGM film (%E = 26.5 ± 1.8, WVTR = 5.4 × 10^{-12}) [11]. These results indicated that additional film emulsion may decrease the homogeneity of the film matrix structure lowering the tensile strength and increasing %E, while increasing WVTR may have come from reducing the mass transfer resistance to water molecules [14,15]. Nevertheless, these changed physical properties improved film flexibility, and more moisture passing through the material could result in more rapid release of the antimicrobial agents in the blended film when the film is in contact with the surface of the food [10]. In addition, the film in this study showed a higher tensile strength and WVTR while %E was lower when compared to commercial plastic films which are more flexible and water resistant, such as low-density polyethylene (tensile strength = 14.9 ± 0.49 MPa, %E = 1073 ± 27.67, and WVTR = 4.98 × 10^{-13} g·cm^{-2}·h^{-1}) [16] and oriented polypropylene (tensile strength = 28 ± 1.05 MPa, %E = 780 ± 18.24, WVTR = 4.27 × 10^{-13}) [17].

Table 3. Mechanical property of antimicrobial film.

Mechanical Properties	Values
Thickness (mm)	0.074 ± 0.018
Tensile Strength (MPa)	68.08 ± 7.54
Elongation (%)	33.56 ± 7.38
WVTR (g·cm^{-2}·h^{-1})	$4.44 \times 10^{-3} \pm 1.38 \times 10^{-4}$

3.2. Microbial Assessment of Fresh Cut Vegetable during Storage

Due to the importance of microbial assessment in real food systems, two types of vegetables were selected for use in this study, baby cos lettuce (Figures 2 and 3) and spring onion (Table 4).

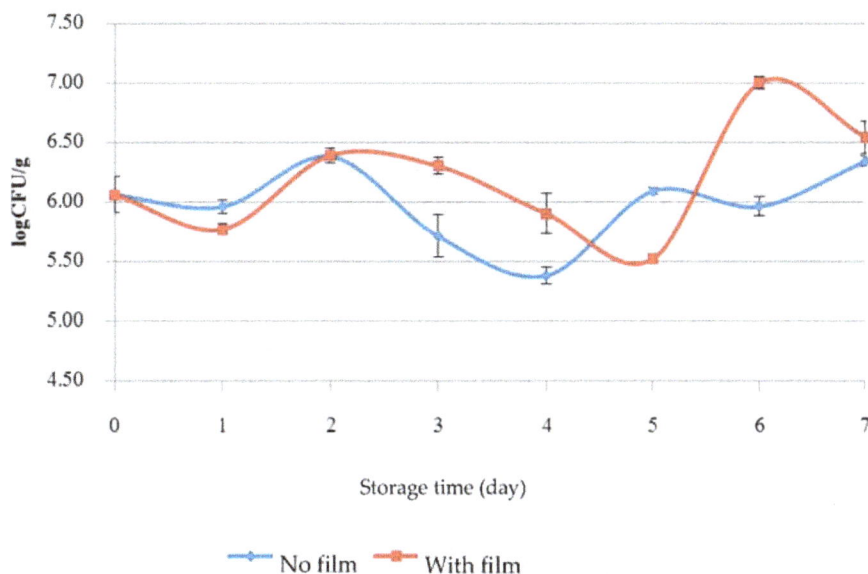

Figure 2. Number of microorganisms by total plate count as colony forming units (CFU) recovered from baby cos lettuce during 7 days of cold storage.

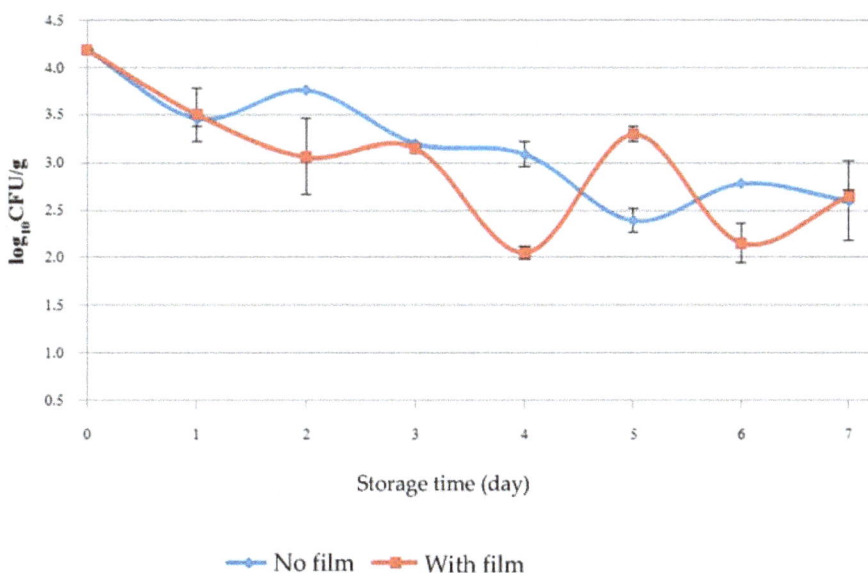

Figure 3. Number of coliforms as colony forming units (CFU) recovered from baby cos lettuce during 7 days of cold storage.

Table 4. Number of microorganisms as colony forming units (CFU) recovered from spring onion after 24 h.

Storage Time (Day)	Treatment	Total Plate Count ($\log_{10}$ CFU/g)	Coliform ($\log_{10}$ CFU/g)
0	Fresh cut vegetable	5.79 ± 0.15 [b]	4.92 ± 0.12 [b]
1	No film	9.05 ± 0.95 [a]	8.16 ± 0.41 [a]
1	Film placed inside the package	8.69 ± 1.29 [a]	7.85 ± 0.21 [a]

Different superscript letters (a,b) indicate mean values in the same column differ significantly ($p \leq 0.05$).

KGM/SB film was not effective by total plate count analysis when used with low temperature storage at $4 \pm 1\,^\circ$C (Figure 2). This may be because low temperature suppressed the evaporation of the bioactive volatile compound SB oil to the surrounding air space. It is consistent with the results of Suppakul et al. [11] who studied efficacy of antimicrobial films containing constituents of SB oil used in cheese and found that lower temperatures suppressed the evaporation of bioactive volatile compound, while at a higher temperature the release rate of the bioactive compound from the film onto the food surface was increased. The mass transfer rate from the film surface to the air is dominant, because of convection and the larger concentration gradient than that in the film interior [14]. Any reduction in coliform CFU may be assumed due to the effect of cold storage more than the effect of antimicrobial film (Figure 3). *E. coli* was not found in the samples during storage for 7 days. Antimicrobials must directly contact the surface of food to exert the greatest antimicrobial effect [18]. Therefore, items with irregular shapes or rough surfaces like baby cos lettuce where KGM/SB film cannot directly contact the surfaces would result in bacterial cells that would be less exposed to antimicrobial compound.

For spring onion, after 24 h storage at ambient temperature, the KGM/SB film did not reduce microbial activity significantly when compared to no film (Table 4). Similar results were observed by Cano et al. [19] who studied incorporation of neem oil in starch-PVA films and found that the film did not improve the antimicrobial properties. Even more, the film seemed to promote the early growth of bacteria during the storage period. Ponce et al. [20] also found that the antibacterial activities of olive and rosemary oil emulsions were reduced when they were applied in the coating material. These results are probably due to the dispersion effect of the active compounds and the interactions among the oils and films. The low effectiveness of antimicrobial film with a real food system as in this study may occur for many reasons such as strong entrapment of the emulsion in the film structure, inhibiting diffusion to the film surface where microbial growth occurs [14], or essential oil emulsion strong interaction with hydroxyl groups of the polymers, thus limiting their diffusion to the film surface and their antimicrobial effects [19]. Moreover, the coliform bacteria that was used in this study were gram-negative which are generally more resistant to growth inhibition and the killing effects of various antibiotics and antimicrobial agents due to their strong surface hydrophilicity acting as a permeability barrier. The surface also possesses divalent cations that could stabilize the lipopolysaccharide association within the membrane, and may prevent active compounds from reaching the cytoplasmic membrane [11].

4. Conclusions

Sweet basil oil incorporated into konjac glucomannan film exhibited antimicrobial activity at a concentration of 4% (v/v) (volume-to-area ratio of 0.325 mL·cm^{-2}) and a concentration of 6% (v/v), (volume-to-area ratio of 0.455 mL·cm^{-2}). The 4% (v/v) (volume-to-area ratio of 0.325 mL·cm^{-2}) film was clear with a smooth surface, a thickness of 0.074 mm, a tensile strength 6.81 MPa, elongation of 33.56%, and a water vapor transmission rate of 4.44×10^{-3} g·mm^{-2}·h^{-1}, respectively. In this preliminary study on application of the antimicrobial film to fresh-cut baby cos lettuce and spring onion in packages, KGM/SB film did not show an antimicrobial effect. However, there is a need for further study of the controlled release properties of KGM/SB film to enhance potential antimicrobial properties of the film against spoilage microorganisms and to extend shelf life of fresh-cut vegetables.

Acknowledgments: The authors would like to thank Chulalongkorn University for financial supporting of the study.

Author Contributions: "Chaleeda Borompichaicharkul, Panya Saeheng and Panuwat Eamsakulrat conceived and designed the experiments; Panya Saeheng and Panuwat Eamsakulrat performed the experiments; Chaleeda Borompichaichartkul, Orachorn Mekkerdchoo, Panya Saeheng and Panuwat Eamsakulrat analyzed the data; Chaleeda Borompichaichartkul and Orachorn Mekkerdchoo wrote the paper". Authorship must be limited to those who have contributed substantially to the work reported.

Conflicts of Interest: The authors declare no conflict of interest.

References

1. American Standard for Testing and Materials. *Standard Test Method for Water Vapor Transmission of Material*; ASTM E95-96; Annual Book of ASTM Standard; American Society for Testing and Materials Publishing: Philadelphia, PA, USA, 1995.

2. American Standard for Testing and Materials. *Standard Test Method for Tensile Properties of Thin Plastic Sheeting*; ASTM D882; Annual Book of ASTM Standard; American Society for Testing and Materials Publishing: Philadelphia, PA, USA, 2001.

3. Davé, V.; McCarthy, S.P. Review of konjac glucomannan. *J. Environ. Polym. Degrad.* **1997**, *5*, 237–241.

4. Dave, V.; Sheth, M.; McCarthy, S.P.; Ratto, J.A.; Kaplan, D.L. Liquid crystalline, rheological and thermal properties of konjac glucomannan. *Polymer* **1998**, *39*, 1139–1148. [CrossRef]

5. Nakano, M.; Takikawa, K.; Arita, T. Release characteristics of dibucaine dispersed in konjac gels. *J. Biomed. Mater. Res.* **1979**, *13*, 811–819. [CrossRef] [PubMed]

6. Mekkerdchoo, O.; Patipasena, P.; Borompichaichartkul, C. Liposome encapsulation of antimicrobial extracts in pectin film for inhibition of food spoilage microorganisms. *Asian J. Food Agro-Ind.* **2009**, *2*, 817–838.

7. Ponce, A.; Fritz, R.; del Valle, C.; Roura, S. Antimicrobial activity of essential oils on the native microflora of organic Swiss chard. *Lebensm. Wiss. Technol.* **2003**, *36*, 679–684. [CrossRef]

8. Arunwatcharin, P. Pathogens and spoilage microorganism's inhibition ability of Thai herb essential oils. In Proceeding of the 45th Conference on Agriculture and Home Economics, Agro-Industry, Kasetsart University, Bangkok, Thailand, 30 January–2 February 2007; pp. 508–515. (In Thai)

9. Chacon, P.A.; Buffo, R.A.; Holley, R.A. Inhibitory effects of microencapsulated allyl isothiocyanate (AIT) against *Escherichia coli* O157:H7 in refrigerated, nitrogen packed, finely chopped beef. *Int. J. Food Microbiol.* **2006**, *107*, 231–237. [CrossRef] [PubMed]

10. Li, B.; Kennedy, J.F.; Peng, J.L.; Yie, X.; Xie, B.J. Preparation and performance evaluation of glucomannan-chitosan-nisin ternary antimicrobial blend film. *Carbohydr. Polym.* **2006**, *65*, 488–494. [CrossRef]

11. Suppakul, P.; Sonneveld, K.; Bigger, S.W.; Miltz, J. Efficacy of polyethylene-based antimicrobial films containing principal constituents of basil. *LWT-Food Sci. Technol.* **2008**, *41*, 779–788. [CrossRef]

12. Kowalczyk, D.; Gustaw, W.; Świeca, M.; Baraniak, B. A study on the mechanical properties of pea protein isolate films. *J. Food Process. Preserv.* **2014**, *38*, 1726–1736. [CrossRef]

13. Pranoto, Y.; Rakshit, S.K.; Salokhe, V.M. Physical and antimicrobial properties of alginate-based edible film incorporated with garlic oil. *Food Res. Int.* **2005**, *38*, 267–272. [CrossRef]

14. Ma, Q.; Zhang, Y.; Critzer, F.; Davidson, P.M.; Zivanovic, S.; Zhong, Q. Physical, mechanical, and antimicrobial properties of chitosan films with microemulsions of cinnamon bark oil and soybean oil. *Food Hydrocoll.* **2016**, *52*, 533–542. [CrossRef]

15. Bonilla, J.; Atarés, L.; Vargas, M.; Chiralt, A. Effect of essential oils and homogenization conditions on properties of chitosan-based films. *Food Hydrocoll.* **2012**, *26*, 9–16. [CrossRef]

16. Manikantan, M.R.; Varadharaju, N. Preparation and properties of linear low density polyethylene based nanocomposite films for food packaging. *Indian J. Eng. Mater. Sci.* **2012**, *19*, 54–66.

17. Hyun, K.; Chong, W.; Koo, M.; Chung, I.J. Physical properties of polyethylene/silicate nanocomposite blown films. *J. Appl. Polym. Sci.* **2003**, *89*, 2131–2136. [CrossRef]

18. Guo, M.; Jin, T.Z.; Wang, L.; Scullen, O.J.; Sommers, C.H. Antimicrobial films and coatings for inactivation of *Listeria innocua* on ready-to-eat deli turkey meat. *Food Control* **2014**, *40*, 64–70. [CrossRef]

19. Cano, A.; Cháfer, M.; Chiralt, A.; González-Martínez, C. Physical and antimicrobial properties of starch-PVA blend films as affected by the incorporation of natural antimicrobial agents. *Foods* **2015**. [CrossRef]

20. Ponce, A.G.; Roura, S.I.; del Valle, C.E.; Moreira, M.R. Antimicrobial and antioxidant activities of edible coatings enriched with natural plant extracts: In vitro and in vivo studies. *Postharvest Biol. Technol.* **2008**, *49*, 294–300. [CrossRef]

20

Clean Foods: The Inhibitory Effect of Five Natural Juices on the Browning of Apple Slices during Drying

Anna Hubackova and Jan Banout *

Department of Sustainable Technologies, Faculty of Tropical AgriSciences, Czech University of Life Sciences, Prague 16521, Czech Republic; a.hubackova@gmail.com

* Correspondence: banout@ftz.czu.cz

Abstract: Enzymatic browning, which accompanies food preservation processes based on drying, is a common obstacle to obtaining marketable and consumer-appealing products. In this paper, we investigate the application of one of the methods commonly used in order to counter the browning process, namely dipping. Dipping involves a soaking of the foods in liquids or solutions in order to suppress the activity of polyphenoloxidase (PPO) enzymes, either by physically preventing oxygen from interacting with it, or by dramatically decreasing water activity and thus slowing down water-dependent reactions. In this study, juices from selected fresh fruits with high ascorbic acid content were used as natural preservatives with antibrowning effects. The juices were prepared from fruit of the following species: *Sorbus aucuparia, Diospyros kaki, Hippophae rhamnoides, Actinidia deliciosa,* and *Rosa canina.* The effect of selected juices on color change was tested on dried "Idared" apple slices and was compared to slices of freshly cut apples (standard). The browning index of all sample groups showed significant differences between treatment groups and the standard except for samples treated with *Rosa canina* juice. The effect of the juices was also evaluated via a sensory panel, where color change, degree of browning, change of taste, and overall acceptability of the resulting color and taste were evaluated. Results showed that the best antibrowning effect was achieved by macerate from fruits of *Rosa canina*. The results of this study showed that dipping in some plant juices has the potential of complementing or replacing the sulphite-based approach, which is the current method of choice of the food industry.

Keywords: natural preservants; *Rosa canina*; drying; browning

1. Introduction

Fruit is an important part of human nourishment, as it contains a number of beneficial and essential compounds, including saccharides, vitamins, minerals, fiber, and natural pigments [1,2]. To secure a fruit supply throughout the year, it is often necessary to preserve it [2]. One of the oldest techniques of preserving foods is by lowering the water activity in foods, for example, by drying [3]. However, through drying, fruit may lose much of its natural texture and color [4–6]. Color change is caused by both enzymatic and non-enzymatic browning [7]. One of the methods of suppressing enzymatic browning is by use of ascorbic acid [2]. Thus, the main objective of this research was to investigate the effect of natural juices with ascorbic acid for suppressing enzymatic browning of apple (*Malus domestica* Borkh.) slices during dehydration.

2. Experimental Section

Natural fruit juices of five species containing high amounts of vitamin C (*Sorbus aucuparia* (mountain-ash), *Diospyros kaki* (Japanese persimmon), *Hippophae rhamnoides* (sea buckthorn),

Actinidia deliciosa (kiwifruit), *Rosa canina*, (dog-rose)) and a 1% solution of ascorbic acid (AA) were used in this study. Except for the juice from *Rosa canina*, all juices were prepared in a juicer and diluted with an equal amount of distilled water. The juice from fruits of *Rosa canina* was prepared by macerating the fruits in the dark at 18 °C.

Vitamin C content in the juices was measured by high-performance liquid chromatography (HPLC). Samples of juices were centrifuged in 1.5-mL Eppendorf test tubes at 14,500 rpm and then filtered through a PTFE filter with a 0.45-μm pore size. The samples were diluted with water at 20:1 and then analyzed via HLPC with a diode array detector (Dionex Summit, Dionex Inc., Sunnyvale, CA, USA). The chromatography conditions were as follows: isocratic elution with a mobile phase of 0.1% TFA, and a flow rate of 0.8 mL/min on a Gemini C-18, 250 × 4.6 mm, a particle size of 5 μm (Phenomenex, Inc., Torrance, CA, USA) column with a C-18 pre-column. Column temperature was 25 °C, and the injection volume was 20 μL. The program duration was 7.5 min with elution time of the ascorbic acid at 6.2 min. Sample ascorbic acid content was measured at 243 nm and quantified by external calibration using a linear calibration for ascorbic acid of 0.004–1.000 mg/mL.

A single juice or ascorbic acid solution was applied to five slices of "Idared" apple, cold stored prior to use, by soaking the slices for 10 min. The apples were purchased at a local fruit market in Prague, Czech Republic. The treated apple slices, as well as untreated slices, were then dried in a home fruit dryer. The dryer temperature was set to 57 °C with a constant air flow rate of 1.0 m/s. The drying was continued until a constant mass was obtained.

After drying, the degree of browning and the color change were measured on a colorimeter MiniScan® XE Plus. The results were reported using the *L*–a*–b** system. Freshly cut apples were measured as a standard, and untreated dried samples measured as a control. After that, the samples treated with the fruit juices and 1% ascorbic acid were measured.

Parameters such as browning index (*BI*), hue angle (h°), and saturation index (C*) were calculated. The *BI* represents purity of brown color and is generally accepted as a significant parameter in color change. It is calculated as [8]

$$BI = \frac{100\,(x - 0.31)}{0.17} \quad x = \frac{a + 1.75L}{5.645L + a - 3.012b}$$

The *BI* assumes values between 0 and 100, where 0 corresponds to white and 100 corresponds to black. Hue angle (h°) is defined as the angle between the hypotenuse and 0° on the *a** (bluish-green/red-purple) axis, and is calculated from the arctangent of *b*/a**. The hue angle value corresponds to whether the object is red, orange, yellow, green, blue, or violet. Saturation index C* represents the hypotenuse of a right triangle created by joining points (0, 0), (*a**, *b**), and (*a**, 0). The greater the C* value, the purer the color appears. The total color difference Δ*E** was calculated as Δ*E** = (Δ*L**2 + Δ*a**2 + Δ*b**2)1/2.

Besides colorimetry, the organoleptic properties of the apple slices were evaluated by 14 trained panelists. During panel tasting, participants filled out a questionnaire about the perceived color change, degree of browning, taste, and overall attractiveness of presented dried apples. The degree of browning and color change was evaluated in comparison with a standard of freshly cut apples. All tests were done in triplicate.

3. Results and Discussion

3.1. Ascorbic Acid Content in Juices

Ascorbic acid content was determined for all juices by HPLC except for *Diospyros kaki*, where this method failed. The average amount of AA in juice from *Diospyros kaki* was taken from the literature [9]. The highest amount of AA was found in the juice from *Sorbus aucuparia* and *Hippophae rhamnoides* (Table 1). About one third that amount was found in the juice from *Rosa canina* and *Actinidia deliciosa*.

Table 1. Ascorbic acid content in selected fruit juices. Values are mean ± standard deviation.

Natural Juices	Ascorbic Acid (mg/mL)
Sorbus aucuparia	1.79 ± 0.20
Diospyros kaki	0.43 ± 0.08
Hippophae rhamnoides	1.65 ± 0.23
Actinidia deliciosa	0.50 ± 0.08
Rosa canina	0.57 ± 0.22

3.2. Colorimetric Measurements

Color is one of the most important parameters that determine the quality of raw materials and processed products. This characteristic affects consumer acceptability of the product [10]. The process of drying caused significant changes in L^*, a^* and b^* values compared to the standard (Table 2). The freshly cut standard had values indicating a light shade of green and yellow, similar to results described by Vega-Gálvez et al. [11]. According to the ΔE^* values (or total color difference), the samples treated with *Rosa canina* were the closest (smallest ΔE^*) to the standard in color. In contrast, the *Sorbus aucuparia*-treated samples showed the greatest difference. Furthermore, none of the treatments achieved a greener shade than the standard. All of the treatments showed an increase in the yellow color component with the most prominent difference measured with *Sorbus aucuparia* samples. The smallest difference in the yellow component was found with the *Hippophae rhamnoides* and *Actinidia deliciosa* samples. A small amount of blue color was found with the ascorbic acid and *Rosa canina* samples.

Table 2. Colorimetric data for freshly cut apple slices (standard), and samples treated with natural juices, ascorbic acid (AA) solution, or left untreated. Values are mean ± standard deviation.

Treatments	L^*	a^*	b^*	ΔL^* [z]	Δa^* [y]	Δb^* [x]	ΔE^* [w]
Standard	86.78 ± 0.98	−0.29 ± 0.33	25.71 ± 1.37				
Sorbus aucuparia	82.08 ± 2.50	7.27 ± 2.02	40.98 ± 3.11	−4.70 ± 2.50	7.56 ± 2.02	15.27 ± 2.11	18.11 ± 2.79
Diospyros kaki	80.46 ± 1.28	5.56 ± 0.98	31.44 ± 1.10	−6.32 ± 2.55	5.85 ± 1.88	5.73 ± 2.46	10.52 ± 2.89
Hippophae rhamnoides	77.48 ± 2.21	7.82 ± 2.19	27.15 ± 2.22	−9.30 ± 2.21	8.11 ± 2.19	1.44 ± 2.22	13.15 ± 2.92
Actinidia deliciosa	79.34 ± 1.54	6.21 ± 2.62	27.46 ± 2.35	−7.44 ± 2.54	6.50 ± 2.62	1.75 ± 2.35	10.82 ± 2.47
Rosa canina	84.29 ± 2.77	4.41 ± 1.91	23.18 ± 2.03	−2.05 ± 2.77	4.70 ± 1.91	−2.53 ± 2.03	8.51 ± 2.02
1% solution of AA	74.09 ± 1.71	11.56 ± 2.21	25.28 ± 1.85	−12.7 ± 1.71	11.85 ± 2.21	−0.43 ± 1.85	17.76 ± 2.04
Untreated samples	77.78 ± 2.72	8.19 ± 2.11	31.89 ± 2.07	9.00 ± 2.72	8.48 ± 2.11	6.18 ± 2.07	13.98 ± 2.82

[z] $+\Delta L^*$ = sample is lighter than the standard, and $-\Delta L^*$ = sample is darker than the standard; [y] $+\Delta a^*$ = sample is redder than standard, and $-\Delta a^*$ = sample is greener than standard; [x] $+\Delta b^*$ = sample is yellower than standard, and $-\Delta b^*$ = sample is bluer than standard; [w] ΔE^* = the total color difference.

3.3. Hue Angle, Saturation, and the Browning Index

Values of L^*, a^*, and b^* (Table 2) were used for the calculation of h°, C^*, and BI (Table 3). According to the results of analysis of variance of BI (F-test, $p \leq 0.005$), there were statistically significant differences between the means of BI of the different treatments. The Box-and-Whisker plot (Figure 1) shows the specific differences among the treatments. It is obvious that only the *Rosa canina* treatment approached the standard in quality. These results are in agreement with Zocca et al. [12], where there was significant inhibition by *Rosa canina* extracts against polyphenol oxidase (PPO). The highest degree of browning was observed with apples treated with juices from *Sorbus aucuparia*. Based on h° and C^*, the closest values to the standard were also found with apple slices treated with *Rosa canina* juice.

Table 3. Hue angle (h°), saturation index (C*), and browning index (BI) of apple slices treated with selected fruit juices, ascorbic acid (AA), or left untreated. Values are mean ± standard deviation.

Treatments	h°	C*	BI
Standard	90.65 ± 3.02	25.71 ± 2.23	33.59 ± 2.36
Sorbus aucuparia	79.94 ± 2.36	41.62 ± 4.36	73.98 ± 3.93
Diospyros kaki	79.98 ± 3.47	31.93 ± 2.14	54.16 ± 4.45
Hippophae rhamnoides	73.93 ± 4.02	28.25 ± 1.89	51.08 ± 6.35
Actinidia deliciosa	77.26 ± 3.69	28.15 ± 1.62	48.58 ± 4.03
Rosa canina	79.23 ± 2.78	23.60 ± 1.44	36.66 ± 8.11
1% solution of AA	65.42 ± 3.56	27.80 ± 2.40	53.37 ± 4.89
Untreated samples	75.60 ± 4.25	32.93 ± 3.32	60.06 ± 3.97

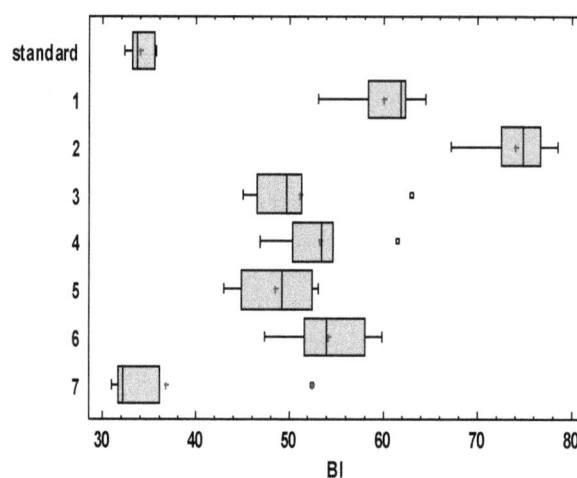

Figure 1. Box-and-Whisker Plot. Y-axis: 1 = untreated samples; 2 = *Sorbus aucuparia*; 3 = *Hippophae rhamnoides*; 4 = 1% solution of AA; 5 = *Actinidia deliciosa*; 6 = *Diospyros kaki*; and 7 = *Rosa canina*. BI = browning index.

3.4. Organoleptic Analyses

Consumer sensory analyses revealed that perceived color change, compared with the standard, was rated as "moderate" for samples treated with *Rosa canina* juice and ascorbic acid solution (Table 4). In the samples treated with *Actinidia deliciosa*, *Hippophae rhamnoides*, and *Diospyros kaki* juices, as well as the untreated samples, color change was rated as "medium," and, for samples treated with *Sorbus aucuparia*, color change was "significant."

Table 4. Sensory evaluation results of apple slices treated with selected fruit juices, ascorbic acid (AA), or left untreated. Values are mean ± standard deviation.

Treatment	Color Change versus Standard	Browning versus Standard	Overall Appearance
Sorbus aucuparia	2.6 ± 0.50 [z]	2.1 ± 0.86	2.1 ± 0.36 [y]
Diospyros kaki	2.3 ± 0.61	2.3 ± 0.61	2.0 ± 0.28
Hippophae rhamnoides	2.1 ± 0.95	3.0 ± 0.25	2.6 ± 0.42
Actinidia deliciosa	1.9 ± 0.47	2.0 ± 0.39	1.9 ± 0.31
Rosa canina	1.0 ± 0.10	1.0 ± 0.16	1.8 ± 0.14
1% solution of AA	1.3 ± 0.36	1.6 ± 0.51	1.9 ± 0.41
Untreated samples	2.3 ± 0.61	1.8 ± 0.7	1.7 ± 0.37

[z] Values for color change compared with the standard and browning compared with the standard: 1 = moderate; 2 = medium; 3 = significant; [y] Overall appearance: 1 = very good; 2 = satisfactory; and 3 = unacceptable.

Samples treated with *Rosa canina* juice displayed the lowest degree of browning compared to the standard, while there was a "medium" degree of browning with the AA solution, with juice from *Actinidia deliciosa*, *Sorbus aucuparia*, and *Diospyros kaki*, and with those that were left untreated. The greatest browning was observed in the *Hippophae rhamnoides* samples.

Overall appearance was not compared with the standard; rather, it focused on the overall acceptability by the consumer. The best appearances were given to samples treated with 1% AA solution, *Rosa canina* juice, and to the untreated samples, while samples treated with *Actinidia deliciosa*, *Diospyros kaki*, and *Sorbus aucuparia* exhibited a "satisfactory" appearance. Samples treated with juice from *Hippophae rhamnoides* averaged between "satisfactory" and "unacceptable".

The fourth aspect evaluated via sensory analysis was taste, specifically sweetness, bitterness, and sourness. The least sweet were the *Rosa canina*-treated samples (Table 5). A sweet taste was more pronounced in the samples treated with juices from *Hippophae rhamnoides*, *Diospyros kaki*, *Actinidia deliciosa*, and the AA solution. Untreated samples were rated as the sweetest. Most samples showed only a "mild" degree of bitterness; only *Sorbus aucuparia* caused the samples to be rated as "medium" with regard to bitterness. Sourness was the least pronounced in samples treated with *Diospyros kaki* juice and in untreated samples. Samples treated with *Actinidia deliciosa*, *Hippophae rhamnoides*, *Sorbus aucuparia*, and *Rosa canina* juices showed a "medium" degree of sourness, whereas those treated with the AA solution were rated as the most sour.

Table 5. Degustation of apple slices treated with selected fruit juices, ascorbic acid (AA), or left untreated. Values are mean ± standard deviation.

Treatment	Sweetness	Bitterness	Sourness	Overall Flavor
Sorbus aucuparia	1.8 ± 0.16 [z]	1.5 ± 0.25	2.1 ± 0.15	2.3 ± 0.34 [y]
Diospyros kaki	2.0 ± 0.32	1.3 ± 0.47	1.4 ± 0.63	2.0 ± 0.36
Hippophae rhamnoides	1.9 ± 0.24	1.4 ± 0.50	1.8 ± 0.40	1.5 ± 0.41
Actinidia deliciosa	2.0 ± 0.47	1.3 ± 0.44	1.6 ± 0.46	2.1 ± 0.23
Rosa canina	1.5 ± 0.33	1.3 ± 0.37	2.1 ± 0.64	1.8 ± 0.34
1% solution of AA	1.6 ± 0.26	1.4 ± 0.63	2.6 ± 0.30	2.3 ± 0.52
Untreated apples	2.1 ± 0.28	1.1 ± 0.36	1.5 ± 0.26	2.0 ± 0.46

[z] Sweetness, bitterness and sourness values: 1 = mild; 2 = medium; 3 = significant; [y] Overall flavor: very good, satisfactory, unacceptable.

Samples treated with the juice from *Hippophae rhamnoides* achieved the highest rating for overall flavor. The *Rosa canina*, *Diospyros kaki*, *Actinidia deliciosa*, *Sorbus aucuparia,* ascorbic acid, and untreated samples received mostly "satisfactory" ratings.

4. Conclusions

In this study, the influence of selected fruit juices on enzymatic browning in dried apple slices was evaluated. Those treated with a diluted *Rosa canina* juice showed the smallest total color difference and the lowest degree of browning, as determined from colorimetric data. The sensory evaluation data suggested that using the *Rosa canina* extract resulted in a product possessing the most appealing organoleptic properties. We conclude that, based on this study, using plant-based food preservatives is a viable and efficient alternative to the currently used techniques.

Acknowledgments: This research was supported by the Internal Grant Agency of the Faculty of Tropical AgriSciences, Czech University of Life Sciences Prague, project no. 20155017.

Author Contributions: Anna Hubackova—experimental procedure and measurements, data collection and analyses, writing of the manuscript. Jan Banout—experimental design and research management, data evaluation, contribution to manuscript preparation mainly the chapter Results and Discussion, ensuring financial sources for the research.

Conflicts of Interest: The authors declare no conflicts of interest.

References

1. Nour, V.; Trandafir, I.; Ionica, M.E. Compositional characteristics of fruits of several apple (*Malus domestica* borkh.) cultivars. *Not. Bot. Hortic. Agrobot. Cluj* **2010**, *38*, 228–233.
2. Rahman, M.S. *Handbook of Food Preservation*; CRC Press: Boca Raton, FL, USA, 2007.
3. Prvulovic, S.; Tolmac, D. *Convection Drying in the Food Industry*; International Commission of Agricultural Engineering: Beograd, Serbia, 2007.
4. Pilližota, V.; Šubarić, D. Control of enzymatic browning of foods. *Food Technol. Biotechnol.* **1998**, *36*, 219–227.
5. Raymond, G.G. Reporting of objective color measurements. *HortScience* **1992**, *27*, 1254–1255.
6. Lozano, J.E. *Fruit Manufacturing: Scientific Basis, Engineering Properties and Deteriorative Reactions of Technological Importance*; Washington State University: Bahia Blanca, Argentina, 2006.
7. Özglu, H.; Bayindirli, A. Inhibition of enzymic browning in cloudy apple juice with selected antibrowning agents. *Food Control* **2002**, *13*, 213–221. [CrossRef]
8. Mohammad, A.; Rafiee, S.; Emam-Djomeh, Z.; Keyhani, A. Kinetic models for color changes in kiwifruit slices during hot air drying. *World J. Agric. Sci.* **2008**, *4*, 376–383.
9. Souza, E.L.; Souza, A.L.K.; Tiecher, A.; Girardic, C.L.; Nora, L.; da Silva, J.A.; Argenta, L.C.; Rombaldi, C.V. Changes in enzymatic activity, accumulation of proteins and softening of persimmon (*Diospyros kaki* Thunb.) flesh as a function of pre-cooling acclimatization. *Sci. Hortic.* **2011**, *127*, 242–248. [CrossRef]
10. Fijalkowska, A.; Nowacka, M.; Wiktor, A.; Sledz, M.; Witrowa-Rajcher, D. Ultrasound as a pretreatment method to improve drying kinetics and sensory properties of dried apple. *J. Food Proc. Eng.* **2015**. [CrossRef]
11. Vega-Gálvez, A.; Ah-Hen, K.; Chacana, M.; Vergara, J.; Martínez-Monzó, J.; García-Segovia, P.; Lemus-Mondaca, R.; Scala, K. Effect of temperature and air velocity on drying kinetics, antioxidant capacity, total phenolic content, color, texture and microstructure of apple (var. *Granny Smith*) slices. *Food Chem.* **2012**, *132*, 51–59. [CrossRef] [PubMed]
12. Zocca, F.; Lomolino, G.; Lante, A. Dog rose and pomegranate extracts as agents to control enzymatic browning. *Food Res. Int.* **2011**, *44*, 957–963. [CrossRef]

21

Rehydration and Textural Properties of Dried Konjac Noodles: Effect of Alkaline and Some Gelling Agents

Rarisara Impaprasert [1,*], Sujika Piyarat [1], Natnicha Sophontanakij [1], Nattiya Sakulnate [1], Suwit Paengkanya [2], Chaleeda Borompichaichartkul [3] and George Srzednicki [4]

[1] Department of Microbiology, King Mongkut's University of Technology Thonburi, Bangkok 10140, Thailand; ps.sujika@gmail.com (S.P.); p.pearl@windowslive.com (N.S.); pimwida@hotmail.com (N.S.)

[2] Division of Energy Technology, School of Energy, Environment and Materials, King Mongkut's University of Technology Thonburi, Bangkok 10140, Thailand; suwit-energy@hotmail.com

[3] Department of Food Technology, Chulalongkorn University, Bangkok 10330, Thailand; chaleedab@hotmail.com

[4] Food Science and Technology, University of New South Wales, Sydney, NSW 2052, Australia; georgesrz@yahoo.com

* Correspondence: rarisara.imp@mail.kmutt.ac.th

Abstract: Konjac glucomannan flour, which mainly consists of glucomannan, is an indigestible dietary fiber. Therefore, it has been broadly used as low-calorie food ingredient in various kinds of foods, beverages, and pharmaceutical products. In this study, the production of dried konjac noodles was evaluated by studying the effects of alkalinity using limewater versus calcium hydroxide and the gelling agent sodium alginate on textural properties of konjac noodles. Drying and rehydration conditions were studied to evaluate the optimum conditions for producing dried konjac noodles. By considering the springiness and cohesiveness of the konjac noodles, the results indicated that using 3% konjac glucomannan flour with limewater and an incubation time of 30 min were the most suitable conditions. In addition, hot air drying at 80 °C for 55 min and soaking in hot water for 9 min were the optimum drying and rehydration conditions.

Keywords: konjac glucomannan flour; limewater; calcium hydroxide; sodium alginate; springiness; cohesiveness; water uptake; hot air drying

1. Introduction

Konjac glucomannan (KGM) flour is high in dietary fiber. KGM flour is a hydrocolloidal polysaccharide obtained from the tubers of various species of *Amorphophallus*. The proximate composition of konjac glucomannan flour obtained from *Amorphophallus muelleri* contains 86.8% glucomannan, 0.2% other carbohydrates, 3% protein, 0.1% lipids, 3.3% ash, and 6.6% moisture content [1]. The main component of KGM is a non-ionic, high molecular weight polysaccharide of mannose and glucose, with a small number of branched side chains, connected by β-1,4-glycosidic linkages with a mannose: glucose molar ratio of approximately 1.6 to 4:1. Acetyl groups along the glucomannan backbone contribute to solubility properties and are located every 9 to 19 sugar units [2]. The outstanding characteristics of KGM are its indigestible dietary fiber, water-absorbing ability, and high viscosity. Thus, KGM has been given much attention in food industries for centuries, being broadly used as a food, a food additive, and in traditional Chinese medicine. Studies on KGM applications have been presented in many fields, such as pharmaceuticals, biotechnology, fine chemicals, and other areas [3].

Functional uses of KGM as a food and a food additive have been approved for many purposes such as a carrier, emulsifier, stabilizer, thickener, gelling agent, glazing agent, and humectant in many kinds of foods and beverages [2]. For food applications, KGM is also used as a fat replacement in low-fat meat products [4]. It has recently been marketed in capsule form, as a drink mix, and in food products for the treatment of obesity and diabetes. The potential use of KGM as a prebiotic has also been suggested [5].

As for the health benefits of KGM, it is a soluble fiber which can increase transit time of food and prolong gastric emptying time, delay glucose diffusion in the intestinal lumen, reduce body weight, decrease the ingestion of foods that increase cholesterol and glucose concentrations, reduce the postprandial rise in plasma glucose, suppress hepatic cholesterol synthesis, and increase the fecal elimination of cholesterol containing bile acids [6].

Konjac glucomannan-based foods were originally developed in Japan and China over several decades. They are a popular health food in world markets since they are a great source of water-soluble dietary fiber and very low calories. A unique characteristic of KGM foods is that they are translucent and gelatinous, with no flavor of their own. KGM foods can be made into many different styles and shapes, with examples at http://www.konjacfoods.com. One of the most popular styles is as a noodle. In order to make KGM foods, KGM is first dissolved in water where it becomes a highly viscous pseudoplastic solution, even at low concentrations [7]. However, it does not form a gel in water. Gelation of KGM takes place when it loses its acetyl groups in the presence of alkaline coagulants (e.g., calcium hydroxide). Thermo-irreversible and heat-stable gels are created by the deacetylation reaction [8]. KGM can also become a gel by synergistic interactions with other plant/algal hydrocolloids such as starch, carrageenan, furcellaran, xanthan and gellan gum [9]. Sodium alginate is one of the hydrocolloids which can interact with KGM to form a strong gel network structure [10,11]. However, there are no scientific reports about noodles derived from KGM and sodium alginate.

Currently, konjac noodles are found in wet forms. In order to preserve the stability of konjac gel, it needs to be packaged in alkaline solution and stored at a low temperature (4–5 °C). The drawbacks of wet KGM noodles are their bitterness and astringent taste, unpleasant odor, high transportation cost, short shelf life, and the requirement for refrigeration even before opening. There is, however, no literature dealing with the use konjac glucomannan flour as a main ingredient in the noodle making process. Therefore, the aim of this research was to evaluate the effects of alkaline solution (limewater and calcium hydroxide) and a gelling agent (sodium alginate with calcium chloride) on textural and rehydration properties of KGM noodles for the possible reduction in use of alkaline solution and its replacement with a gelling agent. The drying temperature and rehydration conditions were also studied to evaluate the optimum conditions for producing dried KGM noodles.

2. Experimental Section

2.1. Material

Commercial konjac glucomannan flour (>98% glucomannan) was purchased from Yunnan Lvyuan Bio-technology Co., Ltd. (Kunming, China). Limewater was prepared by dissolving traditional Thai red lime paste (a mixture of calcium hydroxide, turmeric, and salt) in distilled water until the solution was saturated (pH 12). Calcium hydroxide, sodium alginate, and calcium chloride (AR grade) were purchased from CTi & Science Co., Ltd. (Bangkok, Thailand).

2.2. Production of Konjac Noodles by Using Alkaline Solution

KGM flour was dissolved in alkaline solution to prepare a 3% (w/w) solution. The two alkaline solutions used in this study were limewater and 0.1 M calcium hydroxide. The final pH values of the alkaline solutions were 12.56 and 12.73 for limewater and calcium hydroxide, respectively. Based on preliminary work, the solution of KGM and limewater was incubated for 15, 20, 25, 30, and 35 min, while the KGM-calcium hydroxide solution was incubated for 0.5, 1.0, 1.5 and 2 h, all at room temperature (35 °C). Then, the KGM-alkaline solutions were extruded through a 2 mm diameter-opening of a syringe into boiling water and held for 1 min. After that, the konjac noodles were placed in 4 °C water for 1 min, and then placed in a sieve to drain the excess water.

2.3. Production of Konjac Noodles by Using Sodium Alginate and Calcium Chloride

KGM flour was dissolved in 0.25%, 0.5%, 1.0%, 1.25%, and 1.5% sodium alginate solution to prepare a 3% (w/w) solution. Then, KGM-sodium alginate solution was extruded through a 2 mm diameter-opening of a syringe into a 1% calcium chloride solution for a minute. The resulting KGM noodles were placed on sieve to drain the excess water.

2.4. Drying Experiments

The KGM noodles were analyzed for initial moisture content by using a hot air oven (Programmable Laboratory Oven, Fisher Scientific Worldwide, United States) at 105 ± 2 °C for 3 h [12]. For the drying experiment, 100 ± 0.5 g of konjac noodles were placed on a stainless tray and placed in the hot air oven. The air temperature of the dryer was set at 60, 70, or 80 °C. The weight of konjac noodles and tray were determined every 10 min until they reached a constant weight (the equilibrium moisture content). All drying experiments were performed in triplicate. The moisture content change of konjac noodles during the drying process was expressed as a moisture ratio, calculated by the following Equation (1) to describe the drying curve [13].

$$\text{Moisture ratio} = \frac{M - M_e}{M_o - M_e} \qquad (1)$$

where M = moisture content (% dry weight basis (d.b.)) at each weighing time, M_0 = initial moisture content (% d.b.), M_e = equilibrium moisture content (% d.b.).

2.5. Rehydration Properties

The water uptake percentage of the dried konjac noodles was determined according to Zhou et al. [14] with some modifications. Dried konjac noodles (10 ± 0.5 g) were placed in a beaker containing 400 mL of ambient temperature water, hot water (98 °C), or boiling water for 3, 6, 9, or 12 min. The rehydrated konjac noodles were then drained on a sieve for 5 min before weighing. The water uptake percentage was calculated by following Equation (2) [15].

$$\text{Water uptake (\%)} = \frac{\text{weight of rehydrated konjac noodles}}{\text{weight of dried konjac noodles}} \times 100 \qquad (2)$$

2.6. Texture Profile Analysis

Texture profile analyses (TPA) were performed by using a TA.XT Plus Stable Micro Systems Texture Analyser (Stable Micro Systems, model TA.XT. Plus, Surrey, UK) with the Texture Expert software. The TPA parameters were evaluated both on fresh and rehydrated konjac noodles. According to Jimenez-Colmenero et al. [9], Zhou et al. [14], and Kaur et al. [15] with some modifications, each strand of konjac noodles had a cross-sectional area of approximately 2 mm and length of 40 mm. A set of four strands was placed parallel on the flat metal plate. The samples were axially compressed twice to 50% of their original sample height. A 2 kg load cell with a cylinder P/36R probe was used at 2, 0.8, and 0.8 mm/s for pre-test, test, and post-test speeds, respectively. The trigger type is auto with a trigger force of 0.5 g. Ten replicate samples were tested. Some of the TPA parameters obtained from the force-time curves such as cohesiveness and springiness were determined. Tensile strength and % elongation were measured by using the A/SPR-Spaghetti/Noodle Rig following Stable Micro Systems guidelines at 2 mm/s with a 2 kg load cell.

2.7. Microstructure of Konjac Noodles

Cross-sections of each konjac noodle sample were examined by cutting the konjac noodles and mounting on a stub. The cross-sectional surface of all samples was observed from the top under a scanning electron microscope (JEOL, model JSM-6400LV, Tokyo, Japan) at an accelerating voltage of 15 kV in high vacuum mode. Magnifications of $30\times$, $5000\times$ and $20,000\times$ were used for observing the cross-sections.

2.8. Statistical Analysis

Except for the TPA analyses which were replicated 10 times, all other analyses were replicated 3 times. The differences between treatment means were established using analysis of variance (ANOVA) and Duncan's New Multiple Range test at a confidence level of $p = 0.05$ using the SPSS Statistical Analysis Program for Windows (SPSS Inc., Chicago, IL, USA).

3. Results and Discussion

3.1. Production of Konjac Noodles

In general, when KGM is dissolved in water, it become a viscous solution and cannot form a gel. However, KGM gel can be produced by deacetylation using alkaline solution or in synergistic interactions with other hydrocolloids such as sodium alginate, κ-carrageenan, or xanthan gum [4,5,7,8,10,11]. The texture of alkaline-KGM gel is stronger than with other hydrocolloid-KGM gels. In this study, the konjac noodles, which were produced by using 0.1 M limewater with 15, 20, 25, 30, 35 min of incubation time, 0.1 M calcium hydroxide with 0.5, 1.0, 1.5 and 2 h of incubation time, and 0.25%, 0.5%, 1.0%, 1.25%, and 1.5% sodium alginate solution/1% calcium chloride, were evaluated by considering the TPA of springiness and cohesiveness values (Table 1).

Table 1. Some textural profile analysis (TPA) parameters of fresh konjac noodles.

KGM Sample	Springiness	Cohesiveness
Commercial product	0.87 ± 0.03 [abc]	0.75 ± 0.02 [a]
Limewater/15 min incubation	0.89 ± 0.05 [abc]	0.70 ± 0.02 [bc]
Limewater/20 min incubation	0.89 ± 0.03 [ab]	0.70 ± 0.02 [bc]
Limewater/25 min incubation	0.88 ± 0.04 [abc]	0.69 ± 0.03 [bc]
Limewater/30 min incubation	0.90 ± 0.05 [a]	0.74 ± 0.02 [a]
Limewater/35 min incubation	0.84 ± 0.05 [c]	0.68 ± 0.02 [cd]
Calcium hydroxide/0.5 h incubation	N/A	N/A
Calcium hydroxide/1 h incubation	N/A	N/A
Calcium hydroxide/1.5 h incubation	0.84 ± 0.06 [c]	0.71 ± 0.03 [b]
Calcium hydroxide/2 h incubation	0.86 ± 0.02 [abc]	0.66 ± 0.03 [e]
0.25% sodium alginate	N/A	N/A
0.5% sodium alginate	N/A	N/A
1% sodium alginate	0.85 ± 0.05 [c]	0.66 ± 0.03 [de]
1.25% sodium alginate	0.85 ± 0.05 [c]	0.63 ± 0.04 [f]
1.5% sodium alginate	0.85 ± 0.05 [bc]	0.71 ± 0.04 [b]

Means ± standard deviation. Different letters in the same column indicate significant differences ($p \leq 0.05$). N/A = not available (no gel formation).

Springiness is a measure of how well the product physically responds after it has been deformed during a first compression, while cohesiveness is how well the product withstands a second deformation relative to how it behaved after the first deformation, which is related to the strength of hydrogen bonding among KGM molecules. Noodles are considered good quality if they exhibit high springiness and cohesiveness values.

The springiness values of konjac noodles produced from all treatments showed no significant difference from the commercial product, although some variation among treatments was evident, while cohesiveness values from some treatments did differ from the commercial product. Also, 1.25% sodium alginate had the lowest cohesiveness. The results generally showed higher cohesiveness values when using alkaline solution compared with sodium alginate. This may due to the loss of acetyl groups from KGM chains by adding alkaline solution, called deacetylation [16–21]. The effects of eliminating acetyl groups produced junction zones through hydrogen bonding between calcium ions and deacetylated anions ($R–CH_2O^-$), between water molecules and hydroxyl groups (OH^-) of deacetylated KGM, and by van der Waals, charge transfer and hydrophobic interactions, forming a strong, elastic, and thermally stable gel network structure [11,22,23] to provide high springiness and cohesiveness of samples. While KGM and alkaline solution are cross-linked by several bonds, the springiness and cohesiveness of konjac noodles from KGM and sodium alginate came from only gelation of divalent cations (like Ca^{2+}) and polysaccharides to form an "egg box" model structure [24]. Thus, using alkaline solution showed better cohesiveness values than sodium alginate. It may be possible to use calcium ions (Ca^{2+}) from calcium hydroxide to absorb CO_2 and precipitate calcium carbonate ($CaCO_3$) [25]. Thus, the strength of konjac noodle structure may result from Ca^{2+} from limewater or calcium hydroxide solution which reacted with CO_2 in the environment of the samples to produce crystals of $CaCO_3$ in the noodle matrix to make a strong and elastic gel. It can be concluded that the different types of alkaline solutions and incubation times significantly influenced the gel strength of KGM noodles as observed in springiness and cohesiveness values. The optimum conditions for producing the highest springiness and cohesiveness of KGM noodles was 3% KGM with limewater, incubated for 30 min.

Figure 1 shows that konjac noodles produced from 3% KGM and 1% sodium alginate were transparent, while noodles from the other methods were opaque as a result of using alkaline solution.

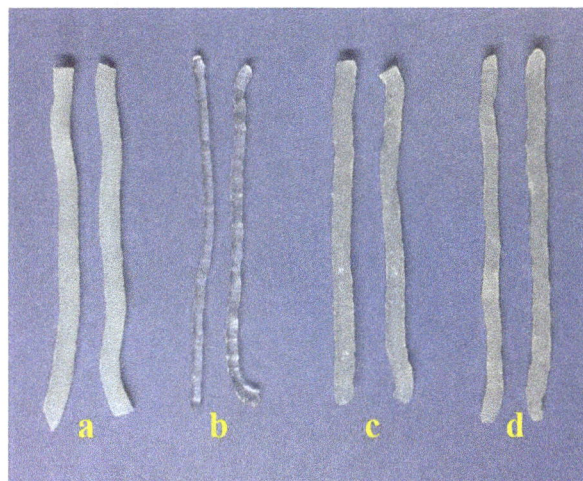

Figure 1. Konjac noodles from (**a**) commercial product, or produced from (**b**) 3% KGM and 1% sodium alginate; (**c**) 3% KGM and $Ca(OH)_2$ with 2 h incubation time; and (**d**) 3% KGM and limewater with 30 min incubation time.

Scanning electron micrographs of cross-sectioned konjac noodles revealed the inner structure of the noodles (Figure 2). The konjac noodles formed strong and elastic hydrogels when heated with alkali by deacetylation and disruption of the hydrogen bonding between the glucomannan chain and water molecules [26]. Thus, a reticular structure was formed. Only konjac noodle produced from 3% KGM and sodium alginate showed a different structure, a ionically cross-linked by divalent Ca^{2+} cations. The results also indicated that the structures of the noodles tended to be denser when increasing the concentration of sodium alginate.

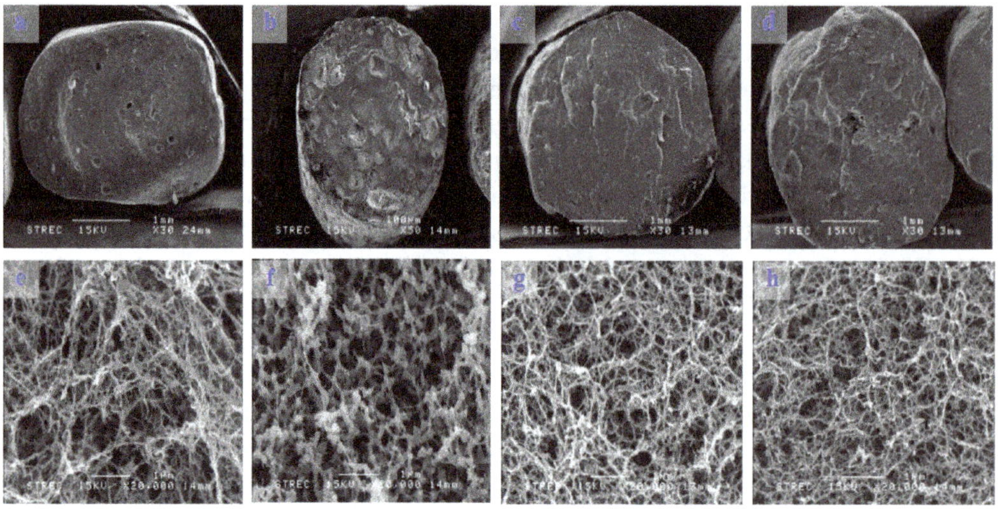

Figure 2. Scanning electron micrographs (SEM) of konjac noodle from a commercial process (**a**,**e**); from 3% KGM and 1% sodium alginate (**b**,**f**); from 3% KGM and $Ca(OH)_2$ with a 2 h incubation time (**c**,**g**); and from 3% KGM and limewater with 30 min incubation time (**d**,**h**). Magnifications of 30× (**a**–**d**) and 20,000× (**e**–**h**) were used.

3.2. Drying Experiments

In order to study the drying time to reduce moisture content of konjac noodles from the initial moisture content of 1.24% ± 22% d.b. (moisture ratio = 1) to the target moisture content 10%–12% d.b. (moisture ratio = 0.01), konjac noodles were dried by using hot air at 60, 70, or 80 °C. The drying patterns among the temperatures were determined from the mass loss of the samples. The drying progress of konjac noodles subjected to hot air drying treatments (60–80 °C) is shown in Figure 3.

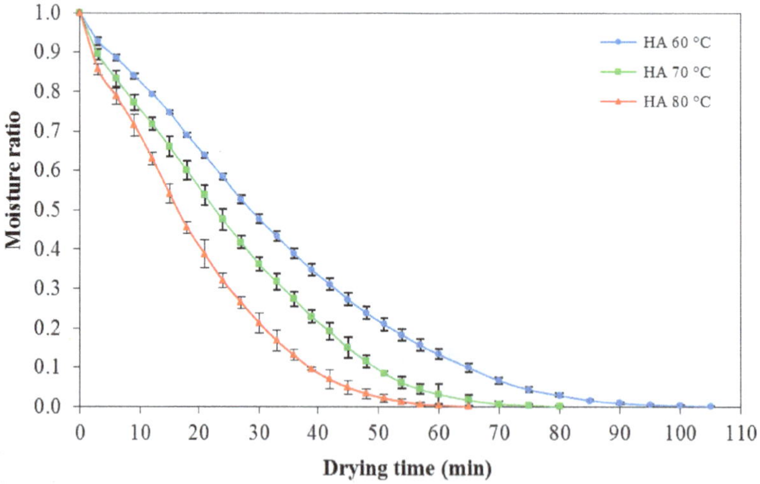

Figure 3. Moisture ratio versus drying time of konjac noodles using hot air at 60, 70, or 80 °C.

The results indicated that drying of konjac noodles using hot air at 60, 70, and 80 °C required 85, 65, and 55 min, respectively, to reduce moisture content to 10%–12% d.b. Thus, drying at 80 °C was chosen for the following experiment since it had the shortest drying time.

3.3. Rehydration Properties, Texture Profile Analysis, and Microstructure Observation

The dried konjac noodles obtained from the previous experiment (3% KGM in limewater, incubation for 30 min, drying at 80 °C for 55 min) were used to study the water absorption properties of the samples by analyzing the water uptake percentage, the tensile strength, and the elongation percentage of rehydrated konjac noodles to determine the most suitable rehydration conditions. Tensile strength is the maximum amount of stress (force per unit area) that is required for pulling the noodle to the breaking point.

The water uptake percentage of the samples tended to increase with increasing the rehydration time and water temperature (Figure 4). In general, the dried noodle has a porous structure that is created during dehydration. The pores serve as channels for water to enter the noodles, leading to rehydration [27]. The higher the water temperature that was used, the more rapidly water was adsorbed.

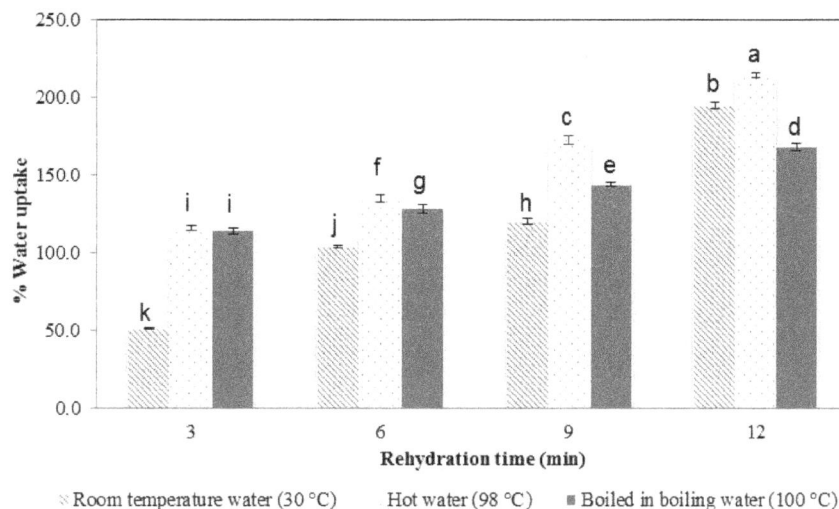

Figure 4. Water uptake percentage as affected by duration and water temperature of rehydrated konjac noodles created from 3% KGM with limewater and incubation for 30 min. Means with different letters are significantly different ($p \leq 0.05$).

Scanning electron micrographs showed that the drying process had a large impact on the reticular structure of konjac noodles (Figure 5). Moreover, the higher water temperature for rehydration also affected their reticular structure. When the reticular structure was destroyed, the rehydration properties decreased. Thus, these two factors were extremely important for the rehydration of the konjac noodle.

When konjac noodles were boiled in hot water, the water migrated from the noodle surface toward the center, affecting the structure of the noodles (Figure 5). Therefore, the noodle textural traits of tensile strength and elongation changed, with the sample becoming softer on rehydration with a longer hydration time in hot water.

In the early stages of rehydration, the konjac noodles had absorbed less water (Figure 4) and were tough and showed a higher tensile strength (Figures 6 and 7). By the latter stages of rehydration, the konjac noodles had absorbed more water and showed a softer texture and lower tensile strength. Thus, rehydration temperature and time affected the strength of the samples. Elongation percentage also tended to decrease with time (Figures 6 and 7). The most suitable rehydration method was soaking in hot water for 9 min which provided the optimum values of tensile strength, elongation, and water uptake percentage.

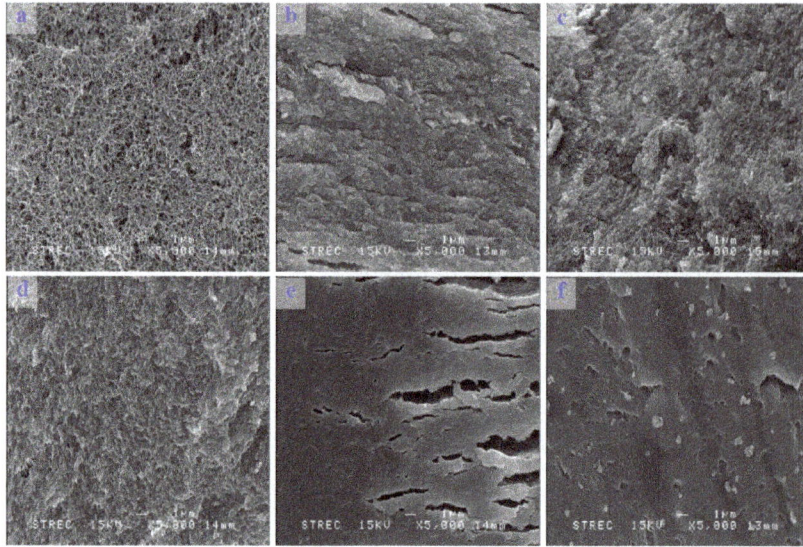

Figure 5. Scanning electron micrographs (SEM) of (**a**) a konjac noodle from 3% KGM and limewater with 30 min incubation; (**b**) a dried konjac noodle; (**c**) a rehydrated konjac noodle in boiling water for 3 min; (**d**) 6 min; (**e**) 9 min; and (**f**) 12 min. Magnification was at 5000×.

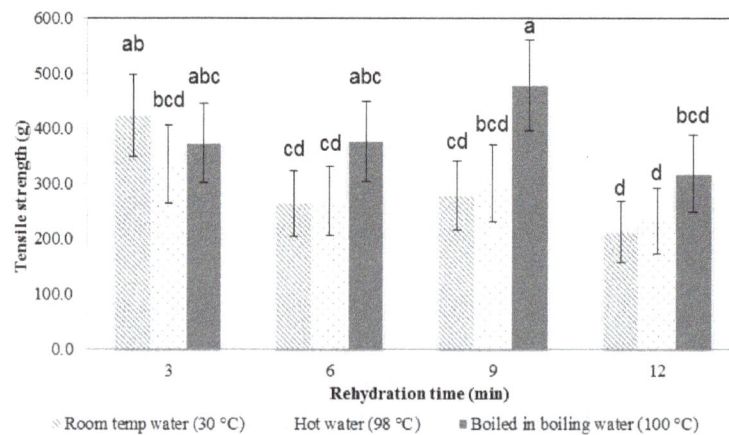

Figure 6. Tensile strength of rehydrated konjac noodles from 3% KGM with limewater and incubation for 30 min. Means with different letter are significantly different ($p \leq 0.05$).

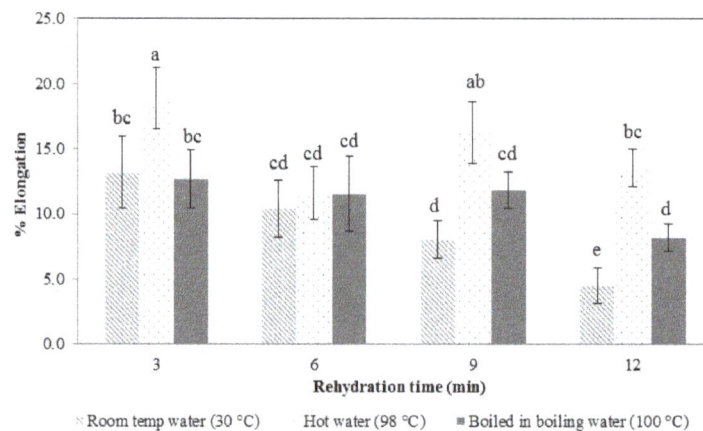

Figure 7. Elongation percentage of rehydrated konjac noodles from 3% KGM with limewater and incubation for 30 min. Means with different letters are significantly different ($p \leq 0.05$).

4. Conclusions

This study showed that springiness values of konjac noodles produced from all treatments showed some variation. Higher cohesiveness values resulted when using alkaline solution compared with sodium alginate. However, the springiness and cohesiveness values were comparable to a commercial product. The optimum method for producing the highest springiness and cohesiveness of KGM noodles was 3% KGM with limewater and incubation for 30 min. Hot air drying at 80 °C for 55 min was the optimum drying condition for producing dried konjac noodles since it resulted in the shortest drying time. The most suitable rehydration procedure was soaking in hot water for 9 min which provided optimum values of tensile strength, elongation, and water uptake percentage. The drying process and rehydration conditions had significant impacts on the structure and rehydration properties of konjac noodles.

Acknowledgments: Authors would like to thanks Department of Microbiology, King Mongkut's University of Technology Thonburi for the research facilities. This research is supported by Faculty of Science, King Mongkut's University of Technology Thonburi Grant No. SCI58-001.

Author Contributions: This work was a product of the combined effort of all authors. Sujika Piyarat, Natnicha Sophontanakij, and Nattiya Sakulnate jointly performed the experiments, gathered and analyzed the data. Rarisara Impaprasert designed the experiment and wrote the manuscript with the help of Suwit Paengkanya. Chaleeda Borompichaichartkul and George Srzednicki revised and improved the manuscript.

Conflicts of Interest: The authors declare no conflict of interest.

References

1. Impaprasert, R.; Borompichaichartkul, C.; Srzednicki, G. A new drying approach to enhance quality of konjac glucomannan extracted from *Amorph. Muelleri. Dry. Technol. Int. J.* **2014**, *32*, 851–860. [CrossRef]
2. Konjac Flour. Available online: http://www.fao.org/gsfaonline/additives/details.html?id=10 (accessed on 15 August 2015).
3. Zhang, Y.Q.; Xie, B.J.; Gan, X. Advance in the applications of konjac glucomannan and its derivatives. *Carbohyd. Polym.* **2005**, *60*, 27–31. [CrossRef]
4. Tye, R.J. Konjac flour: Properties and applications. *Food Technol.* **1991**, *45*, 86–92.
5. Chua, M.; Baldwin, T.C.; Hocking, T.J.; Chan, K. Traditional uses and potential health benefits of *Amorphophallus konjac* K Koch ex N.E.Br. *J. Ethnopharmacol.* **2010**, *128*, 268–278. [CrossRef] [PubMed]
6. Shah, B.R.; Li, B.; Wang, L.; Liu, S.; Li, Y.; Wei, X.; Weiping, J.; Zhenshun, L. Health benefits of konjac glucomannan with special focus on diabetes. *Bioact. Carbohydr. Diet. Fiber* **2015**, *5*, 179–187. [CrossRef]
7. Dave, V.; McCarthy, S.P. Review of Konjac Glucomannan. *J. Environ. Polym. Degr.* **1997**, *5*, 237–243.
8. Nishinari, K.; Zhang, H. Recent advances in the understanding of heat set gelling polysaccharides. *Trends Food Sci. Technol.* **2004**, *15*, 305–312. [CrossRef]
9. Jimenez-Colmenero, F.; Cofrades, S.; Herrero, A.M.; Solas, M.T.; Ruiz-Capillas, C. Konjac gel for use as potential fat analogue for healthier meat product development: Effect of chilled and frozen storage. *Food Hydrocoll.* **2013**, *30*, 351–357. [CrossRef]
10. Harding, S.E.; Smith, I.H.; Lawson, C.J.; Gahlerd, R.J.; Wood, S. Studies on macromolecular interactions in ternary mixtures of konjac glucomannan, xanthan gum and sodium alginate. *Carbohydr. Polym.* **2011**, *83*, 329–338. [CrossRef]
11. Wang, J.; Liu, C.; Shuai, Y.; Cui, X.; Nie, L. Controlled release of anticancer drug using graphene oxide as a drug-binding effector in konjac glucomannan/sodium alginate hydrogels. *Colloids Surf. B Biointerfaces* **2014**, *113*, 223–229. [CrossRef] [PubMed]
12. Association of Official Analytical Chemists. *Official Methods of Analysis*, 18th ed.; Association of Official Analytical Chemists International: Washington, DC, USA, 2006.
13. Chen, J.Y.; Zhang, H.; Miao, Y. The effect of quantity of salt on the drying characteristics of fresh noodles. *Agric. Agric. Sci. Procedia.* **2014**, *2*, 207–211. [CrossRef]
14. Zhou, Y.; Cao, H.; Hou, M.; Nirasawa, S.; Tatsumi, E.; Foster, T.J.; Cheng, Y. Effect of konjac glucomannan on physical and sensory properties of noodles made from low-protein wheat flour. *Food Res. Int.* **2013**, *51*, 879–885. [CrossRef]

15. Kaur, A.; Singh, N.; Kaur, S.; Katyal, M.; Virdi, A.S.; Kaur, D.; Ahlawat, A.K.; Singh, A.H. Relationship of various flour properties with noodle making characteristics among durum wheat varieties. *Food Chem.* **2015**, *188*, 517–526. [CrossRef] [PubMed]

16. Huang, L.; Takahashi, R.; Kobayashi, S.; Kawase, T.; Nishinari, K. Gelation behavior of native and acetylated konjac glucomannan. *Biomacromolecules* **2002**, *3*, 1296–1303. [CrossRef] [PubMed]

17. Maekaji, K. Mechanism of gelation of konjac mannan. *Agric. Biol. Chem.* **1974**, *38*, 315–321. [CrossRef]

18. Williams, M.A.; Foster, T.J.; Martin, D.R.; Norton, I.T.; Yoshimura, M.; Nishinari, K. A molecular description of the gelation mechanism of konjac mannan. *Biomacromolecules* **2000**, *1*, 440–450. [CrossRef] [PubMed]

19. Yoshimura, M.; Nishinari, K. Dynamic viscoelastic study on the gelation of konjac glucomannan with different molecular weights. *Food Hydrocoll.* **1999**, *13*, 227–233. [CrossRef]

20. Zhang, H.; Yoshimura, M.; Nishinari, K.; Williams, M.A.K.; Foster, T.J.; Norton, I.T. Gelation behaviour of konjac glucomannan with different molecular weights. *Biopolymers* **2001**, *59*, 38–50. [CrossRef]

21. Solo-de-Zaldívar, B.; Herranz, B.; Borderias, J. First steps in using glucomannan to make thermostable gels for potential use in mince fish restructuration. *Int. J. Food Eng.* **2012**, *8*, 1–10. [CrossRef]

22. Lapasin, R.; Pricl, S. *Rheology of Industrial Polysaccharides: Theory and Applications*; Aspen Publishers: Gaithersburg, MD, USA, 1999.

23. Herranz, B.; Tovar, C.A.; Solo-de-Zaldívar, B.; Borderias, A.J. Effect of alkalis on konjac glucomannan gels for use as potential gelling agents in restructured seafood products. *Food Hydrocoll.* **2012**, *27*, 145–153. [CrossRef]

24. Grant, G.T.; Morris, E.R.; Rees, D.A.; Smith, P.J.C.; Thom, D. Biological interactions between polysaccharides and divalent cations: The Egg-Box model. *FEBS Lett.* **1973**, *32*, 195–198. [CrossRef]

25. Chuajiw, W.; Nakano, M.; Takatori, K.; Kojima, T.; Wakimoto, Y.; Fukushima, Y. Effects of amine, amine salt and amide on the behaviour of carbon dioxide absorption into calcium hydroxide suspension to precipitate calcium carbonate. *J. Environ. Sci.* **2013**, *25*, 2507–2515. [CrossRef]

26. Luo, X.G.; He, P.; Lin, X.Y. The mechanism of sodium hydroxide solution promoting the gelation of Konjac glucomannan (KGM). *Food Hydrocoll.* **2013**, *30*, 92–99. [CrossRef]

27. Jang, A.; Kim, H.; Shim, J.; Lee, S.K.; Lee, S. Correlation of thermal conductivity of instant noodles with their textural property for rehydration study. *J. Texture Stud.* **2016**, *47*, 87–91. [CrossRef]

22

Evaluation of Some Postharvest Storage Approaches on Essential Oil Characteristics of Fresh Organic Damask Rose (*Rosa damascena* Mill.) Flowers

Maryam Mirzaei [1], Nima Ahmadi [1,*], Fatemeh Sefidkon [2], Abdolali Shojaeiyan [1] and Alireza Mazaheri [3]

[1] Department of Horticultural Sciences, Tarbiat Modares University, Tehran 336-14115, Iran; maryammirzaeem@gmail.com (M.M.); shoja_a@yahoo.com (A.S.)

[2] Research Institute of Forests and Rangelands, Tehran 13185-116, Iran; sefidkon@rifr-ac.ir

[3] R & D Department, Golkaran Essential Oils and Herbal Extracts Company, Kashan 1178, Iran; mazaheri_58@yahoo.com

* Correspondence: ahmadin@modares.ac.ir

Abstract: Damask rose (*Rosa damascena* Mill.) is an economically important species in the Rosaceae family for rose oil and rose water production, obtained exclusively from freshly-gathered Damask rose flowers. Iran is famous for producing organic rose essential oil, mainly used in the perfume industry and for folk medicines due to its healing properties. Due to their high perishability, postharvest handling of the flowers prior to oil extraction are a key point in rose essential oil production. In this study, postharvest handling conditions of Damask rose flowers were evaluated for the effect on yield and quality of the extracted essential oils. Organically-grown flowers were stored under different conditions included packaging in polyethylene bags and immersing into water containers, and then held at $25 \pm 2\,°C$ or $4 \pm 1\,°C$ for 1, 2, or 3 days. Identification of the chemical composition of the essential oils was performed by GC and GC/MS. The results indicated that petal storage in water resulted in the highest essential oil content and quality, and storage in polyethylene bags resulted in the greatest loss, compared to those from unstored petals. The results provide a reference for interested groups such as producers and consumers who are concerned about Damask rose flower storage and subsequent oil extraction.

Keywords: essential oil characteristics; organic damask rose; post-harvest handling

1. Introduction

Rosa damascena Mill., commonly known as the oil-bearing rose, is one of the most important Rosa species cultivated for production of rose oil and rose water [1]. Rose oil, one of the most expensive base materials in the flavor and perfume industries, is characterized by a high percentage of monoterpene alcohols including citronellol, nerol, geraniol, linalool, and phenyl ethyl alcohol. These components are the main contributors to the perfume value of rose oil [2]. According to the international standard of rose essential oil [3], rose oil requires 20%–34% citronellol, 15%–22% geraniol, and 8%–15% nonadecane as the main compounds [3]. As harvested Damask rose flowers are highly perishable, postharvest handling of the flowers prior to oil extraction is a vital step in rose essential oil production. Due to the huge volume of Damask rose flowers delivered to the factories during the harvest period, it often takes a long time before flowers are eventually extracted and distilled. Excessive temperatures inside the stacks of harvested flowers caused by their high respiration rate results in losses of essential oil quantity and quality [2]. Providing optimal storage conditions throughout postharvest handling could reduce losses of yield and quality of essential oils [4,5]. The high value of the essential oil of Damask rose is an important reason for finding proper storage techniques for the flower crop after harvest. Therefore, the objectives of this study were to determine the effects of ambient and reduced temperatures as well as different storage durations and surrounding mediums on the yield and chemical components of organic Damask rose flowers throughout a postharvest handling period of 3 days.

2. Experimental Section

2.1. Plant Material

Organically-grown fresh Damask roses petals were handpicked from a Rosarium located in Mashhad-e Ardehal, Kashan, Iran, and transported immediately to the laboratory. The essential oils obtained from unstored petals (0 day storage duration), which were hydrodistilled shortly after harvest, were considered as the reference point in order to determine the effect of storage conditions on essential oil content and oil composition.

2.2. Storage Conditions

The petals were stored under the following conditions: packaged using sealed polyethylene (PE) bags (60 μm thick), or immersed in water inside sealed plastic containers (20 cm × 25 cm × 30 cm), and then held at 25 °C (25 ± 2 °C, 45% ± 5% RH) or 4 °C (4 ± 1 °C, 90% ± 5% RH) for 1, 2 and 3 days.

2.3. Essential Oil Extraction

The essential oil was extracted from 250 g petals by hydrodistillation with 750 mL of distilled water using a 4 L Clevenger-type apparatus for each experimental replication. The distillation time was 1.5 h and the oil content was measured as percentage ($v/w\%$). In the case of the water storage method, the water used for storage was also hydrodistilled.

2.4. Characterization of Essential Oil Components

2.4.1. Gas Chromatography (GC)

The essential oils were analyzed on a Thermo-UFM (Ultra-fast model, Milan, Italy) gas chromatograph using a 10 m × 0.1 mm capillary column PH-5 (nonpolar), with an inner surface with a stationary phase of 5% dimethyl siloxane phenyl. The oven temperature was held at 60 °C for 3 min, programmed to 285 °C at the rate of 80 °C/min, and detector (FID) and injection chamber temperatures were 280 °C. Helium was used as the carrier gas with inlet pressure of 0.5 mL/min.

2.4.2. Gas Chromatography-Mass Spectrometry (GC-MS)

The GC-MS analysis was performed using a Varian 3400 gas chromatograph connected to a mass spectrometer model Saturn II, using an ion trap system having ionization energy of 70 eV, with a DB-5 column (semi-polar), with dimensions of 30 m × 0.25 mm, and a film thickness 0.25 mm. Gas pressure was 2.46 kg/cm^2, column temperature was 40 to 250 °C with a rate of increase of 3 °C/min, and the injection chamber temperature and the transfer line were set at 260 °C and 270 °C, respectively. The carrier gas was helium with a linear velocity of 31.5 cm/s, and a split ratio of 1:60. A scan time of 1 s with a mass range 40–300 were used. Spectra identification were carried out using retention indices and by injection of normal hydrocarbons (C7-C25) under the same conditions and confirmed with the Wiley 275-L library and the data reported in the literature [6–8].

2.5. Data Analysis

The experiment was carried out as a factorial with the factors of storage duration, petal storage medium, and storage temperature in a randomized complete block design (RCBD). There were three replications of essential oil extraction from the petals for every treatment combination. Analysis of data from the essential oil content was performed using the general linear model (GLM) of SAS 9.2 [9]. Mean values were compared at the 5% ($p \leq 0.05$) and 1% ($p \leq 0.01$) levels of significance using Fisher's Least Significant Difference (LSD) test. There were three replications of every treatment combination. Component analyses of the oils were performed by injecting a mix of the three replications of each treatment combination on both GC and GC/MS.

3. Results and Discussion

3.1. Essential Oil Contents

The petal essential oil content was significantly affected by storage condition and duration, but not temperature alone. There was a significantly greater content of essential oil stored in water at 4 °C and 25 °C at 1, 2 and 3 day compared to petals stored in PE bags (Figure 1). Storage of flowers in water recorded a significant increase ($p \leq 0.01$) of 16%, 19% and 23% in essential oil content after 1, 2, and 3 day, respectively, over the unstored petals with an average value of 0.069 ($v/w\%$). This increase could be explained by considering the limitation of O_2 inside the containers along with the presence of water that may have caused fermentation, as it has been indicated that rose petals produce more essential oil with a fermentation period after harvest [10]. The mean essential oil content obtained from petals stored in PE for 1 day were close to the value at harvest (unstored petals) but were significantly lower after 2 or 3 day of storage. Cold-stored flowers in water contained higher oil content than those in PE at the same storage temperature, which might be explained by the ability of water to maintain volatile oil content of the petals during the storage period, while the permeability of the PE bags might result in losses from volatilization of essential oils, particularly with the extension of the storage period.

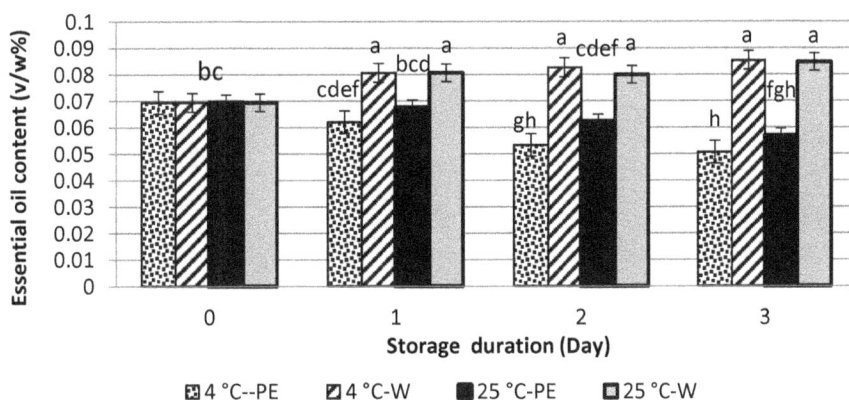

Figure 1. Mean essential oil content ($v/w\%$, fresh weight basis) of organic Damask rose under different storage conditions. PE: (Polyethylene bag); W: (Water).

3.2. The Essential Oil Components

The changing patterns of the essential oil components under the storage conditions indicated that the chemical profile of rose oil throughout storage of petals in PE bags was affected more than storage in water under either 25 °C or 4 °C storage conditions, with a higher degradation of geraniol in favor of citronellol at 25 °C (Table 1). Storage of petals in water impacted the maintenance of the major essential oil compounds. The component analyses of oils from 2 and 3 day samples indicated that storage of the petals in the cold water (4 °C-W) maintained the rose oil characteristics better than the other handling method, with approximately 38% lower levels of long-chain aliphatic hydrocarbons after 1 day of storage, which resulted in improvement of the oil quality. A possible reason for the rise of monoterpene alcohols might be ascribed to the proper conditions for the action of enzymes, such as β-glucosidase involved in hydrolysis and liberation of monoterpene alcohols from their glycosidic forms [11].

Table 1. Damask rose essential oil components under different storage conditions.

Factors of Variance		Compound	Phenyl Ethyl Alcohol	cis-Rose Oxide	Dihydrolinalool	Citronellol	Neral	Geraniol	Geranial	Citronellyl Acetate	E-β-Damascone	n-Heptadecane	E,E-Farnesyl Acetate	Nonadecane	n-Eicosane	n-Heneicosane	n-Tricosane
Time	Storage conditions	RI	1109	1113	1135	1228	1240	1255	1269	1355	1416	1693	1728	1902	2003	2102	2298
0 day	Unstored petals		1.3	0.3	0.8	35.5	13.3	16.8	3.1	0.4	0.2	2.3	1.1	15.6	0.7	2.6	0.8*
1 day	4°C-PE		1.0	0.3	0.8	42.5	3.9	13.0	0.3	-	-	2.3	0.2	17.9	1.4	8.1	1.2
	4°C-W		1.3	0.3	0.9	38.2	21.5	13.4	2.9	0.2	0.3	2.5	0.9	9.1	0.5	1.0	0.3
	25°C-PE		1.6	0.4	1.3	47.4	16.5	9.8	5.8	-	0.3	2.6	0.4	8.8	0.3	4.1	0.8
	25°C-W		1.1	0.7	0.9	42.3	11.4	12.6	2.0	0.1	0.4	2.9	0.7	20.5	1.0	6.3	0.9
2 days	4°C-PE		0.9	0.4	0.8	45.6	12.5	10.8	1.6	-	0.4	2.6	0.5	18.9	2.1	12.3	2.2
	4°C-W		1.7	0.5	1.4	48.2	15.6	8.1	6.2	-	0.3	2.8	0.7	7.3	0.4	0.6	0.8
	25°C-PE		1.2	0.3	0.5	36.5	18.7	10.0	2.0	0.2	0.1	2.6	0.5	12.8	0.4	1.7	0.5
	25°C-W		1.1	1.1	1.1	43.4	10.2	11.8	1.2	-	0.6	3.3	0.4	22.2	1.1	5.0	1.1
3 days	4°C-PE		1.0	0.3	0.7	47.4	7.4	9.0	0.6	-	-	2.9	-	20.7	2.7	14.3	4.1
	4°C-W		0.7	0.4	1.0	35.6	16.4	15.4	3.4	-	0.4	3.1	-	7.6	0.6	2.2	0.4
	25°C-PE		1.9	0.6	0.8	36.5	12.8	5.2	5.9	-	0.2	3.4	0.8	12.6	0.4	3.6	0.7
	25°C-W		1.0	1.5	1.4	46.7	9.2	10.0	0.9	-	0.8	3.9	0.1	23.6	1.2	4.6	1.3

RI: Retention Index; * Relative percentage of each component in relation to total essential oil; W: water; PE (polyethylene).

4. Conclusions

As the water used for the storage of the flowers was then used in the hydrodistillation, water, particularly at low temperatures, might be an effective storage method for oil-bearing rose petals after harvest to maintain essential oil content and component quantities close to the content of freshly-harvested flowers. These results are useful for choosing the best storage techniques for Damask rose flowers prior to processing.

Acknowledgments: The authors would like to express their appreciation to Barij Essence Company for their support and for providing facilities for this study.

Author Contributions: Maryam Mirzaei, Nima Ahmadi, Fateme Sefidkon and Abdolali Shojaeiyan designed the research; Maryam Mirzaei, Nima Ahmadi and Fateme Sefidkon performed the research; Nima Ahmadi, Fateme Sefidkon, Alireza Mazaheri and Abdolali Shojaeiyan contributed analysis tools/materials/reagents; Maryam Mirzaei analyzed the data; Maryam Mirzaei and Nima Ahmadi wrote the paper.

Conflicts of Interest: The authors declare no conflict of interest.

References

1. Widrlechner, M.P. History and utilization of *Rosa damascena*. *Econ. Bot.* **1981**, *35*, 42–58. [CrossRef]
2. Baydar, H.; Baydar, N. The effects of harvest date, fermentation duration and Tween 20 treatment on essential oil content and composition of industrial oil rose (*Rosa damascena* Mill.). *Ind. Crops Prod.* **2005**, *21*, 251–255. [CrossRef]
3. Oil of Rose (*Rosa* × *damascena* Mill.). International Standards for Business, Government and Society. Available online: http://www.iso.org (accessed on 10 August 2015).
4. Najafian, S. Storage conditions affect the essential oil composition of cultivated Balm Mint Herb (Lamiaceae) in Iran. *Ind. Crops Prod.* **2014**, *52*, 575–581. [CrossRef]
5. Kazaz, S.; Sabri, E.; Baydar, H.; Dilmacunal, T.; Koyuncu, M.A. Cold storage of oil rose (*Rosa damascena* Mill.) flowers. *Sci. Hortic.* **2010**, *126*, 284–290. [CrossRef]
6. Adams, R.P. *Identification of Essential Oils by Ion Trap Mass Spectroscopy*; Academic Press: New York, NY, USA, 1989; p. 302.
7. Shibamoto, T. *Retention Indices in Essential Oil Analysis in Capillary Gas Chromatography in Essential Oil Analysis*; Sandra, P., Bicci, C., Eds.; Verlag: New York, NY, USA, 1987; pp. 259–274.
8. McLafferty, F.W.; Stauffer, D.B. *The Wiley/NBS Registry of Mass Spectral Data*; Wiley and Sons: New York, NY, USA, 1989.
9. SAS Institute Inc. *SAS Online Doc® 9.2*; SAS Institute Inc.: Cary, NC, USA, 2008.
10. Menary, R.; Mactavish, H.S. Rural industries research and development corporation. In *Boronia Extraction: Increasing Yield and Quality: A Report for the Rural Industries*; Research and Development Corporation: Hobart, Australia, 2000.
11. Baldermann, S.; Yang, Z.; Sakai, M.; Fleischmann, P.; Watanabe, N. Volatile constituents in the scent of roses. *Floric. Ornam. Biotechnol.* **2009**, *3*, 89–97.

23

Shelf Life of Tropical Canarium Nut Stored under Ambient Conditions

David A. Walton [1,*], **Bruce W. Randall** [1], **Matthew Poienou** [2], **Tio Nevenimo** [2], **John Moxon** [3] and **Helen M. Wallace** [1]

[1]   Genecology Research Centre, Faculty of Science, Health, Education and Engineering, University of the Sunshine Coast, Maroochydore DC, QLD 4558, Australia; brandall@usc.edu.au (B.W.R.); hwallace@usc.edu.au (H.M.W.)

[2]   National Agriculture Research Institute, Islands Regional Centre, P.O. Box 204, Kokopo 613, Papua New Guinea; mathew.poienou@nari.org.pg (M.P.); tio.nevenimo@nari.org.pg (T.N.)

[3]   Cocoa Board of Papua New Guinea, P.O. Box 532, Kokopo 613, Papua New Guinea; pm@ppap.cocoaboard.org.pg

*   Correspondence: dwalton3@usc.edu.au

Abstract: There is a need to develop alternative crops to improve the food security and prosperity of developing countries. The tropical nut *Canarium indicum* (canarium nut) is increasingly used as a shade tree for cocoa and has potential for commercialization as a sustainable crop that will improve food security and livelihoods in Melanesia and East Asia. There is no information on canarium nut shelf life characteristics. Canarium kernels may be prone to rancidity, due to a high content of unsaturated fatty acids. Kernels at 5.4% moisture content were vacuum-packed with a domestic vacuum-packaging system and stored for six months in Papua New Guinea and for nine months in Southeast Queensland, Australia at both ambient temperatures (22 to 31 °C and 22 to 25 °C, respectively) and under refrigeration. Nuts were analysed for changes in peroxide values and free fatty acids (FFAs) over the storage periods that might indicate development of rancidity. Peroxide values indicated very low levels of oxidation in all treatments. Free fatty acids were at low levels but increased significantly during storage at ambient temperatures. The results suggested that vacuum-packed Canarium nuts can be stored safely under ambient tropical conditions for six months with daytime temperatures around 31 °C, and for nine months at 25 °C. Increasing FFA levels at ambient temperatures indicate caution about longer storage time at ambient temperatures. Storage under refrigeration greatly prolonged shelf life.

Keywords: nuts; *Canarium indicum*; canarium nut; shelf life; peroxide values; free fatty acids; food security

1. Introduction

Only 103 domesticated species of the world's 250,000 species of flowering plants provide 90% of the world's food [1]. A diversification away from over-reliance on staple food crops will be important as part of the process of achieving security of food production for an increasing world population in the future [2,3]. There is potential for expansion of species currently used as indigenous food crops. Nuts in particular are increasing in importance as food crops, with total world tree nut production for 2014–2015 at ~3.5 million metric tons (kernels) [4]. Edible nuts in the form of raw or roasted kernels [5,6] are increasingly recommended for inclusion in healthy diets due to evidence of beneficial effects on blood cholesterol, coronary heart disease, diabetes, sudden death, inflammation, thrombosis and vascular reactivity [7–10]. More than 80% of the world trade in tree nuts is comprised of just four species, walnuts, hazelnuts, pistachios, and almonds [4].

Many other species of edible nuts around the world have potential to be developed and produce a commercial return, for example, macadamia and Brazil nuts are case studies. Macadamia originated in the sub-tropical rainforest of eastern Australia, and is a rare example of a nut from the tropics grown commercially in plantations [11–17]. Brazil nuts have also become an important industry for countries along the Amazon River such as Brazil, Peru and Colombia, however, Brazil nuts are harvested from the forest floor rather than plantations [18]. In Papua New Guinea, the Solomon Islands and Vanuatu, potential exists for domesticating a variety of new tree nuts that have been used in traditional cultures [19–21].

The genus *Canarium* contains tree nuts with beneficial properties, including *Canarium odontophyllum*, *C. ovatum* and *C. album* [22–26]. *Canarium ovatum* (Pili) is cultivated commercially in the Philippines, producing 23,000 metric tons of edible nuts in 2011 [19,27]. *Canarium indicum* L. (Burseraceae) is a tropical nut species indigenous to lowland rainforests of eastern Indonesia, Papua New Guinea (PNG), the Solomon Islands and Vanuatu, producing nutritious nuts and high quality timber [20,28]. Canarium nuts have been an important food for the people of Papua New Guinea for thousands of years [29,30]. *Canarium indicum* has great potential for commercialization and is in a unique position to develop as a sustainable industry, adding to food security in Melanesia [20,31,32]. Traditionally, the nut has been mainly harvested from the ground in forests, similar to the Brazil nut, but large numbers of trees are now being planted as shade trees in cocoa plantations in Papua New Guinea [20]. In traditional culture, surplus nuts are typically stored in woven baskets above a fireplace. This practice is not suitable for a commercial product.

In other crops postharvest handling methods have been developed to reduce waste in the postharvest chain and maintain quality and nutritional attributes, [12–17,33]. Nuts are prone to rancidity because of high content of unsaturated fatty acids. Rancidity is one of the most important defects of this food [34,35]. Canarium nut has very high oil content of around 67 to 75%, with 40% oleic acid (monounsaturated, with one double bond) and 14% linoleic acid (polyunsaturated, with two double bonds) [36]. The high degree of unsaturation of canarium oil, especially linoleic acid, creates potential for oxidative rancidity to develop [37,38]. Hydrolytic rancidity can also occur if moisture content of foods is not reduced and maintained sufficiently [39,40]. It is the first step in the deteriorative rancidity pathway [39]. Fatty acids generated from hydrolysis are more prone to oxidation than intact triglycerides [38,41] and are the preferred substrate for respiration, generating heat and more water [42]. Tropical conditions with high temperatures and humidity may favour development of rancidity [43,44]. There is some knowledge of culture and postharvest processing of canarium nuts [45–49], however more research is needed to identify suitable postharvest treatments.

We conducted trials to investigate postharvest shelf life of *Canarium indicum* nuts stored as whole nuts and as kernels. Canarium kernels were packed using a domestic vacuum packaging system and stored under ambient and refrigerated conditions to investigate the shelf life characteristics of canarium under conditions likely to prevail in the postharvest chain in a tropical climate. For comparison, kernels were also stored in a laboratory under ambient laboratory conditions that may approximate conditions sometimes pertaining within a building in the tropics. Knowledge of shelf life of canarium kernels will provide information on suitable storage conditions to maintain quality of nuts in conditions encountered in developing countries.

2. Materials and Methods

2.1. General Methods

Canarium indicum fruit of up to two weeks post-abscission age was gathered from the ground in plantations in East New Britain, PNG. Fruit was depulped in PNG by a traditional method of soaking fruit in water, then allowing it to ferment in wet hessian sacks for two to three days until the pulp (exocarp) was soft enough to remove from the nuts. Sacks were maintained wet by external application of water as required. Three shelf-life experiments were conducted: (1) shelf-life of kernels stored in-shell as whole nuts at three different initial moisture contents (16.1% ± 0.3%, 10.3% ± 0.2% and 5.4% ± 0.1%) at mean laboratory temperatures of 22.2 ± 0.04 °C minimum and 24.5 ± 0.04 °C maximum; (2) shelf-life of kernels dried to an initial moisture content of 5.4% ± 0.1 stored at both ambient tropical temperatures (mean minimim 22.0 ± 0.01 °C to mean maximum 31.2 ± 0.26 °C) and under refrigeration (4 °C); and (3) shelf-life of kernels dried to an initial moisture content of 5.4% ± 0.1% stored at both laboratory temperature as above and under refrigeration (4 °C). Nuts were dried in a Memmert fan-forced laboratory oven (Memmert Gbh & Co. KG, Schwabach, Germany). The moisture content (wet basis (w.b.)) of nuts for shelf life studies was determined by drying three replicates of five nuts at 105 °C for 24 h to remove all free water [50] Moisture content was determined by the formula: Moisture content (w.b.) = (initial weight − final weight) × 100 / initial weight.

All moisture contents were calculated on a wet basis (w.b.). For shelf-life of kernels, whole nuts were dried for 3 days at 38 °C followed by 3 days at 45 °C to achieve nut moisture content of 5.4% ± 0.1% (w.b.). Nuts were cracked by hand with a modified TJ's™ macadamia nutcracker to release the kernels [45]. Kernels in testa were vacuum-packed for storage using a Foodsaver®Vac 440 domestic vacuum-sealer (Sunbeam Corporation, Botany, Australia) in polyethylene packaging film (Foodsaver VSO520, Sunbeam Corporation, Botany, Australia). At assessment kernels were blanched by pouring hot water at 100 °C over kernels in testa and allowing them to stand for 1.5 min before removing the testa. Kernels were then re-dried for 15 min at 60 °C to remove moisture absorbed during blanching. Shelf-life was assessed by peroxide value (PV) for oxidative rancidity and free fatty acids (FFA) for hydrolytic rancidity.

2.2. Shelf Life of Canarium Kernels Stored as Nuts (Nut-in-Shell)

Nuts at three different initial moisture levels were stored in open-mesh polypropylene bags under laboratory conditions (25 °C) for 11 months. Initial moisture contents were 16.1% ± 0.3%, 10.3% ± 0.2% and 5.4% ± 0.1% (w.b.). For each moisture content, there were five replicates each of 10 nuts. Moisture contents were determined again after six weeks ambient drying in the laboratory. Kernels from nuts oven-dried to 5.4% moisture content (w.b.) as above were used to obtain initial and final values for PV and FFA.

2.3. Shelf Life of Canarium Kernels Stored at Ambient Tropical Temperatures (22 °C to 31 °C)

Canarium nuts at 5.4% moisture content (w.b.) were cracked using a "TJ" macadamia cracker [45]. The kernels obtained were vacuum packed and stored for six months under ambient tropical conditions (mean minimum 22.0 ± 0.01 °C to mean maximum 31.2 ± 0.26 °C and under refrigeration (4 °C). Each temperature treatment consisted of five replicates of 10 kernels. Free fatty acids were determined at the start of the experiment, and PV and FFA analyses were conducted after one month, after three months and after six months of storage.

2.4. Shelf-Life of Canarium Kernels Stored at Laboratory Temperature

Canarium nuts at 5.4% moisture content (w.b.) were cracked as above [45]. The kernels obtained were vacuum packed and stored for nine months under laboratory conditions (22.2 ± 0.04 °C minimum and 24.5 ± 0.04 °C maximum) to examine Canarium shelf life at a lower temperature than tropical, and under refrigeration (4 °C). Each temperature treatment consisted of five replicates each of 10 kernels. FFA were determined at the start of the experiment and PV and FFA analyses were conducted after 1, 3, 6 and 9 months of storage.

Following testa removal, each replicate of ten kernels was crushed twice in a garlic press and mashed in a mortar and pestle with n-pentane. The sample was then stirred in a covered beaker with pentane for half to one hour. The liquid was then transferred to test tubes and centrifuged at 3000 rpm for 3 min to separate the oil and solvent from the nut residue. The supernatant liquid was then decanted and the pentane stripped from the oil in a Buchi Rotavapor (BÜCHI Labortechnik AG, Flawil, Switzerland).

2.5. Peroxide Values and Free Fatty Acids

Peroxide Values were determined according to AOAC Official Method 965.33 modified as follows: smaller samples of oil were used (1 g instead of 5 g); 0.01N $Na_2S_2O_3$ was used instead of 0.1 N $Na_2S_2O_3$; and $Na_2S_2O_3$ titration was accomplished with a 100 µL HPLC syringe instead of a burette. Peroxide value (expressed as milliequivalents of peroxide per kilogram of sample) was calculated according to the formula:

$$PV\ (meq\ perkg) = \frac{(S - B) \times N \times 1000}{sample\ wt\ (g) \times 1000} \qquad (1)$$

where S = sample titration (µL); B = Blank titration; and N = normality of $Na_2S_2O_3$.

Free fatty acids were determined using AOAC method 940.28, modified as follows: smaller samples of oil were used (1 g instead of 5 g); 0.1 N NaOH was used instead of 0.25 N; and NaOH titration was accomplished with a 100 µL HPLC syringe instead of a burette. Free fatty acids were calculated as follows:

$$\%\ FFA\ (as\ oleic) = \frac{(\mu L)alkali \times N\ of\ alkali \times 28.2\ mg}{sample\ wt\ (g) \times 1000} \qquad (2)$$

2.6. Statistical Analysis

Data for PV and FFA for kernels stored for 11 months as nut-in-shell were analysed for significant difference by a non-parametric Mann-Whitney U test. Data for PV and FFA for the kernel shelf life studies (stored at tropical temperatures (22 °C to 31 °C) and stored under laboratory conditions (constant 25 °C) and under refrigeration (4 °C) were analysed with a two-way ANOVA with time and temperature of storage as factors. Where significant differences were detected, means were compared using Duncan's Multiple Range test.

3. Results

3.1. Shelf Life of Canarium Kernels Stored as Nut-in-Shell at Constant Temperature (25 °C)

Nut-in-shell stored at three different initial moisture contents (w.b.) (Moist: 16.1% ± 0.3%; Partly dry: 10.3% ± 0.2%; Dry: 5.4% ± 0.1%) equilibrated to ~7.4% ± 0.44% moisture content after six weeks in an air conditioned laboratory at 25 °C. After 11 months the PV and FFA values of these kernels stored as whole nuts were low (Table 1). However, there was a significant increase (p = 0.008) in FFA from 0.11 to 0.31 during storage.

Table 1. Mean peroxide [1] and free fatty acid values (SE) for canarium kernels stored in nut-in-shell for 11 months at laboratory temperatures (~25 °C). Means in the same column with different superscripts are significantly different (p = 0.008).

Treatment	Peroxide Value	Free Fatty Acids
Initial value	NA [1]	0.11 (0.003) [a]
After 11 months storage at 25 °C	0.31 (0.032)	0.31 (0.29) [b]

[1] Peroxide Values not available. Different superscript letters in the same column indicate significant difference.

3.2. Shelf Life of Canarium Kernels Stored at Ambient Tropical Temperatures (22 °C to 31 °C)

Peroxide values for vacuum-packed kernels stored at tropical temperatures (22 °C to 31 °C) for six months were very low (≤0.61) and there were no significant differences in PV at three months and six months compared with the one-month values (Table 2). All FFAs were also low (0.14 to 0.29), but increased significantly compared with the initial (control, 0.16) after storage at ambient temperature for three months, and were further significantly increased after six months to c. 0.29 (p <0.05) (Figure 1). In contrast, kernels stored under refrigeration showed no significant increase in FFA levels (Figure 1). There were no significant interactions between the time and temperature of kernel storage.

Table 2. Peroxide value [1] means (SE) for vacuum-packed canarium kernels stored at ambient tropical temperatures (mean max. 31.2°C, mean min. 22 °C) and under refrigeration for six months, or at ambient conditions (25°C) and under refrigeration in a laboratory for nine months.

Temperature	Location	1 Month	3 Months	6 Months	9 Months
Tropical	Ambient	0.41 (0.07)	0.40 (0.04)	0.48 (0.07)	
	Refrigerated	0.61 (0.03)	0.42 (0.04)	0.38 (0.03)	
Laboratory	Ambient	NA [1]	0.38 (0.06)	0.25 (0.03)	0.23 (0.01)
	Refrigerated	NA [1]	0.29 (0.03)	0.26 (0.02)	0.19 (0.03)

[1] NA: Peroxide Values not available.

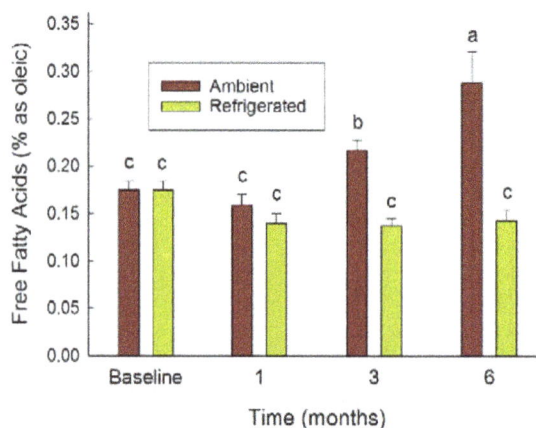

Figure 1. Free fatty acids (% as oleic) for *Canarium indicum* kernels stored at tropical temperatures (22 °C to 31 °C) and refrigerated temperatures (4.0 °C) for six months. Values are means (SE), different superscript letters on columns indicate significant difference between means ($p < 0.05$).

3.3. Shelf Life of Canarium Kernels Stored at Constant Temperature (22–25 °C)

Vacuum-packed canarium kernels did not deteriorate markedly during the nine months of storage and all PVs were very low (≤0.38 meq per kg, Table 2). There were no significant differences between the three, six and nine month treatments or between refrigerated and ambient treatments (Table 2). There were no significant interactions between the time and temperature of kernel storage.

Free fatty acid values were generally low (≤0.2) during laboratory storage for nine months; however, there was a significant upward trend in FFA at ambient temperatures ($p < 0.001$, Figure 2).

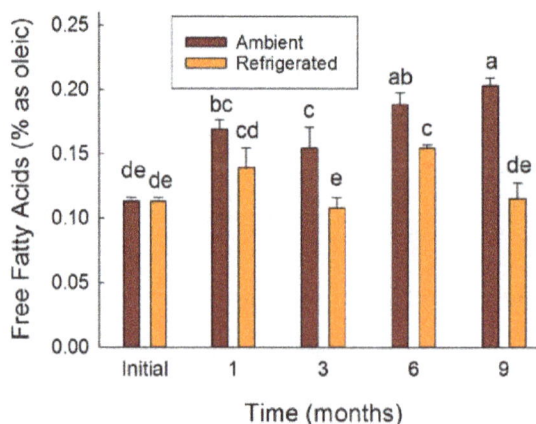

Figure 2. Free fatty acids (% as oleic) for canarium kernels stored at ambient laboratory temperatures (25 °C) and under refrigeration (4.0 °C) for nine months. Values are means (SE), different superscript letters on columns indicate significant difference between means ($p < 0.05$).

4. Discussion

Nuts require appropriate storage because of their high oil content and seasonal cropping nature to maintain eating quality and to make them available out of season [10,51,52]. This study showed that freshly harvested, vacuum packed canarium kernels can be stored for six months at ambient tropical temperatures of 22 to 31 °C, or for nine months at ambient temperature ~25 °C without development of rancidity as indicated by PV and FFA values. The development of unacceptable taste or rancidity is characteristic of nuts that have been stored in inappropriate conditions or stored for too long [53]. Lipid oxidation and hydrolysis during storage are the most common causes of deterioration in the sensory and nutritional quality of nuts [42,54,55]. Rancidity is the most serious defect of edible nuts and success for a new nut crop such as canarium will be greatly affected by the ability to prevent rancidity developing.

We found the PVs for kernels stored for three months, six months and nine months at constant temperature of 25 °C were very low ($\leq$0.38 meq per kg). Similarly, when kernels were stored at ambient tropical temperatures (22 °C to 31 °C) for six months, PVs were low (max. 0.61 meq per kg). We consider that the rancidity indications for macadamias are appropriate for canarium because the two nuts have very similar oil content, of approximately 75% [56]. The Australian Macadamia Industry Quality Handbook has a standard maximum PV of 5.0 [57,58], although other industry sources prefer a maximum of 3.0 [59]. Recognized PV standards for walnuts are less than 3.0, and for almonds less than 2.0 [41]. All PVs were well below all these standards, indicating high quality in terms of oxidative rancidity.

Factors limiting oxidation in this study could be (1) vacuum packing limited oxygen availability; (2) the packaging film limited oxygen availability [52,60,61]; and (3) storing in testa probably helped enhance shelf life, as for almonds [42]. The simple packaging system used enabled safe storage of canarium nuts for six months without refrigeration under ambient, lowland tropical conditions in PNG. Further shelf life studies are needed to determine the storage time limits under tropical conditions using similar and alternative packaging materials.

Free fatty acid values showed only limited hydrolysis of vacuum-packed kernels during nine months of storage at 25 °C. The highest value recorded, ~0.3%, was lower than the general standard of 0.4% recommended for hazelnuts and the nut industry in general [55] and for macadamias of <0.5% [62], although it is the maximum desirable level proposed by McConachie [63]. Although FFA values were low, the highest values indicate caution about the potential for development of hydrolytic rancidity during longer term storage. If macadamia kernels develop FFA over 0.15% in 12 weeks of storage they are considered to have a dramatically reduced shelf life [64]. Free fatty acids for the whole nut (nut-in-shell) storage trial, all ambient temperature treatments, and the six month tropical refrigerated treatment increased to over 0.15%, possibly indicating a limited shelf life after 11 months storage at 25 °C in-shell, nine months storage of kernels at constant 25 °C or six months storage at 22 to 31 °C. However, the current results compare very favourably with the accepted FFA standard for walnuts of <1.5% [41]. All refrigerated kernels at both sites maintained excellent FFA levels, demonstrating the value of refrigerated storage. There was some variability in FFA values in the refrigerated treatment stored for nine months, which could be due to inherent variability of samples, e.g., time on the ground before sampling, variable storage time before processing and possible effects of the traditional rotting method (retting) of depulping the fruit in the source country, PNG. Further canarium shelf life studies will be conducted including shelf life of in-shell kernels, an accelerated shelf life study and shelf life of roasted kernels.

The moisture content of the stored kernels was probably a factor in the free fatty acids increases as hydrolytic cleavage of triglycerides can occur when the moisture content is above the critical monolayer level at which enzymes are activated [42]. Blanching to remove the testa of canarium may deactivate or reduce enzymes [65]. Kernel moisture content at the beginning of storage was approximately three percent, sufficient to provide some free water for hydrolysis and suggesting the desirability of a drying regime to achieve lower kernel moisture content for ambient storage, e.g., the recommended moisture content for storage of macadamia kernels is 1.5%. Hydrolysis typically can increase by a factor of eight as water activity increases from 0.3 to 0.5 [40]. Temperature was also a factor, as under all ambient conditions hydrolysis of oils increased as time elapsed, but under refrigeration, hydrolysis was not sufficient to significantly increase FFA. Hydrolysis of nuts appears to be related to storage at excessively high ambient temperature [42,57,66]. Free fatty acids are very important for nuts, as they contribute off-flavours [67–69]. Free fatty acids from hydrolysis are also important for the oxidation pathway as they are more prone to oxidation than intact triglycerides [38,41,42].

Quality of canarium kernels was maintained at satisfactory levels as determined by PV and FFA when kernels were stored in-shell at 25 °C for 11 months (PV, 0.31 meq per kg; FFA, 0.31 % as oleic). This is similar to hazelnuts, which were stored for eight months in-shell at ambient conditions ranging from 10 °C to 26 °C without substantial deterioration as determined by PV and FFA, but after 12 months storage under these conditions, FFA of 0.47% was above the acceptable limit for hazelnuts [55]. Our research shows that canarium kernels can be safely stored in-shell for 11 months at mean laboratory temperatures ranging from 22.2 °C min. to 24.5 °C max. while awaiting further processing, provided the initial nut-in-shell moisture content is no higher than five percent. The inexpensive vacuum seal used is probably affordable technology for a modest operation in a developing country; however, electric ovens as in the current study may not be practical. A further problem is that the electricity network may be unreliable. There may be potential to use solar drying technology.

5. Conclusions

Canarium nut kernels vacuum-packed using simple domestic equipment were stored for nine months at 25 °C or for six months at 31 °C without PV and FFA values indicating rancidity developing. Free fatty acids that developed during these storage periods indicate caution for further storage under these conditions. Storage under refrigeration greatly prolonged shelf life.

Author Contributions: David Walton contributed perimental design and data analysis, and conducted laboratory experiments. Bruce Randall contributed data analysis and manuscript editing. Matthew Poienou conducted tropical experiment. Tio Nevenimo supervised tropical experimentation. John Moxon contributed experimental design and tropical collaboration in project. Helen Wallace contributed experimental design, manuscript editing, and was the project leader.

Conflicts of Interest: The authors declare no conflict of interest.

References

1.   Prescott-Allen, R.; Prescott-Allen, C. How many plants feed the world. *Conserv. Biol.* **1990**, *4*, 365–374. [CrossRef]
2.   Mayes, S.; Massawe, F.J.; Alderson, P.G.; Roberts, J.A.; Azam-Ali, S.N.; Hermann, M. The potential for underutilized crops to improve security of food production. *J. Exp. Bot.* **2012**, *63*, 1075–1079. [CrossRef] [PubMed]
3.   Kahane, R.; Hodgkin, T.; Jaenicke, H.; Hoogendoorn, C.; Hermann, M.; Keatinge, J.D.H.; Hughes, J.D.; Padulosi, S.; Looney, N. Agrobiodiversity for food security, health and income. *Agron. Sustain. Dev.* **2013**, *33*, 671–693. [CrossRef]
4.   INC. Global Trade of Tree Nuts Increased in 2013 by 36% over 2012. Available online: https://www.nutfruit. org/en/global-statistical-review_13653 (accessed on 19 October 2015).
5.   Fallico, B.; Arena, E.; Zappala, M. Roasting of hazelnuts. Role of oil in colour development and hydroxymethylfurfural formation. *Food Chem.* **2003**, *81*, 569–573. [CrossRef]
6.   Kodad, O.; Company, R.S.I. Variability of oil content and of major fatty acid composition in almond (*prunus amygdalus batsch*) and its relationship with kernel quality. *J. Agric. Food Chem.* **2008**, *56*, 4096–4101. [CrossRef] [PubMed]
7.   Garg, M.L.; Blake, R.J.; Wills, R.B.H.; Clayton, E.H. Macadamia nut consumption modulates favourably risk factors for coronary artery disease in hypercholesterolemic subjects. *Lipids* **2007**, *42*, 583–587. [CrossRef] [PubMed]

8. Griel, A.E.; Cao, Y.; Bagshaw, D.D.; Cifelli, A.M.; Holub, B.; Kris-Etherton, P.M. A macadamia nut-rich diet reduces total and ldl-cholesterol in mildly hypercholesterolemic men and women. *J. Nutr.* **2008**, *138*, 761–767. [PubMed]

9. Orem, A.; Yucesan, F.B.; Orem, C.; Akcan, B.; Kural, B.V.; Alasalvar, C.; Shahidi, F. Hazelnut-enriched diet improves cardiovascular risk biomarkers beyond a lipid-lowering effect in hypercholesterolemic subjects. *J. Clin. Lipidol.* **2013**, *7*, 123–131. [CrossRef] [PubMed]

10. Askari, G.; Yazdekhasti, N.; Mohammadifard, N.; Sarrafzadegan, N.; Bahonar, A.; Badiei, M.; Sajjadi, F.; Taheri, M. The relationship between nut consumption and lipid profile among the iranian adult population; Isfahan healthy heart program. *Eur. J. Clin. Nutr.* **2013**, *67*, 385–389. [CrossRef] [PubMed]

11. Trueman, S.J. The reproductive biology of macadamia. *Sci. Hortic.* **2013**, *150*, 354–359. [CrossRef]

12. Walton, D.A.; Wallace, H.M. The effect of mechanical dehuskers on the quality of macadamia kernels when dehusking macadamia fruit at differing harvest moisture contents. *Sci. Hortic.* **2015**, *182*, 119–123. [CrossRef]

13. Walton, D.A.; Randall, B.W.; Le Lagadec, M.D.; Wallace, H.M. Maintaining high moisture content of macadamia nuts-in-shell during storage induces brown centres in raw kernels. *J. Sci. Food Agric.* **2013**, *93*, 2953–2958. [CrossRef] [PubMed]

14. Walton, D.A.; Wallace, H.M. Postharvest dropping of macadamia nut-in-shell causes damage to kernel. *Postharvest Biol. Technol.* **2008**, *49*, 140–146. [CrossRef]

15. Walton, D.A.; Wallace, H.M. Delayed harvest reduces quality of raw and roasted macadamia kernels. *J. Sci. Food Agric.* **2009**, *89*, 221–226. [CrossRef]

16. Walton, D.A.; Wallace, H.M. Dropping macadamia nuts-in-shell reduces kernel roasting quality. *J. Sci. Food Agric.* **2010**, *90*, 2163–2167. [CrossRef] [PubMed]

17. Walton, D.A.; Wallace, H.M. Quality changes in macadamia kernel between harvest and farm-gate. *J. Sci. Food Agric.* **2011**, *91*, 480–484. [CrossRef] [PubMed]

18. Yang, J. Brazil nuts and associated health benefits: A review. *Lwt-Food Sci. Technol.* **2009**, *42*, 1573–1580. [CrossRef]

19. Evans, B. *Edible Nut Tres in Solomon Islands: A Variety Collection of Canarium, Terminalia and Barringtonia*; Australian Centre for International Agricultural Research: Canberra, Australia, 1999; p. 96.

20. Nevenimo, T.; Moxon, J.; Wemin, J.; Johnston, M.; Bunt, C.; Leakey, R.R.B. Domestication potential and marketing of *canarium indicum* nuts in the Pacific: 1. A literature review. *Agrofor. Syst.* **2007**, *69*, 117–134. [CrossRef]

21. Pauku, R.L.; Lowe, A.J.; Leakey, R.R.B. Domestication of indigenous fruit and nut trees for agroforestry in the Solomon Islands. *For. Trees Livelihoods* **2010**, *19*, 269–287. [CrossRef]

22. Duan, W.J.; Tan, S.Y.; Chen, J.; Liu, S.W.; Jiang, S.B.; Xiang, H.Y.; Xie, Y. Isolation of anti-hiv components from *canarium album* fruits by high-speed counter-current chromatography. *Anal. Lett.* **2013**, *46*, 1057–1068. [CrossRef]

23. Azlan, A.; Prasad, K.N.; Khoo, H.E.; Abdul-Aziz, N.; Mohamad, A.; Ismail, A.; Amom, Z. Comparison of fatty acids, vitamin E and physicochemical properties of *canarium odontophyllum* miq. (dabai), olive and palm oils. *J. Food Compos. Anal.* **2010**, *23*, 772–776. [CrossRef]

24. Shakirin, F.H.; Prasad, K.N.; Ismail, A.; Yuon, L.C.; Azlan, A. Antioxidant capacity of underutilized malaysian *canarium odontophyllum* (dabai) miq. Fruit. *J. Food Compos. Anal.* **2010**, *23*, 777–781. [CrossRef]

25. Chew, L.Y.; Prasad, K.N.; Amin, I.; Azrina, A.; Lau, C.Y. Nutritional composition and antioxidant properties of *canarium odontophyllum* mig. (dabai) fruits. *J. Food Compos. Anal.* **2011**, *24*, 670–677. [CrossRef]

26. Kikuchi, T.; Watanabe, K.; Tochigi, Y.; Yamamoto, A.; Fukatsu, M.; Ezaki, Y.; Tanaka, R.; Akihis, T. Melanogenesis inhibitory activity of sesquiterpenes from *canarium ovatum* resin in mouse b16 melanoma cells. *Chem. Biodivers.* **2012**, *9*, 1500–1507. [CrossRef] [PubMed]

27. Imperial, R. Pili (*canarium ovatum* engl.): A Promising Indigenous Crop of the Bicol Region in the Phillipines for Food and Nutrition. Available online: http://www.fao.org/fileadmin/templates/rap/files/meetings/2012/120531_iib3.pdf (accessed on 19 October 2015).

28. Thompson, L.; Evans, B. *Canarium indicum* var. *Indicum* and c. *Harveyi* (canarium nut) report 2004. In *Species Profiles for Pacific Island Agroforestry*; Permanent Agriculture Resources (PAR): Holualoa, HI, USA, 2006; pp. 209–228.

29. Matthews, P.J.; Gosden, C. Plant remains from waterlogged sites in the Arawe Islands, West New Britain province, Papua New Guinea: Implications for the history of plant use and domestication. *Econ. Bot.* **1997**, *51*, 121–133. [CrossRef]

30. Yen, D.E. The origins of subsistence agriculture in oceania and the potentials for future tropical food crops. *Econ. Bot.* **1993**, *47*, 3–14. [CrossRef]

31. Nevenimo, T.I.O.; Johnston, M.; Binifa, J.; Gwabu, C.; Angen, J.; Moxon, J.; Leakey, R. Domestication potential and marketing of *canarium indicum* nuts in the pacific: Producer and consumer surveys in Papua New Guinea (East New Britain). *For. Trees Livelihoods* **2008**, *18*, 253–269. [CrossRef]

32. Bunt, C.; Leakey, R. Domestication potential and marketing of *canarium indicum* nuts in the Pacific: Commercialization and market development. *For. Trees Livelihoods* **2008**, *18*, 271–289. [CrossRef]

33. Kitinoja, L.; Saran, S.; Roy, S.K.; Kader, A.A. Postharvest technology for developing countries: Challenges and opportunities in research, outreach and advocacy. *J. Sci. Food Agric.* **2011**, *91*, 597–603. [CrossRef] [PubMed]

34. Dominguez, I.L.; Azuara, E.; Vernon-Carter, E.J.; Beristain, C.I. Thermodynamic analysis of the effect of water activity on the stability of macadamia nut. *J. Food Eng.* **2007**, *81*, 566–571. [CrossRef]

35. Wall, M.M. Functional lipid characteristics, oxidative stability, and antioxidant activity of macadamia nut (*macadamia integrifolia*) cultivars. *Food Chem.* **2010**, *121*, 1103–1108. [CrossRef]

36. 36 Leakey, R.; Fuller, S.; Treloar, T.; Stevenson, L.; Hunter, D.; Nevenimo, T.; Binifa, J.; Moxon, J. Characterization of tree-to-tree variation in morphological, nutritional and medicinal properties of *canarium indicum* nuts. *Agrofor. Syst.* **2008**, *73*, 77–87. [CrossRef]

37. StAngelo, A.J. Lipid oxidation in foods. *Crit. Rev. Food Sci. Nutr.* **1996**, *36*, 175–224. [CrossRef] [PubMed]

38. Robards, K.; Kerr, A.F.; Patsalides, E. Rancidity and its measurement in edible oils and snack foods—A review. *Analyst* **1988**, *113*, 213–224. [CrossRef] [PubMed]

39. Davies, C. Lipolysis in lipid oxidation. In *Understanding and Measuring the Shelf-Life of Foods*; Steele, R., Ed.; Woodhead Publishing Limited: Cambridge, UK, 2004; pp. 142–156.

40. Esse, R.; Saari, A. Shelf-life and moisture management. In *Understanding and Measuring the Shelf-Life of Foods*; Steele, R., Ed.; Woodhead Publishing Limited: Cambridge, UK, 2004; pp. 24–41.

41. 41 Buransompob, A.; Tang, J.; Ma, R.S.; Swanson, B.G. Rancidity of walnuts and almonds affected by short time heat treatments for insect control. *J. Food Process. Preserv.* **2003**, *27*, 445–464. [CrossRef]

42. Lin, X.; Wu, J.; Zhu, R.; Chen, P.; Huang, G.; Li, Y.; Ye, N.; Huang, B.; Lai, Y.; Zhang, H.; et al. California almond shelf life: Lipid deterioration during storage. *J. Food Sci.* **2012**, *77*, C583–C593. [CrossRef] [PubMed]

43. Labuza, T.P. Kinetics of lipid oxidation in foods. *CRC Crit. Rev. Food Technol.* **1971**, *2*, 355–405. [CrossRef]

44. Robards, K.; Kerr, A.F.; Patsalides, E. Rancidity and its measurement in edible oils and snack foods: A review. *Analyst* **1988**, *113*, 213–224. [CrossRef] [PubMed]

45. Wallace, H.M.; Poienou, M.; Randall, B.; Moxon, J. Postharvest cracking and testa removal methods for *canarium indicum* nuts in the pacific. *Acta Hortic.* **2010**, *880*, 499–502. [CrossRef]

46. Walton, D.A.; Randall, B.W.; Wallace, H.M.; Poienou, M.; Moxon, J. Maturity indices of *canarium indicum* (burseraceae) nut. *Acta Hortic.* **2016**, *1109*, 17–22. [CrossRef]

47. Walton, D.A.; Randall, B.W.; Wallace, H.M.; Poienou, M.; Moxon, J. A roasting study for the tropical nut *canarium indicum* (burseraceae). *Acta Hortic.* **2016**, *1109*, 43–48. [CrossRef]

48. Randall, B.W.; Walton, D.A.; Zekele, P.; Gua, B.; Pakau, R.; Wallace, H.M. Selection of the tropical nut *canarium indicum* for early fruiting, nut-in-shell size and kernel size. *Acta Hortic.* **2016**, *1109*, 169–174. [CrossRef]

49. Wallace, H.; Randall, B.; Grant, E.; Jones, K.; Walton, D.; Poienou, M.; Nevenimo, T.; Moxon, J.; Pauku, R. Processing methods for canarium nuts in the pacific. *Acta Hortic.* **2016**, *1128*, 145–149. [CrossRef]

50. Braga, G.C.; Couto, S.M.; Hara, T.; Neto, J. Mechanical behaviour of macadamia nut under compression loading. *J. Agric. Eng. Res.* **1999**, *72*, 239–245. [CrossRef]

51. Fourie, P.C.; Basson, D.S. Predicting occurrence of rancidity in stored nuts by means of chemical-analyses. *Lebensm. Wiss. Technol.* **1989**, *22*, 251–253.

52. Mexis, S.F.; Badeka, A.V.; Riganakos, K.A.; Karakostas, K.X.; Kontominas, M.G. Effect of packaging and storage conditions on quality of shelled walnuts. *Food Control* **2009**, *20*, 743–751. [CrossRef]

53. Kaijser, A.; Dutta, P.; Savage, G. Oxidative stability and lipid composition of macadamia nuts grown in New Zealand. *Food Chem.* **2000**, *71*, 67–70. [CrossRef]

54. Zajdenwerg, C.; Branco, G.F.; Alamed, J.; Decker, E.A.; Castro, I.A. Correlation between sensory and chemical markers in the evaluation of Brazil nut oxidative shelf-life. *Eur. Food Res. Technol.* **2011**, *233*, 109–116. [CrossRef]

55. Ghirardello, D.; Contessa, C.; Valentini, N.; Zeppa, G.; Rolle, L.; Gerbi, V.; Botta, R. Effect of storage conditions on chemical and physical characteristics of hazelnut (*corylus avellana* l.). *Postharvest Biol. Technol.* **2013**, *81*, 37–43. [CrossRef]

56. Trueman, S.J.; Richards, S.; McConchie, C.A.; Turnbull, C.G.N. Relationships between kernel oil content, fruit removal force and abscission in macadamia. *Aust. J. Exp. Agric.* **2000**, *40*, 859–866. [CrossRef]

57. AMS. *Macadamia Industry Quality Handbook*; Australian Macadamia Society: Lismore, Australia, 2008.

58. O'Hare, S.R.; Quinlan, K.; Vock, N. *Macadamia Growers Handbook*; Department of Primary Industries and Fisheries: Brisbane, Queensland, Australia, 2004; p. 214.

59. Himstedt, S. Oil Content and Other Componeents as Indicators of Quality and Shelf Life of Macadamia Kernels. Doctoral Dissertation, University of Queensland, Brisbane, Australia, October 2002.

60. Bakkalbasi, E.; Yilmaz, O.M.; Javidipour, I.; Artik, N. Effects of packaging materials, storage conditions and variety on oxidative stability of shelled walnuts. *Lebensm. Wiss. Technol.* **2012**, *46*, 203–209. [CrossRef]

61. Shakerardekani, A.; Karim, R. Effect of different types of plastic packaging films on the moisture and aflatoxin contents of pistachio nuts during storage. *J. Food Sci. Technol.* **2013**, *50*, 409–411. [CrossRef] [PubMed]

62. Mason, R.L.; Nottingham, S.M.; Reid, C.E.; Gathambiri, C. The quality of macadamia kernels stored in simulated bulk retail dispensers. *Food Aust.* **2004**, *56*, 133–139.

63. McConachie, I. Fats and oils as applied to macadamia nuts: A very basic chemistry lesson. *Aust. Macadamia Soc. News Bull.* **1996**, *23*, 34.

64. Mason, R.L.; Himstedt, S.; Nottingham, S.; McConchie, C.; Meyers, N. *Oil Content and Other Components as Indicators of Quality and Shelf Life of Macadamia*; Project No. MC97010; Horticulture Australia Limited: Sydney, Australia, 2003.

65. Saltveit, M.E. Wound induced changes in phenolic metabolism and tissue browning are altered by heat shock. *J. Food Sci. Technol.* **2000**, *21*, 61–69. [CrossRef]

66. Cavaletto, C. Macadamia nuts. In *Handbook of Tropical Foods*; Chan, I., Harvey, T., Eds.; Marcel Dekker: New York, NY, USA, 1983; pp. 361–397.

67. Dela Cruz, A.; Cavaletto, C.; Yamamoto, Y.; Ross, E. Factors affecting macadamia stability: 2. Roasted kernels. *Food Chem.* **1966**, *20*, 123–124.

68. Das, I.; Shah, N.G.; Kumar, G. Properties of walnut influenced by short time microwave treatment for disinfestation of insect infestation. *J. Stored Prod. Res.* **2014**, *59*, 152–157. [CrossRef]

69. Narasimhan, S.; Rajalakshmi, D.; Chand, N.; Mahadeviah, B.; Indiramma, A.R. Palm oil quality in different packaging materials—Sensory and physicochemical parameters. *JAOCS* **2001**, *78*, 257–264. [CrossRef]

Permissions

List of Contributors

Ali Farhadi
Agriculture and Natural Resource Research Center of Isfahan, Iran
Department of Horticulture Sciences, Agriculture Faculty, Ferdowsi University of Mashhad, Iran

Hossain Aroeii and Hossain Nemati
Department of Horticulture Sciences, Agriculture Faculty, Ferdowsi University of Mashhad, Iran

Reza Salehi
Department of Horticultural Sciences, Campus of Agriculture & Natural Resources, University of Tehran, Karaj, Iran

Francesco Giuffrida
Department of Agriculture, Food and Environment, Catania University, Catania 95100, Italy

Uwe Schindler and Lothar Müller
Leibniz Centre for Agricultural Landscape Research (ZALF), Institute of Landscape Hydrology, Eberswalder St. 84, Muencheberg D15374, Germany

Scott Schaeffer and Amit Dhingra
Molecular Plant Science Program, Washington State University, Pullman, WA 99164, USA
Department of Horticulture, Washington State University, Pullman, WA 99164, USA

Christopher Hendrickson and Rachel Fox
Department of Horticulture, Washington State University, Pullman, WA 99164, USA

Sisir Mitra
Section Tropical and Subtropical Fruits, International Society for Horticultural Science, Faculty of Horticulture, Bidhan Chandra Krishi ViswaVidyalaya, Mohanpur, B-12/48, Kalyani, Nadia, West Bengal 741252, India

Hidangmayum Devi
 Indian Council of Agriculture Research complex for NEH Region, Tripura Centre, Lembucherra, West Tripura 799210, India

Holy Ranaivoarisoa, Solofoniaina Ravoninjiva and Sylvain Ramananarivo
Agro Management, Développement Durable et Territoires, Ecole Doctorale Gestion des Ressources Naturelles et Développement, Ecole Supérieure des Sciences Agronomiques, University of Antananarivo, Antananarivo 101, Madagascar

Romaine Ramananarivo
Ecole Supérieure de Management et d'Informatique Appliquée, Antananarivo 101, Madagascar

Amarat Simonne
Food Safety and Quality Research Laboratory, Department of Family, Youth and Community Sciences, Institute of Food and Agricultural Sciences, University of Florida, Gainesville, FL 32603, USA

Monica Ozores-Hampton
Horticultural Sciences Department at SWFREC, Institute of Food and Agricultural Sciences, University of Florida, Immokalee, FL 34142, USA

Danielle Treadwell
Horticultural Sciences Department, Institute of Food and Agricultural Sciences, University of Florida, Gainesville, FL 32611, USA

Lisa House
Food and Resource Economics Department, University of Florida, Gainesville, FL 32611, USA

Uwe Schindler, Gunnar Lischeid and Lothar Müller
Leibniz Centre for Agricultural Landscape Research (ZALF), Institute of Landscape Hydrology, Eberswalder St. 84, Muencheberg D15374

Audrius Sasnauskas, Vidmantas Bendokas, Rasa Karkleliene, Danguolė Juškevičciene, Tadeušas Šikšnianas, Dalia Gelvonauskienė, Rytis Rugienius, Danas Baniulis, Sidona Sikorskaitė-Gudžiūnienė, Ingrida Mažeikienė, Audrius Radzevičius, Nijolė Maročkienė, Eugenijus Dambrauskas and Vidmantas Stanys
Institute of Horticulture, Lithuanian Research Centre of Agriculture and Forestry, Babtai LT-54333, Lithuania

Frank Eulenstein, Marion Tauschke, Axel Behrendt, Uwe Schindler and Marcos A. Lana
Leibniz Centre for Agricultural Landscape Research (ZALF), Müncheberg 15374, Germany

Jana Monk
AgResearch Ltd., Christchurch 8140, New Zealand

Shaun Monk
Grasslanz Technology Ltd., Palmerston North 4442, New Zealand

Niranjani P. K. Semananda, James D. Ward and Baden R. Myers
School of Natural and Built Environments, University of South Australia, Mawson Lakes Campus, Adelaide 5095, Australia

Hector Valenzuela
Department of Plant and Environmental Protection Sciences, University of Hawaii at Manoa, 3190 Maile Way No. 307, Honolulu, HI 96822, USA

Frank Eulenstein
Leibniz-Zentrum für Agrarlandschaftsforschung (ZALF) Müncheberg, Eberswalder Straße 84, Müncheber 15374, Germany
Department Agro-Chemistry, Kuban State Agrarian University, Krasnodar 350044, Russia

Marcos Lana, Marion Tauschke and Axel Behrend
Leibniz-Zentrum für Agrarlandschaftsforschung (ZALF) Müncheberg, Eberswalder Straße 84, Müncheber 15374, Germany

Askhad Sheudzhen
Department Agro-Chemistry, Kuban State Agrarian University, Krasnodar 350044, Russia

Sandro Schlindwein
Departamento de Engenharia Rural, Universidade Federal de Santa Catarina, Florianópolis 88034-000, Brazil

Edgardo Guevara and Santiago Meira
Department of Crop Production INTA—Instituto Nacional de Tecnologia Agropecuaria, Pergamino 2700, Argentina

Khalid M. Al-Absi
Department of Plant Production, Faculty of Agriculture, Mútah University, Al-Karak 61710, Jordan

Douglas D. Archbold
Department of Horticulture, University of Kentucky, Lexington, KY 40546, USA

Ana Patrícia Costa
Faculty of Sciences, Universidade do Porto, Porto 4169-007, Portugal

Isabel Pôças
Linking Landscape, Environment, Agriculture and Food, Instituto Superior de Agronomia, Universidade de Lisboa, Lisboa 1349-017, Portugal
Geo-Space Sciences Research Centre, Universidade do Porto, Porto 4169-007, Portugal

Mário Cunha
Faculty of Sciences, Universidade do Porto, Porto 4169-007, Portugal
Geo-Space Sciences Research Centre, Universidade do Porto, Porto 4169-007, Portugal
CITAB—Center for the Research and Technology of Agro-Environmental and Biological Sciences, Universidade de Trás-os-Montes e Alto Douro, Quinta de Prados, Ap. 1013, Vila Real 5001-801, Portugal

Tamby Ramanankonenana, Jules Razafiarijaona, Sylvain Ramananarivo and Romaine Ramananarivo
Agro-Management, Développement Durable et Territoires, Ecole Doctorale Gestion des Ressources Naturelles et Développement, Ecole Supérieure des Sciences Agronomiques, University of Antananarivo, Antananarivo 101, Madagascar

Eric Simonne
Horticultural Sciences Department, Institute of Food and Agricultural Sciences, University of Florida, Gainesville, FL 32611, USA

Robert Hochmuth
Suwannee Valley Agricultural Extension Center, Florida Cooperative Extension Service, Live Oak, FL 32060, USA

Kobra Khalaj, Nima Ahmadi and Mohammad Kazem Souri
Department of Horticultural Sciences, Tarbiat Modares University, Tehran 336-14115, Iran

Maria do Rosá rio Cameira and Mariana Mota
Department of Biosystems Engineering, Instituto Superior de Agronomia, University of Lisbon, Tapada da Ajuda, 1349-017 Lisboa, Portugal
Linking Landscape, Environment, Agriculture and Food (LEAF), Instituto Superior de Agronomia, University of Lisbon, Tapada da Ajuda, 1349-017 Lisboa, Portugal

Panya Saeheng, Panuwat Eamsakulrat and Chaleeda Borompichaichartkul
Department of Food Technology, Faculty of Science, Chulalongkorn University, Bangkok 10330, Thailand

Orachorn Mekkerdchoo
Division of Fermentation Technology, Faculty of Agro-Industry, King Mongkut's Institute of Technology Ladkrabang, Bangkok 10520, Thailand

Anna Hubackova and Jan Banout
Department of Sustainable Technologies, Faculty of Tropical AgriSciences, Czech University of Life Sciences, Prague 16521, Czech Republic

Rarisara Impaprasert, Sujika Piyarat, Natnicha Sophontanakij and Nattiya Sakulnate
Department of Microbiology, King Mongkut's University of Technology Thonburi, Bangkok 10140, Thailand

Suwit Paengkanya
Division of Energy Technology, School of Energy, Environment and Materials, King Mongkut's University of Technology Thonburi, Bangkok 10140, Thailand

Chaleeda Borompichaichartkul
Department of Food Technology, Chulalongkorn University, Bangkok 10330, Thailand

George Srzednicki
Food Science and Technology, University of New South Wales, Sydney, NSW 2052, Australia

Maryam Mirzaei, Nima Ahmadi and Abdolali Shojaeiyan
Department of Horticultural Sciences, Tarbiat Modares University, Tehran 336-14115, Iran

Fatemeh Sefidkon
Research Institute of Forests and Rangelands, Tehran 13185-116, Iran

Alireza Mazaheri
R & D Department, Golkaran Essential Oils and Herbal Extracts Company, Kashan 1178, Iran

David A. Walton, BruceW. Randall and Helen M. Wallace
Genecology Research Centre, Faculty of Science, Health, Education and Engineering, University of the Sunshine Coast, Maroochydore DC, QLD 4558, Australia

Matthew Poienou and Tio Nevenimo
National Agriculture Research Institute, Islands Regional Centre, P.O. Box 204, Kokopo 613, Papua New Guinea

John Moxon
Cocoa Board of Papua New Guinea, P.O. Box 532, Kokopo 613, Papua New Guinea

Index